# FARMERS, SCIENTISTS AND PLANT BREEDING
## Integrating Knowledge and Practice

# Contents

# Contributors

**Marianne Bänziger** is senior scientist at the International Maize and Wheat Improvement Research Centre (CIMMYT) and stationed in Zimbabwe. Her research on the improvement of crops for abiotic stress tolerance spans more than 10 years. She coordinates CIMMYT's global programme 'Maize for Sustainable Production in Stress Environments', and also leads the maize stress breeding network in southern Africa.

**Salvatore Ceccarelli** received a PhD in Applied Genetics at the Institute of Genetics of the Faculty of Agriculture of the University of Milano (Italy) and has been associate professor of plant genetic resources and full professor of genetics and plant breeding at the University of Perugia (Italy). He has been a barley breeder with the International Centre for Agricultural Research in the Dry Areas (ICARDA) since 1984, and is currently the manager of the barley improvement programme that is targeted to all developing countries. He is author or co-author of 77 refereed journal articles, 38 conference papers, 17 non-refereed papers, one book on agricultural genetics (in Italian), and chapters in six books. He has been an invited speaker at about 18 international conferences or workshops.

**David A. Cleveland** is an associate professor in the Environmental Studies Program and the Department of Anthropology at the University of California, Santa Barbara, and is Co-Director of the Centre for People, Food and Environment. He has carried out research with farmers and agricultural scientists on sustainable,

small-scale agriculture and development policy in southern
Mexico, West Africa and south-western USA. His current
research is on comparing the knowledge and practice of
indigenous farmers and formal scientists, focusing on plant
breeding, and including the roles of biotechnology, crop genetic
resource conservation and intellectual property rights.

**Donald N. Duvick** is a retired plant breeder and affiliate professor
of plant breeding at Iowa State University, with experience in
breeding most of the major field crops in the USA and abroad.
He was employed in the private sector, where he was both a
breeder and an administrator, and has also served as trustee
and on advisory committees for the International Agricultural
Research Centres, CIMMYT and the International Rice Research
Institute (IRRI), on the External Evaluation Panel for the USAID
Cowpea/Bean CRSP, the US Department of Agriculture's (USDA)
National Plant Genetic Resources Board, the USDA's Plant
Gene Expression Centre Advisory Council, and the Steering
Committee of the Keystone International Dialogue on Plant
Genetic Resources.

**David Frossard** is a cultural anthropologist studying peasant
challenges to chemical-intensive 'green revolution' rice farming
in South-east Asia and elsewhere. He has taught anthropology
at the University of California, Irvine (where he received his
PhD degree in 1994), the Colorado School of Mines, Golden,
Colorado, and the International College of China Agricultural
University, Beijing.

**Stefania Grando** received a PhD in Productivity of Crop Plants at the
Institute of Plant Breeding, Faculty of Agriculture, University of
Perugia, Italy. She joined the barley improvement project at
ICARDA in 1987, and is currently primarily responsible for
research and training on stabilizing and enhancing productivity
of barley in Africa and Yemen with emphasis on breeding
varieties for poor farming communities.

**Krishna D. Joshi** is a plant breeder. He worked with Lumle
Agricultural Centre before joining Local Initiatives for
Biodiversity Research and Development (LI-BIRD), Pokhara,
Nepal. He is involved in a number of research projects on
participatory plant breeding, participatory varietal selection,
*in situ* crop conservation and participatory scaling up. He has
long experience in developing and scaling up of participatory
crop improvement approaches both in marginal and high
potential production systems of Nepal.

**Shawn McGuire** studied biology, and later development studies. His
interests are in collaborative breeding and genetic resources

management, focusing especially on farmers' seed systems and on the institutional aspects of formal research. He is a PhD candidate with the Technology and Agrarian Development Group at Wageningen University, investigating farmers' sorghum seed systems and institutional change in Ethiopia.

**Julien de Meyer** is a Junior Professional Officer seconded by the Swiss Agency for Development and Cooperation to the maize stress breeding programme at CIMMYT in Zimbabwe. For the past 3 years, he has been responsible for the Mother–Baby Pilot Project in Zimbabwe that was to develop a more cost-effective and appropriate maize variety testing scheme for stress environments in southern Africa.

**Ram B. Rana** is a socio-economist at LI-BIRD. He is currently involved in research on socio-economic and cultural contexts of farmer decision making processes in plant genetic resources management on farm. He is also involved in research on the lives of shepherds of migratory sheep and goats in the high Himalayas.

**Deepak K. Rijal** is an agroecologist at LI-BIRD. In recent years he has been working with on-farm management of plant genetic resources involving farming communities in contrasting agroecological and sociocultural settings in Nepal. He leads this research in the area of adding values and benefits through non-breeding means.

**Humberto Ríos Labrada** is a plant breeder with experience in low-input agriculture. He was a lecturer on Research Methodology and Plant Breeding at the Higher Pedagogical Institute for Technical and Professional Education, Cuba. Currently, he is coordinator of the participatory plant breeding programme of the National Institute of Agricultural Sciences, La Habana, Cuba.

**Jürg Schneider** is an anthropologist currently working as a senior scientific adviser for development cooperation at the International Affairs Division of the Swiss Agency for the Environment, Forests and Landscape. Previously, he has worked in Indonesia for several years conducting research on agricultural development, local knowledge and genetic resource conservation. He has also been teaching anthropology at Swiss universities.

**Melinda Smale,** at the time Chapter 3 was completed, was Senior Economist, International Plant Genetic Resources Institute (IPGRI), and Visiting Research Fellow, International Food Policy Research Institute (IFPRI). Many of the concepts and results discussed here are based on her work with colleagues at CIMMYT, where she began to study economic aspects of crop genetic diversity in 1994. In the 1980s and early 1990s, she conducted research on technology adoption in Somalia, Pakistan and Malawi, working with CIMMYT and Volunteers in Technical Assistance.

**Steven E. Smith** is a quantitative geneticist and associate professor at the University of Arizona. His research concentrates on the improvement of forage plants in both rangeland and cropland environments. This involves selection for stress tolerance in forage crops, investigation of the physiological–morphological basis of forage productivity and survival, rapid assessment of genetic variation and development of improved selection strategies.

**Daniela Soleri** is an ethnoecologist interested in the genetic implications of 'indigenous' and 'scientific' knowledge and practices in agriculture with the objective of using both of these for developing practical approaches to improve the well-being of poor agricultural communities.

**Bhuwon Sthapit** was a plant breeder with Lumle Agricultural Centre, Nepal Agricultural Research Council and LI-BIRD. Currently he is a scientist specializing in *in situ* crop conservation, home gardens and participatory approaches with IPGRI, he has been working in Asia Pacific and the Oceania region for over 4 years, and has been a focal point of participatory plant breeding (PPB) at IPGRI. He has long experience in the field of PPB approaches in Nepal and Vietnam.

**Madhu Subedi** is a plant breeder. He worked with Lumle Agricultural Centre and Nepal Agricultural Research Council before joining LI-BIRD. He was involved with participatory plant breeding projects before joining the University of Wolverhampton, UK, as a PhD student.

**John Witcombe** is a plant breeder at the University of Wales, Bangor. He is involved in a number of research projects in South Asia addressing participatory plant breeding. Through these projects, he is playing a leading role in research addressing this issue.

**Karl Zimmerer** is geographer and professor at the University of Wisconsin; he has carried out research with Quechua farmers of the Peruvian Andes since 1985 on their management of crop genetic diversity during a period of great change.

# Preface

This book is about knowledge and practice in plant breeding by farmers and scientists. The book's authors, from around the world, include those with expertise in social science, plant breeding, participatory and collaborative development, and project implementation, who have distinguished themselves by their innovative and insightful approaches to understanding farmers' and scientists' plant breeding, and to improving the theoretical, empirical and methodological basis of crop improvement. We asked all of them to address the following questions on the basis of their own research and experience: What is the nature of plant breeding knowledge, in terms of intuition, empiricism and theory? In what ways are farmers' and plant breeders' knowledge and practice different or similar? What do the answers to the preceding questions imply about the potential for further collaboration?

We appreciate the authors' patience with our, in many cases extensive, editorial suggestions in our efforts to ensure that all of the chapters be focused on the central theme, and useful to the wide range of intended readers, from social scientists to plant breeders, from policy makers to project personnel. We encouraged authors to: (i) make explanations understandable to non-specialists; (ii) describe the basis for statements in terms of theory, data and methods; and (iii) be explicit about their assumptions. We specifically asked plant breeders to discuss the theory underlying their approaches and why, even though they share the same basic theory with other plant breeders and with whole programmes that have historically followed more conventional strategies, their application of that theory differs. We thank all of the authors for their contributions; we have learned much in the process of editing and discussing their manuscripts.

The purpose of this book is to encourage further discussion and research on the nature of farmers' and scientists' knowledge and practice in plant breeding, in order to support the development of more environmentally, socially and economically sustainable agriculture. To this end, we have added cross-references in all of the chapters to other chapters in the book, in order to facilitate readers' comparisons of the similarities and differences in methods, assumptions and conclusions.

We hope that this book will be useful not only to those directly involved in plant breeding that brings farmers and scientists closer together, but also to others interested in the comparison between local and scientific knowledge, and the application of this research to the practical problems of improving local and global well-being: policy makers and researchers in agricultural development and plant breeding, social scientists, agricultural scientists and development professionals.

David A. Cleveland and Daniela Soleri
University of California, Santa Barbara, and
Centre for People, Food and Environment
August 2001

# Introduction: Farmers, Scientists and Plant Breeding: Knowledge, Practice and the Possibilities for Collaboration

**1**

## DAVID A. CLEVELAND[1] AND DANIELA SOLERI[2]

[1]*Environmental Studies Program and Department of Anthropology, University of California, Santa Barbara, CA 93106-4160, USA;*
[2]*Centre for People, Food and Environment, Santa Barbara, California, USA and Environmental Studies Program, University of California, Santa Barbara, CA 93106-4160, USA*

## Abstract

Control over management of the world's resources is increasingly contested because of economic, political and biophysical globalization, and increasing demands of a growing population of more than 6 billion. This has led to new interest in indigenous or traditional knowledge in many areas, including agriculture and plant breeding. Farmers were the first plant breeders, beginning with domestication of plants over 12,000 years ago. Modern, scientific plant breeding developed in the last two centuries, and has become increasingly separated from farmers, especially in non-industrial regions. Plant breeding systems consist not only of crop genotypes and growing environments, but also of the social structures in which plant breeding is carried out, and the knowledge of farmers and scientists. Because of the challenge to make plant breeding and agriculture more environmentally, socially and economically sustainable, there is increasing interest in reuniting farmer and scientific plant breeding.

## Globalization, Indigenous and Scientific Knowledge, and Plant Breeding

Population growth, environmental degradation and the integration of physical, biological and sociocultural systems on a global scale, have all increased dramatically in the last few centuries, and especially in the last 50 years. With productive resources becoming scarcer and more contested, attention has focused on the potential value of local knowledge and local systems of resource use and conservation as more environmentally, economically and socially sustainable alternatives to modern, industrial systems. This attention has increased dramatically in recent years, especially since the Convention on Biological Diversity (CBD) in 1992, which asserts the importance of indigenous or local people's knowledge for sustainable development, as well as the rights of these people to this knowledge. As these ideas have achieved wider acceptance, local and indigenous peoples have gained a voice for their viewpoints in international fora (Fowler, 1994; Cleveland and Murray, 1997).

One of the greatest challenges for understanding the promise of local knowledge and practice for helping to solve global problems will be developing the required levels of communication, which will need to be 'broad and deep beyond precedent', will need to take advantage of global communication networks (Ostrom *et al.*, 1999), and that renders the search for a more balanced, indeed more scientific, treatment of disparate knowledge systems inevitable (Nader, 1996: 6–7). Therefore, at the centre of the debate about more sustainable alternatives to conventional, modern agriculture are questions about the comparative environmental sustainability of modern and traditionally based agri-culture, about the degree of similarity between scientists' and farmers' knowledge, about who controls the representation of this knowledge, and about what all of this means for the possibility of collaboration between farmers and scientists (*Diversity*, 1998; Sillitoe, 1998; Dove, 2000).

A sign of the lack of information and of the political importance of this topic is that there is a great deal of disagreement over terminology. 'Indigenous knowledge', 'local knowledge' and 'traditional knowledge' often have conflicting meanings, and these meanings differ among researchers (Ellen and Harris, 2000). We use the terms 'indigenous knowledge' (IK) and farmers' knowledge (FK) to refer to the knowledge of people who are not in the modern, global scientific system. We understand that what is often referred to as IK is not 'indigenous' in the more restricted sense of arising from only local sources (Cleveland and Murray, 1997; Dove, 2000), but use the term in a general way to empha-size the contrast of relatively more local IK with relatively more general (at least in a geographical sense) modern, scientific knowledge (SK).

Disagreement over terminology has not prevented new efforts by social and natural scientists to understand the potential contribution of IK to sustainable development, and new efforts by local peoples and their supporters to try to capture more of the power of SK to serve their own goals.

Small-scale farmers whose well-being and way of life are threat-ened by globalization, opportunistically make use of the possibilities offered by other aspects of globalization to improve their situation (Cleveland, 1998), or simply to be able to remain farmers. Therefore, local groups may define 'indigenous' or 'traditional' agriculture in ways that include industrial agriculture technologies, in part because this serves their larger goal of maintaining their physical and cultural identity; *they localize global SK.* Local communities are increasingly taking the initiative, or working with national and international non-governmental organizations, to gain more control over the process of improving IK (e.g. Millar *et al.*, 2001). For example, Zuni indigenous farmers have learned how to use global positioning system (GPS) technology to map their family farm fields, and this became an important force for resolving land disputes that have impeded the revitalization of indigenous agriculture (Cleveland *et al.*, 1995).

Scientists who perceive negative impacts on the natural environment and on society of the application of SK are also interested in *globalizing IK* for increasing sustainability. Biological scientists have supported this integration in agriculture, advocating 'the development of more ecologically designed agricultural systems that reintegrate features of traditional agricultural knowledge and add new ecological knowledge' (Matson *et al.*, 1997: 508). Interest in cataloguing IK in terms of SK concepts to facilitate its use in more locally appropriate, participatory development has reached the mainstream, though it is often criticized for detaching IK from its local contexts and thus rendering it useless (Sillitoe, 1998). The status of IK as a complement to SK in promoting more sustainable development is still far from certain, and it is difficult to separate empirical evidence from its political contexts (Ellen and Harris, 2000).

As a result, local farmers, project workers, agricultural scientists, social scientists and development policy makers are increasingly asking (implicitly and explicitly), 'Is it possible for scientists and local peoples to collaborate to reach common development goals?' Could the answer be 'No, this idea is just a politically correct fad, doomed to failure because the social and biophysical environments, knowledge and goals of the two groups are so different'? But what if they aren't so different? Perhaps we should first ask, 'In what ways are scientists' and farmers' environments, knowledge and goals similar, as well as different? What are the reasons for differences and similarities?' Approached in this way the answer to questions about the possibility of collaboration might not necessarily always be 'No'. It might also be 'Yes, scientists and local peoples can collaborate to reach common development goals.' Then questions need to be asked about collaboration. 'What form would collaboration take? How would the relative value of local and scientific knowledge be determined for a given situation, and what methods could be used for integrating them? How would scientists and local people talk to each other? Would scientists or local people be in charge?'

This book addresses these questions for the important case of plant breeding. Different chapters deal with very different situations, and focus on different components of the plant breeding system of farmers, of plant breeders or of both. Most are written from the perspective of plant breeders and/or social scientists, although some also take the perspective (through the lens of outsiders) of farmers. The authors of the chapters also have different methodological approaches and theoretical orientations, and are working with data from different and, to some extent, unique situations. Yet the authors of each chapter reflect on their own knowledge and practice, and that of the farmers and scientists they work with, have the plant breeding system as a whole as a reference, and strive to make their methods and assumptions explicit. The chapters provide valuable insights on the importance of understanding the dynamics of farmer and scientist knowledge in assessing the possibilities for collaboration.

We asked the authors to describe the way in which working with farmers, their crop varieties and growing environments has led them to reinterpretations of conventional plant breeding or social theory and to new insights, methods and practices, and to be as explicit as possible about their understanding of farmers' and scientists' knowledge and practice. These requests made of the authors reflect an important assumption on which this book is based: through rigorous empirical research, theory building, self-reflection and cross disciplinary communication, we can gain greater understanding of the details of, and causes for, both general patterns and unique situations, both similarities and differences between farmers and scientific plant breeders, and thus of the possibilities for collaboration between them.

In the remainder of this introduction we discuss the separation between farmer and scientist plant breeding, a broad definition of the plant breeding system, and the current move to bring farmers and scientists more closely together in plant breeding.

## The Development of Plant Breeding and the Separation of Farmers and Scientists

Since the first domestications of wild plants about 12,000 years ago, farmer plant breeders have been responsible for the development of thousands of crop varieties in hundreds of species (Harlan, 1992). Plant breeding as a specialized activity began about 200 years ago in industrial countries (Simmonds, 1979). Modern professional plant breeding developed in the early part of the 20th century based on

Darwin's theory of evolution through selection and the genetic mechanisms of evolution, together with the basic mechanisms of inheritance and expression of the phenotype (via genotype × environment interaction, $G \times E$) discovered by Mendel in 1865 and rediscovered and elaborated by others in the first decades of the 20th century (Simmonds, 1979; Allard, 1999; Duvick, Chapter 8, this volume). For example, Johannsen demonstrated that quantitative traits followed the same principles of inheritance that Mendel demonstrated for qualitative traits; Nilsson-Ehle and East showed that many different genes could affect one character; Turesson found that different genotypes of a species are adapted to a specific range of environmental variables; and Fisher and associates demonstrated that the inheritance of quantitative characters could be analysed statistically (Hill *et al.*, 1998; Allard, 1999).

The crop varieties developed by plant breeders and farmers are often considered to be contrasting, although this is a simplification to which there are exceptions (Evans, 1993; Frankel *et al.*, 1995; Fischer, 1996). The emphasis of most scientific, professional plant breeders (hereafter simply 'plant breeders') has typically been on developing a relatively small number of genetically more uniform *modern varieties* (MVs), adapted to geographically wide, *optimal* (relatively low stress and uniform) growing environments, with high yield and yield stability in these environments. *Farmers' varieties* (FVs) are characterized by narrow geographical adaptation to *marginal* (relatively high stress and variable) growing environments, and high yield stability in those environments from year to year. We use the term FVs here to include landraces, traditional varieties selected by farmers, MVs adapted to farmers' environments by farmer and natural selection, and progeny from crosses between landraces and MVs (sometimes referred to as 'creolized' varieties or 'degenerate' MVs).

Plant breeding by scientists has become increasingly separated from plant breeding by farmers (Simmonds, 1979), as have seed supply systems (Cromwell *et al.*, 1993). Schneider documents the process of decreasing collaboration between wheat farmers and wheat breeders in Switzerland, and suggests political, institutional and technological reasons for this change (Schneider, Chapter 7, this volume). The distance between scientific and farmer plant breeding is especially great in the case of small-scale farmers planting in marginal growing environments with limited access to external inputs, as documented by many of the chapters in this book. In some industrial societies communication between conventional commercial farmers and plant breeders is still important, for example between large-scale maize farmers in the central United States and commercial maize breeders (Duvick, Chapter 8, this volume).

## The Plant Breeding System

The process of formal plant breeding begins with the initial decision about breeding goals and plans, and concludes with the release of a new variety and its subsequent dissemination to farmers (Weltzien *et al.*, 2000). In between are four basic steps: (i) creation of a large amount of genetic diversity through choosing parent germplasm, hybridization (crossing) and recombination in filial generations; (ii) selection of individual plants and populations initially in a limited range of *selection environments*; (iii) evaluation of the 'best' populations resulting from selection across a wider range of *test environments*; and (iv) the choice of varieties for release in the *target environments* on the basis of their potential to out-perform (out-yield) the existing varieties (Simmonds, 1979; Stoskopf *et al.*, 1993).

### Selection vs. choice

To understand plant breeding systems it is important to differentiate between *choice* of populations or varieties, which does not change the genetic make-up of these units, and the *selection* of plants from within populations or varieties, with the potential to change the genetic make-up of these units, and result in new varieties (Cleveland *et al.*, 2000). While this distinction is commonly made in the participatory plant breeding literature (e.g. Witcombe *et al.*, 1996), the terms 'choice' and 'selection' are often not explicitly defined, and may sometimes be used interchangeably.

The *choice* of germplasm (populations and varieties) determines the genetic diversity available within a crop, both as a basis for selection (by farmers and breeders), and for production (by farmers). Farmers and plant breeders make choices between varieties and populations, especially (for plant breeders) in the initial stages of the selection process when choosing germplasm for making crosses, and in the final stages when choosing among populations/varieties generated from those crosses for further testing (Hallauer and Miranda, 1988), planting (farmers) or release (plant breeders). Farmers' choices when saving seed for planting, in seed procurement and in allocating different varieties to different growing environments also affect the genetic diversity of their repertoires of crops and crop varieties, and determine the diversity on which future selection will be based.

Artificial *selection* of plants by farmers and breeders within segregating plant populations can change the genetic make-up of the population and lead to the development of new varieties. Artificial selection is both *indirect*, a result of the environments created in farmers' and plant breeders' fields and store rooms, and *direct*, a result of human selection of planting material. Direct artificial selection can be both *conscious* (based on explicit criteria), the result of decisions to select for certain traits, and *unconscious* (based on implicit criteria), when no conscious decision is made about the trait selected for, as when large seeds are automatically selected because they are easier to handle (Harlan, 1992). There is some confusion over terms in the literature; indirect artificial selection is sometimes defined as 'natural' selection (Simmonds, 1979: 14–15), as the same as conscious selection (Allard, 1999: 19, 26), or as entirely 'unconscious' selection (Poehlman and Sleper, 1995: 9).

## Broadening the definition of plant breeding

While the standard definition of plant breeding emphasizes its biological aspects, it is obvious that the human element is critical, and needs to be explicitly addressed when collaboration between scientists and farmers is a goal. Therefore, the plant breeding system (Fig. 1.1) can be more broadly defined to include not only

- the *biophysical* components of crop populations and growing environments; and

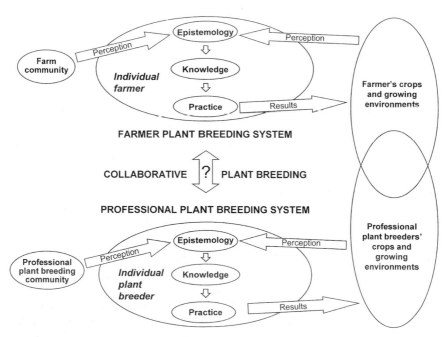

**Fig. 1.1.** The plant breeding system.

- the *practice* of choice and selection between and within crop populations;

but also

- the *social* and *institutional* components (communities of farmers and plant breeders, including social structure, economic and power relations, behaviours and oral/written expressions of knowledge);
- the *knowledge* of individual farmers and plant breeders about their crop populations and growing environments (both conscious and unconscious, including intuition, values, empirical data, theory); and
- *epistemology*, the way knowledge is acquired through the processing of physical stimuli from the outside (affected by sensory perception, brain structure and function, language, technology, practice and pre-existing knowledge).

The chapters in this book consider the plant breeding system in this broad perspective, though each chapter focuses on a limited portion of it.

## The biological basis of plant breeding

The elementary biological model on which plant breeding is based, and as presented in the standard textbooks, is universally accepted among plant breeders (e.g. Simmonds, 1979; Falconer and Mackay, 1996). First, variation in population phenotype ($V_P$) on which choice and selection are based is determined by genetic variation ($V_G$), environmental variation ($V_E$), and variation in genotype × environment (G × E) interaction ($V_{G \times E}$), thus $V_P = V_G + V_E + V_{G \times E}$. Broad sense heritability (H) is the proportion of $V_P$ due to genetic variance ($V_G/V_P$), while narrow sense heritability ($h^2$) is the proportion of $V_P$ due to additive genetic variance ($V_A/V_P$), that is, the proportion of $V_G$ directly transmissible from parents to progeny, and therefore of primary interest to breeders.

Second, response to selection (R) is the difference for the traits measured between the mean of the whole population from which the parents were selected and the mean of the next generation that is produced by planting those selected seeds under the same conditions. R is the product of two different factors, $h^2$ and S ($R = h^2S$), where S is the selection differential, the difference between the mean of the selected group and the mean of whole original population selected from. Expression of S in standard deviation units (the standardized selection differential; Falconer and Mackay, 1996) permits comparison of selections among populations with different amounts or types of variation. The results of selecting for a given trait improve as the proportion of $V_P$ contributed by $V_G$ (especially $V_A$) increases.

The biological relationships described in these simple equations underlie plant breeders' understanding of even the most complex phenomena that they encounter (Cooper and Hammer, 1996; DeLacy *et al.*, 1996). For example, two highly respected English-language plant breeding texts state that the relationship between genotype and phenotype is 'perhaps the most basic concept of genetics and plant breeding' (Allard, 1999: 48), and of $R = h^2S$, that 'If there were such a thing as a fundamental equation in plant breeding this would be it' (Simmonds, 1979: 100).

## The social and individual bases of plant breeding

Much of the current discussion about the nature of knowledge is polarized between objectivist and constructivist camps (e.g. Hull, 1988; Harding, 1998). The assumption at the *constructivist* end of the spectrum is that knowledge is dominated by social forces, including power relationships, and is historically and culturally particular; that is, the process that mediates the acquisition of knowledge (epistemology) is dominated by pre-existing knowledge, including values, acquired through participation in a particular institutional or social setting, often mediated by the social control of technology and information. The assumption at the *objectivist* end of the spectrum is that more and more universal and accurate knowledge of biophysical reality is a valid goal; that is, epistemology is dominated by scientific methods capable of discriminating and eliminating social influences and of ascertaining the true nature of the world outside the individual mind.

In an objectivist approach to plant breeding, science is often seen as increasing the amount and accuracy of objective knowledge about plants and their environments solely through testing of theory-based hypotheses, and applying this knowledge to produce new, more desirable crop varieties. Plant breeders consider themselves to be 'applied evolutionists' (Simmonds, 1979: 27; Allard, 1999: 49) and textbooks document the progressive, science-based development of the profession, which increasingly differentiates their SK from the IK of farmers.

In a constructivist approach, the development, application and results of plant breeding science, including the kinds of crop varieties developed, are often seen to be primarily the result of macro political or economic variables, foremost among them industrial modernism. This is the approach of most social scientists who research or discuss plant breeding. From this viewpoint, the SK of plant breeding unobjectively adopts the values of modernism, yet is imperialistic in its claims to universality, and focused on transforming the climate and environment to fit a predetermined 'ideal plant type', in contrast to IK, which is seen as much more complete and sophisticated in terms of objective reality (e.g. Scott, 1998).

From the viewpoint of the more extreme positions that dominate the 'science' wars (Gould, 2000), IK and SK sometimes seem to be mutually exclusive, providing no rationale or capacity for collaboration between them. However, the real challenge lies not in promoting ideologically based conclusions about IK and SK, but in understanding the complexities that determine knowledge and practice in general, and in a particular situation, in order to support change in a socially desirable direction. (Of course, determining what is socially desirable is itself part of the problem.) It demands, to the extent possible, separating conclusions based on values, for example that local communities should have control over their FVs, from conclusions that can be tested by empirical research, for example that local communities are conserving crop genetic diversity.

As an alternative to objectivist and constructivist views, a 'holistic' model of knowledge assumes that both farmer and scientist knowledge of plant breeding are the result of objective observations of reality and social construction, and both may be composed of empirical data, theory and values (Fig. 1.1; see also Soleri *et al.*, Chapter 2, this volume). This approach is being discussed more and more as an alternative to dichotomous, essentializing definitions of indigenous and scientific knowledge in the social sciences (Bernard, 1998; Schweizer, 1998) and in social studies of science (e.g. Hull, 1988; Harding, 1998; Gould, 2000). Plant breeders may also recognize that their theoretical understanding of plants is limited by the lack of required experimental data, and of the technologies and resources necessary to gather them (Simmonds, 1979; Anderson and Hazell, 1989; Duvick, Chapter 8, this volume), although they do not often consider the extent to which their knowledge may be socially constructed.

## The Move to Reunite Farmer and Scientific Plant Breeding

Participatory plant breeding (PPB) proposes to reverse the historical trend of separation between farmers and plant breeders, bringing them together in the process of developing new crop varieties or improving existing ones. In some ways it is a relatively new approach to crop improvement. While PPB is only a very small part of plant breeding as a whole, it has become a popular component of international agricultural development during the last several years, and the main focus of several global-level international development initiatives (Eyzaguirre and Iwanaga, 1996; Witcombe *et al.*, 1996; McGuire *et al.*, 1999), including the Community Biodiversity Development and Conservation programme (CLADES *et al.*, 1994; http://www.cbdcprogram.org/frame. htm) and the PPB component of the CGIAR's System Wide Programme on Participatory Research and Gender Analysis (SWP PRGA) (CGIAR, 1997; http://www.prgaprogram.org/).

Interest in PPB comes from a convergence of the movement towards sustainable agriculture in professional plant breeding, genetic resources conservation and traditionally based agriculture (Cleveland *et al.*, 1994), and towards participatory research and development (Friis-Hansen and Sthapit, 2000). Frossard (Chapter 6, this volume) describes one of the most prominent examples of farmer-initiated PPB in his chapter on MASIPAG in the Philippines. Several chapters in this book are by plant breeders who describe the motivation for beginning to work with farmers in Syria (Ceccarelli and Grando, Chapter 12), Nepal (Joshi *et al.*, Chapter 10), Cuba (Ríos Labrada *et al.*, Chapter 9) and Zimbabwe (Bänziger and de Meyer, Chapter 11).

## Increasing environmental sustainability

Agriculture and plant breeding, like most human activities, are facing unprecedented challenges at both local and global levels. It is widely agreed that human impact on the Earth's ecosystems threatens the current patterns of biological and sociocultural diversity, and this has focused attention on achieving more sustainable human–environment interaction (Vitousek *et al.*, 1997), including agriculture (Matson *et al.*, 1997). At the same time, the demand for food is increasing, while past approaches to increasing food production are often considered to be inadequate (Evans, 1997; Mann, 1999).

Modern, professional plant breeding (in concert with agronomy) has been extremely successful in meeting increasing demands from a growing human population (Evans, 1993, 1998). However, the benefits of modern plant breeding have not reached many of the limited-resource farming communities that characterize much of the developing world, as documented in many of the chapters of this volume (see McGuire, Chapter 5; Frossard, Chapter 6; Zimmerer, Chapter 4). For example, only about 40% of low-input maize production in the developing world is planted to MVs (Heisey and Edmeades, 1999). The reasons for the failure of modern plant breeding to benefit many farmers include the conventional belief that improving productivity of higher input systems is a more effective way to increase food production and people's well-being than is attention to farmers in marginal environments (Heisey and Edmeades, 1999), or perhaps a failure to understand marginal environments and the farmers who make a living there (Ceccarelli *et al.*, 1994; Bänziger and de Meyer, Chapter 11; Ceccarelli and Grando, Chapter 12; Joshi *et al.*, Chapter 10, this volume). Modern, industrial agriculture also faces the challenge of developing varieties that are adapted to growing environments with fewer external inputs, including artificial fertilizers and pesticides, and irrigation (Duvick, 1992). Cuba's response to forced and dramatic reductions of agricultural inputs in 1989 (Ríos Labrada *et al.*, Chapter 9, this volume) is seen by many as an example of what other industrial agricultural systems will face in the future.

At the same time, the success of modern agriculture has threatened the genetic base on which both modern and traditional agriculture depend, the replacement of many FVs with fewer MVs, and the movement of many farmers in marginal environments out of farming. The FVs grown by farmers contain rich but largely unknown genetic resources that will be essential for developing more sustainable crop varieties of both MVs and FVs (Qualset *et al.*, 1997; Brown, 1999) and, just as importantly, may be critical for even modest FV success in many extant traditionally based systems (Soleri and Smith, 1995).

For all of these reasons, there has been increasing awareness among plant breeders of the need to:

• increase yields and yield stability in marginal environments, both (i) those that have been high-yielding, but where inputs are being reduced to reduce production costs and negative environmental impacts; and (ii) those of many of the world's farmers who have not adopted MVs, but whose FVs have inadequate yields;
• conserve the base of genetic diversity on which all plant breeding depends, and which is threatened by the loss of FVs as the area planted to FVs and the number of farmers growing them declines (Fischer, 1996; Qualset *et al.*, 1997; Heisey and Edmeades, 1999).

From the perspective of an increasing number of scientists, plant breeding with farmers is a way to both increase yields and other desirable production components in marginal environments, while at the same time supporting *in situ* conservation of crop genetic diversity (Witcombe *et al.*, 1996; Qualset *et al.*, 1997; Brown, 1999; Weltzien *et al.*, 2000; Ceccarelli and Grando, Chapter 12; Joshi *et al.*, Chapter 10; Ríos Labrada *et al.*, Chapter 9; Smale, Chapter 3, this volume).

## Increasing social and economic sustainability

As with other areas of development, a major incentive for scientists to work with farmers has been the value of IK for increasing environmental sustainability. However, recognition of the claims by indigenous peoples of rights to natural resources, to manage their own development, and to their IK, implies the need to increase the social and economic sustainability of agriculture as well (Cleveland and Murray, 1997). Several chapters in this book demonstrate the importance for plant breeding of understanding the knowledge and practice of farmers, and the social and political systems within which they are embedded (Chapters 2, 4, 5, 6, 7, 9, this volume).

An important method for achieving this has been 'participatory' research and development although, deciding what 'participation' means in terms of valuing IK, of recognition of rights in IK, and who is to be in control of development has been contentious. Since PPB is still relatively new, there is a wide range of understandings of what it entails, and a wide range of activities in PPB projects (Friis-Hansen and Sthapit, 2000).

Much of PPB to date has emphasized the participation of farmers in plant breeders' work. An important reason for this is that most of this work has been initiated by 'foresighted individuals working at otherwise conventional research stations' with objectives therefore focused on developing new products rather than on the process of farmer plant breeding (Friis-Hansen and Sthapit, 2000: 19). Another major reason for not using farmers' plant breeding experience and theory more extensively in PPB may be that very little is known about them by outsiders, either in farmers' own terms, or in terms of the theory of scientific plant breeding (Brown, 1999; Cleveland et al., 2000). Other ways of characterizing participation in plant breeding include a widely used quantitative taxonomy based on the amount of effort borne by farmers (Biggs, 1989; see Joshi et al., Chapter 10; and Soleri et al., Chapter 2, this volume, for more discussion).

Comparing the economic sustainability of PPB with more conventional approaches can be complex and requires evaluation of the 'participatory' aspect as well as, but separate from, other substantial deviations from the conventional model such as decentralization (Ceccarelli et al., 2000). Ríos Labrada et al. (Chapter 9, this volume) include a basic economic comparison of some aspects of two plant breeding methods in Cuba. Benefit/cost analyses of PPB and other types of plant breeding may yield very different results, and will be an important contribution to understanding the basis of PPB, but this area of research is just beginning to be explored (Simmonds, 1990; Heisey et al., 1997). Smale's chapter in this volume (Chapter 3) is the first attempt by an economist, based on her very extensive empirical and theoretical work with both farmers and plant breeders, to lay out a framework for the economic research in this area.

We have suggested *collaborative plant breeding* (CPB) as an alternative to PPB to emphasize two points that we believe to be critical for the intercultural (farmers and scientists) and interdisciplinary (social and biophysical sciences) nature of collaboration between farmers and plant breeders (Cleveland and Soleri, 1997).

**1.** The knowledge and practice of both farmers and breeders are important, neither should be assumed to be 'better' a priori; their relative merits in terms of contribution to CPB need to be empirically assessed in each situation. At a more fundamental level, successful collaboration requires mutual respect that is based on an understanding of differences and similarities.
**2.** Positive biological and social results in CPB are not necessarily correlated with the amount of physical effort invested by farmers. For example, introgression of alleles conferring disease tolerance into FVs may require very little if any physical work on the part of farmers, yet have major benefits for them.

The term PPB, however, is still used by most, and some may prefer it to CPB because 'collaboration' suggests to them an emphasis on social as opposed to biological goals. There are also other terms that overlap to a greater or lesser extent with CPB, such as 'decentralized breeding', 'farmer crop improvement', 'joint breedership' and these are used for the most part interchangeably (Weltzien *et al.*, 2000: 7), including in most chapters of this book. However, some authors and practitioners imply specific meanings with their use of terms (see Joshi *et al.*, Chapter 10 and Soleri *et al.*, Chapter 2, this volume, for examples). Clearly the actual terminology is in many cases irrelevant, but the assumptions that have become associated with terms do require examination. We believe that the use and discussion of different terms is an important part of the process of clarifying what 'collaboration' or 'participation' means, and what they imply in terms of goals.


## Conclusion

Successfully meeting the challenges for environmental, social and economic sustainability of food production that we face in a globalized 21st century will undoubtedly require new strategies. Based on past experience it seems likely that an important component of these will be new understandings of diverse perspectives and identification and pursuit of shared goals. Increased collaboration between scientific plant breeders and farmers could be vital for achieving those understandings and realizing those goals. However, this may require rethinking past approaches, and more work on the theoretical basis, practical implications and potential contributions of collaboration. An essential ingredient, we believe, will be greater understanding of the knowledge and practice of both farmers and plant breeders, and of the differences and similarities between them. This book is a contribution to that end.

## Acknowledgements

We thank the farmers and scientists in Syria, Cuba, Nepal and especially in Oaxaca, Mexico, whose knowledge and practices helped us to start to recognize and focus on these issues, and continues to do so; Eva Weltzien R. and Catherine Longley for comments on a draft of this chapter; and the National Science Foundation (SES-9977996) for financial support for our research with plant breeders and farmers in Syria, Cuba and Nepal that serves as the background for this chapter. Any inaccuracies or misunderstandings are the sole responsibility of the authors.

## References

Allard, R.W. (1999) *Principles of Plant Breeding*, 2nd edn. John Wiley & Sons, New York.

Anderson, J.R. and Hazell, P.B.R. (eds) (1989) *Variability in Grain Yields: Implications for Agricultural Research and Policy in Developing Countries.* Johns Hopkins University Press, Baltimore, Maryland.

Bernard, H.R. (1998) Introduction: on method and methods in anthropology. In: Bernard, H.R. (ed.) *Handbook of Methods in Cultural Anthropology.* Altamira Press, Walnut Creek, California, pp. 9–36.

Brown, A.H.D. (1999) The genetic structure of crop landraces and the challenge to conserve them *in situ* on farms. In: Brush, S.B. (ed.) *Genes in the Field: On-Farm Conservation of Crop Diversity.* Lewis Publishers, Boca Raton, Florida; IPGRI, Rome; IDRC, Ottawa, pp. 29–48.

Ceccarelli, S., Erskine, W., Hamblin, J. and Grando, S. (1994) Genotype by environment interaction and international breeding programmes. *Experimental Agriculture* 30, 177–187.

Ceccarelli, S., Grando, S., Tutwiler, R., Bahar, J., Martini, A.M., Salahieh, H., Goodchild, A. and Michael, M. (2000) A methodological study on participatory barley breeding. I. Selection phase. *Euphytica* 111, 91–104.

CGIAR (1997) *New Frontiers in Participatory Research and Gender Analysis: Proceedings of the International Seminar on Participatory Research and Gender Analysis for Technology Development.* CGIAR SWP, Cali, Colombia.

CLADES, COMMUTECH, CPRO-DLO, GRAIN, NORAGRIC, PGRC/E, RAFI and
    SEARICE (1994) *Community Biodiversity Development and Conservation
    Programme.* Centre for Genetic Resources, Wageningen, The Netherlands;
    and Centro de Education y Tecnologia, Santiago, Chile (Proposal to DGIS,
    IDRC and SIDA for Implementation Phase I – 1994–1997).
Cleveland, D.A. (1998) Balancing on a planet: toward an agricultural anthro-
    pology for the 21st century. *Human Ecology* 26, 323–340.
Cleveland, D.A. and Murray, S.C. (1997) The world's crop genetic resources
    and the rights of indigenous farmers. *Current Anthropology* 38, 477–515.
Cleveland, D.A. and Soleri, D. (1997) Posting to the Farmer-breeding list serve
    <farmer-breeding-l-postmaster@cgnet.com> sponsored by the CGIAR's
    Systemwide Program on Participatory Research and Gender Analysis,
    http://www.prgaprogram.org/prga/, 27 September 1997.
Cleveland, D.A., Soleri, D. and Smith, S.E. (1994) Do folk crop varieties have a
    role in sustainable agriculture? *BioScience* 44, 740–751.
Cleveland, D.A., Bowannie, F.J., Eriacho, D., Laahty, A. and Perramond, E.P.
    (1995) Zuni farming and United States government policy: the politics of
    cultural and biological diversity. *Agriculture and Human Values* 12, 2–18.
Cleveland, D.A., Soleri, D. and Smith, S.E. (2000) A biological framework for
    understanding farmers' plant breeding. *Economic Botany* 54, 377–394.
Cooper, M. and Hammer, G.L. (1996) Synthesis of strategies for crop improve-
    ment. In: Cooper, M. and Hammer, G.L. (eds) *Plant Adaptation and Crop
    Improvement.* CAB International in association with IRRI and ICRISAT,
    Wallingford, UK, pp. 591–623.
Cromwell, E., Wiggins, S. and Wentzel, S. (1993) *Sowing Beyond the State.*
    Overseas Development Institute, London.
DeLacy, I.H., Basford, K.E., Cooper, M. and Fox, P.N. (1996) Retrospective
    analysis of historical data sets from multi-environment trials – theoretical
    development. In: Cooper, M. and Hammer, G.L. (eds) *Plant Adaptation
    and Crop Improvement.* CAB International in association with IRRI and
    ICRISAT, Wallingford, UK, pp. 243–267.
*Diversity* (1998) Special issue on COP-4, CBD. *Diversity* 14(1&2), 6–32 (N).
Dove, M.R. (2000) The life-cycle of indigenous knowledge, and the case of
    natural rubber production. In: Ellen, R., Parkes, P. and Bicker, A. (eds)
    *Indigenous Environmental Knowledge and Its Transformations: Critical
    Anthropological Perspectives.* Harwood Academic Publishers, Amster-
    dam, pp. 213–251.
Duvick, D.N. (1992) Genetic contributions to advances in yield of U.S. maize.
    *Maydica* 37, 69–79.
Ellen, R. and Harris, H. (2000) Introduction. In: Ellen, R., Parkes, P. and Bicker,
    A. (eds) *Indigenous Environmental Knowledge and Its Transformations:
    Critical Anthropological Perspectives.* Harwood Academic Publishers,
    Amsterdam, pp. 1–33.
Evans, L.T. (1993) *Crop Evolution, Adaptation and Yield.* Cambridge
    University Press, Cambridge.
Evans, L.T. (1997) Adapting and improving crops: the endless task. *Philosophi-
    cal Transactions of the Royal Society of London: Biological Sciences* 352,
    901–906.

Evans, L.T. (1998) *Feeding the Ten Billion: Plants and Population Growth.* Cambridge University Press, Cambridge.

Eyzaguirre, P. and Iwanaga, M. (eds) (1996) *Participatory Plant Breeding. Proceedings of a Workshop on Participatory Plant Breeding, 26–29 July 1995, Wageningen, The Netherlands.* International Plant Genetic Resources Institute, Rome.

Falconer, D.S. and Mackay, T.F. (1996) *Introduction to Quantitative Genetics.* Prentice Hall/Pearson Education, Edinburgh.

Fischer, K.S. (1996) Research approaches for variable rainfed systems – thinking globally, acting locally. In: Cooper, M. and Hammer, G.L. (eds) *Plant Adaptation and Crop Improvement.* CAB International in association with IRRI and ICRISAT, Wallingford, UK, pp. 25–35.

Fowler, C. (1994) *Unnatural Selection: Technology, Politics, and Plant Evolution.* Gordon and Breach, Yverdon, Switzerland.

Frankel, O.H., Brown, A.H.D. and Burdon, J.J. (1995) *The Conservation of Plant Biodiversity.* Cambridge University Press, Cambridge.

Friis-Hansen, E. and Sthapit, B. (2000) Concepts and rationale of participatory approaches to conservation and use of plant genetic resources. In: Friis-Hansen, E. and Sthapit, B. (eds) *Participatory Approaches to the Conservation and Use of Plant Genetic Resources.* International Plant Genetic Resources Institute, Rome, pp. 16–19.

Gould, S.J. (2000) Deconstructing the 'science wars' by reconstructing an old mold. *Science* 287, 253–261.

Hallauer, A.R. and Miranda, J.B. (1988) *Quantitative Genetics in Maize Breeding,* 2nd edn. Iowa State University, Ames, Iowa.

Harding, S. (1998) *Is Science Multicultural? Postcolonialisms, Feminisms, and Epistemologies.* Indiana University Press, Bloomington, Indiana.

Harlan, J.R. (1992) *Crops and Man,* 2nd edn. American Society of Agronomy, and Crop Science Society of America, Madison, Wisconsin.

Heisey, P.W. and Edmeades, G.O. (1999) Part 1. Maize production in drought-stressed environments: technical options and research resource allocation. In: CIMMYT (ed.) *World Maize Facts and Trends 1997/98.* CIMMYT, Mexico, DF, pp. 1–36.

Heisey, P.W., Smale, M., Byerlee, D. and Souza, E. (1997) Wheat rusts and the costs of genetic diversity in the Punjab of Pakistan. *American Journal of Agricultural Economics* 79, 726–737.

Hill, J., Becker, H.C. and Tigerstedt, P.M.A. (1998) *Quantitative and Ecological Aspects of Plant Breeding.* Chapman and Hall, London.

Hull, D.L. (1988) *Science as a Process: an Evolutionary Account of the Social and Conceptual Development of Science.* The University of Chicago Press, Chicago.

Mann, C. (1999) Crop scientists seek a new revolution. *Science* 283, 310–314.

Matson, P.A., Parton, W.J., Power, A.G. and Swift, M.J. (1997) Agricultural intensification and ecosystem properties. *Science* 277, 504–509.

McGuire, S., Manicad, G. and Sperling, L. (1999) *Technical and Institutional Issues in Participatory Plant Breeding – Done from the Perspective of Farmer Plant Breeding: a Global Analysis of Issues and of Current Experience.* CGIAR Systemwide Program on Participatory Research and Gender

Analysis for Technology Development and Institutional Innovation. Working Document No. 2, March 1999, Cali, Colombia.

Millar, D., Haverkort, B., van 't Hooft, K. and Hiemstra, W. (2001) Challenging developments: approaches, results and perspectives for endogenous development. *Compass Magazine* 4, 4–7.

Nader, L. (1996) Anthropological inquiry into boundaries, power, and knowledge. In: Nader, L. (ed.) *Naked Science: Anthropological Inquiry into Boundaries, Power and Knowledge*. Routledge, New York, pp. 1–25.

Ostrom, E., Burger, J., Field, C.B., Norgaard, R.B. and Policansky, D. (1999) Revisiting the commons: local lessons, global challenges. *Science* 284, 278–282.

Poehlman, J.M. and Sleper, D.A. (1995) *Breeding Field Crops*, 4th edn. Iowa State University Press, Ames, Iowa.

Qualset, C.O., Damania, A.B., Zanatta, A.C.A. and Brush, S.B. (1997) Locally based crop plant conservation. In: Maxted, N., Ford-Lloyd, B. and Hawkes, J.G. (eds) *Plant Genetic Conservation: the In Situ Approach*. Chapman & Hall, Hants, UK, pp. 160–175.

Schweizer, T. (1998) Epistemology: the nature and validation of anthropological knowledge. In: Bernard, H.R. (ed.) *Handbook of Methods in Cultural Anthropology*. Altamira Press, Walnut Creek, California, pp. 39–87.

Scott, J.C. (1998) *Seeing Like a State: How Certain Schemes to Improve the Human Condition Have Failed*. Yale University Press, New Haven, Connecticut.

Sillitoe, P. (1998) The development of indigenous knowledge: a new applied anthropology. *Current Anthropology* 39, 223–252.

Simmonds, N.W. (1979) *Principles of Crop Improvement*. Longman Group, London.

Simmonds, N.W. (1990) The social context of plant breeding. *Plant Breeding Abstracts* 60, 337–341.

Soleri, D. and Smith, S.E. (1995) Morphological and phenological comparisons of two Hopi maize varieties conserved in situ and ex situ. *Economic Botany* 49, 56–77.

Stoskopf, N.C., Tomes, D.T. and Christie, B.R. (1993) *Plant Breeding Theory and Practice*. Westview Press, Boulder, Colorado.

Vitousek, P.M., Mooney, H.A., Lubchenco, J. and Melillo, J.M. (1997) Human domination of Earth's ecosystems. *Science* 277, 494–499.

Weltzien, E., Smith, M.E., Meitzner, L.S. and Sperling, L. (2000) *Technical and Institutional Issues in Participatory Plant Breeding – From the Perspective of Formal Plant Breeding. A Global Analysis of Issues, Results, and Current Experience*. Working Document No. 3. CGIAR Systemwide Program on Participatory Research and Gender Analysis for Technology Development and Institutional Innovation, Cali, Colombia.

Witcombe, J.R., Joshi, A., Joshi, K.D. and Sthapit, B.R. (1996) Farmer participatory crop improvement. I. Varietal selection and breeding methods and their impact on biodiversity. *Experimental Agriculture* 32, 445–460.

# Understanding Farmers' Knowledge as the Basis for Collaboration with Plant Breeders: Methodological Development and Examples from Ongoing Research in Mexico, Syria, Cuba and Nepal

**2**

DANIELA SOLERI,[1] DAVID A. CLEVELAND,[2]
STEVEN E. SMITH,[3] SALVATORE CECCARELLI,[4]
STEFANIA GRANDO,[4] RAM B. RANA,[5] DEEPAK RIJAL[5]
AND HUMBERTO RÍOS LABRADA[6]

*[1]Centre for People, Food and Environment, Santa Barbara, California, and Environmental Studies Program, University of California, Santa Barbara, CA 93106-4160, USA; [2]Department of Anthropology and Environmental Studies Program, University of California, Santa Barbara, CA 93106-4160, USA; [3]School of Natural Resources, 301 Biosciences East, University of Arizona, Tucson, AZ 85721, USA; [4]Germplasm Program, International Centre for Agricultural Research in Dry Areas (ICARDA), PO Box 5466, Aleppo, Syria; [5]Local Initiatives for Biodiversity Research and Development (LI-BIRD), PO Box 324, Pokhara, Nepal; [6]Instituto Nacional de Ciencias Agricolas (INCA), GP No. 1, San Jose de Las Lajas, La Habana, CP 32700, Cuba*

## Abstract

There has been very little comparative research on farmers' and scientists' theoretical or conceptual knowledge, sometimes leading to reliance on untested assumptions in plant breeding projects that attempt to work with farmers. We propose an alternative approach that is inductive, based on a very basic biological model of plant–environment relationships, and on a holistic model of knowledge. The method we use was developed in Oaxaca, Mexico, and is based on scenarios involving genotype × environment interactions, heritability, and genetic response to selection. It is being modified and applied in a research project with collaborating scientists and farmers in Syria (barley),

Cuba (maize) and Nepal (rice). We are testing the ideas that: (i) farmers' knowledge is complex, and includes conceptual knowledge of genotypes and environments; (ii) farmers' knowledge is both similar to and different from scientists' knowledge; and (iii) a generalizable methodological approach permitting inclusion of farmers' conceptual knowledge in research design and execution can form the basis for enhanced farmer–scientist collaboration for crop conservation and improvement. Results to date suggest that farmers have conceptual knowledge of their genotypes and environments that is congruent with the basic biological model also used by scientists, but that their knowledge is also influenced by the specific, local characteristics of their genotypes and environments, and by their social contexts. Some examples of the practical utility of these research results are given.

## Introduction

There is increasing evidence from social studies of formal scientific knowledge (Hull, 1988) and local or indigenous knowledge (Medin and Atran, 1999) that both may consist of a complex combination of intuition, empiricism and theory, and of verifiable objective observation and social construction, and may, therefore, be similar as well as different (Agrawal, 1995).

For the most part, these findings have not had an impact on social or natural science research in agriculture and agricultural development, including plant breeding. Here it is commonly assumed that scientific knowledge (SK) is theoretical, objectively verifiable and universally generalizable, in contrast to local or indigenous farmer knowledge (FK), which is assumed to be intuitional or empirical and embedded in local social and biophysical contexts. The dominance of these stereotypes means that SK and FK are often considered fundamentally different and incomparable, or FK is considered an inferior version of SK. As a result, most agricultural development projects involving FK have not considered the possibility for collaboration of farmers and scientists based on similarities in their theoretical knowledge. Is an important potential being overlooked? Is it possible that FK about plant genotypes and growing environments is in part conceptual (theoretical)? Are there similarities between FK and SK, and if so could this be an important tool for facilitating collaboration between farmers and scientists for improved production in farmers' fields?

These are the general questions we address in this chapter, based on our work in Mexico beginning in 1996, and in ongoing work begun in 2000 with barley farmers in Syria, maize farmers in Cuba and rice farmers in Nepal. We have been investigating whether farmers have theoretical concepts about their crop genotypes and environments, and to what extent SK and FK are similar because they are based on observations of universal biological relationships between plant genotypes and the environments in which they grow, and different because of differences in the many unique biophysical and social situations in which this knowledge is created. Results suggest that the increased understanding of FK that results from such research can serve to support collaboration between farmers and plant breeders.

We begin by reviewing some of the critical theoretical and practical issues regarding the role of FK in crop improvement. Next is a discussion of the theory and methods we have been using to understand FK, including a holistic model of knowledge, a biological model of plant–environment relations and scenario-based interviews. We then report some of our research findings on FK related to the three main components of the biological model (genotype × environment interaction, heritability and genetic response to plant selection). We conclude with suggestions about the potential importance of our research for collaborative plant breeding (CPB).

## Scientist Knowledge and Plant Breeding

Plant breeding, including both choosing crop varieties and populations and selecting plants, is based on an understanding of plants, environments and the relationship between them (see Cleveland and Soleri, Chapter 1, this volume). There are still many complexities of plant genotype × environment interactions that are not well understood in terms of biological theory (Duvick, Chapter 8, this volume), and about which there continue to be disagreements among plant breeders; for example the effect of selection environment on the range of target environments to which a genotype is adapted (Cleveland, 2001; Soleri and Cleveland, 2001; Bänziger and De Meyer, Chapter 11, this volume; Ceccarelli and Grando, Chapter 12, this volume). However, the fundamentals are well-established and universally accepted by plant breeders: plant phenotype is the result of both genotype and environment, the degree to which a trait is heritable depends on the degree to which it is affected by the genotype vs. the environment, and genetic change in a population due to selection is dependent on the proportion of plants selected and the heritability of the trait selected for (see the section 'A biological model' below, and Cleveland and Soleri, Chapter 1, this volume).

In actual situations, understanding these basic relationships is difficult because a great number of variables affect them, and predicting the results of choice and selection is hampered by the lack of required experimental data, and of the technologies and resources necessary to gather and analyse them. Plant breeders also recognize that their theoretical understanding of plants is limited, and that much plant breeding has been based on intuition and empiricism rather than theory (Simmonds, 1979; Duvick, 1996; Duvick, Chapter 8, this volume; Wallace and Yan, 1998: 320), although intuition and empiricism are likely to be underlain by the basic theoretical understanding of genotype–environment relations (see the section 'A holistic model of knowledge' below).

This *fundamental biological theory* (see section below 'A biological model') on which plant breeding is based is the same no matter where plant breeding is practised. However, the biophysical, economic and sociocultural variables can be quite different; for example between poor farmers' fields in marginal environments and plant breeders' research stations, or between national agricultural policy priorities of large-scale efficiencies and increased inputs and production, and farmers' priorities of reducing risk and optimizing crop production as part of a general household survival strategy. Work under a specific set of circumstances may lead to *interpretation of theory* that is then generalized and broadly applied (for example, when the fundamental theory that as $V_E$ decreases, $h^2$ increases is understood to imply that indirect selection under low $V_E$ is always more efficient than direct selection under high $V_E$), without investigating the implications of those interpretations under all circumstances. Working with farmers often demands re-examining some of these interpretations of theory that form the basis of conventional plant breeding by testing the assumptions (biological, environmental, economic, sociocultural) on which they are based. The results of these tests will have implications for both the interpretation of theory as well as the *methods* and *practices* used.

Thus, in discussions of the role of formal, scientific plant breeding in CPB, it may often be helpful to make a clearer distinction between: (i) fundamental biological theory; (ii) interpretations of fundamental theory; and (iii) methods and practice; (iii) may be very different depending on whether it is based on (i) or (ii), or on different versions of (ii). Many of the disagreements about plant breeding methods for CPB may grow out of disagreements about differences in the interpretation of fundamental biological theory, and disagreements about these interpretations may in turn be based on the belief of proponents that their interpretations of fundamental theory are not based on their unique experiences and assumptions, but rather are part of fundamental theory.

Therefore, especially for those aspects of the biophysical reality of genotypes and environments that are less well understood in terms of plant breeding theory, plant breeders' knowledge may more likely be based on the particular experiences that each one has with the particular environments and crop varieties they work with, and thus may be less generalizable and more apt to be influenced by pre-existing knowledge (including values) specific to the plant breeder's social environment. This means that disagreements among plant breeders could arise even though fundamental genetic and statistical principles remain constant across a range of contexts, because the 'art' of plant breeding is more tied to specific individuals and/or environments (Soleri and Cleveland, 2001).

## Farmer Knowledge and Plant Breeding

There has been growing interest in the potential of FK to make a contribution to agricultural development, both to increase the effectiveness of scientist and farmer research and practice, and to empower farmers. However, very little is known by outsiders about FK of plant breeding, either in farmers' own terms, or in terms of scientific plant breeding (Brown, 1999; McGuire *et al.*, 1999; Cleveland *et al.*, 2000; Weltzien *et al.*, 2000). In the urgency to redress the shortcomings of much formal research by including farmers, most work has been initiated by 'foresighted individuals working at otherwise conventional research stations' and thus having the objectives and professional pressures of such institutions (Friis-Hansen and Sthapit, 2000: 19). These individuals have typically been working in institutions whose interest and expertise is in developing research products, not in experimenting with theory and method for improving participation or in understanding FK. On the other hand, farmers have rarely had the ability, because of their low social status and lack of political power, to take the initiative in working with scientists (see Frossard, Chapter 6, this volume for an important counter example; Schneider, Chapter 7, this volume, for an historical example of how farmer participation was eliminated). Therefore, the possibility that scientific research and development might be improved by learning more about FK related to plant breeding has not typically been considered and, therefore, there was little motivation to learn more about FK.

In addition to these important social and institutional factors, our working definitions of 'knowledge' and of 'participation' based on unexamined assumptions have also contributed to the lack of research on the theoretical content of FK and on comparing FK and SK (with the resulting lack of information in turn reinforcing the effect of these variables).

## Defining farmer knowledge

The lack of empirical information and theoretical analysis has contributed to our using simple, stereotyped definitions of FK (and often of SK as well), and the frequent failure to test the many assumptions underlying these definitions (Scoones and Thompson, 1993; Sillitoe, 1998). We can very roughly divide current views of FK into two categories: those that see FK as fundamentally different from SK, and those that see FK as empirically similar to SK (see Ellen and Harris (2000) and Sillitoe (1998) for reviews).

In the first group, definitions of FK emphasize that it is primarily intuition and skill, socially constructed, and based on the local social and environmental contexts and cultural values. FK and SK are seen as fundamentally different, and attempts to explain FK in scientific terms impede true appreciation of that knowledge (e.g. Selener, 1997). This implies that the role of outsiders should be to empower local people and validate FK in its own terms.

The second major category of definitions of FK emphasizes that it consists primarily of rational empirical knowledge of the environment. Definitions of FK as *economically rational* tend to assume that scientists are more rational. The role of outsiders should be to facilitate the replacement or modernization of small-scale farmers, including replacement of their crop varieties (farmer varieties, FVs) with modern varieties (MVs) (Srivastava and Jaffee, 1993). Definitions of FK as *ecologically rational* emphasize farmers' accurate and, therefore, sustainable ecological knowledge of their environments, supported by many empirical data, especially in ethnotaxonomic studies of plants and animals, while recognizing variation in distribution of cultural knowledge as the result of factors including age, gender, social status and affiliation, kinship, personal experience and intelligence (Berlin, 1992). These definitions generally do not include theoretical content of FK (Medin and Atran, 1999). Here the role of outsiders is to understand how FK can be explained in terms of SK, and can make the application of SK more effective.

Participatory research has usually been based on definitions of the second type. As a result, FK has been used as either a *descriptive* or a *discriminatory* tool in participatory plant breeding (PPB). FK as a *descriptive* tool has most commonly been used. For example, a major survey of 49 PPB projects found that the primary focus was soliciting farmers' descriptions and rankings of selection criteria. For about two-thirds of these projects 'identifying, verifying, and testing of specific selection criteria was the main aim of the research', and 85% obtained farmers' selection criteria for new varieties (Weltzien *et al.*, 2000: 18, 51, 75). The main impact on scientific plant breeding appears to have been 'better understanding of new ideotypes based on farmers' experiences, specific preferences and needs' that will affect priorities of formal plant breeding and the 'process of formal variety development' (Weltzien *et al.*, 2000: 75).

More recently, using FK of crops as a *discriminatory* tool has become more common. This has been important in some PPB work, with farmers asked to choose among varieties already released in other areas (e.g. for rice and chickpea, Joshi and Witcombe, 1996), among new and experimental varieties (e.g. for pearl millet, Weltzien *et al.*, 1998), among segregating populations (e.g. $F_3$ bulks with barley, Ceccarelli *et al.*, 2000), or to select individual plants within segregating populations (e.g. $F_5$ bulks with rice, Sthapit *et al.*, 1996). When such choice or selection is accomplished using actual plants, plant parts or propagules, analysis of results can reveal farmers' implicit criteria that they may not be able to verbalize easily or at all (i.e. it may be unconscious) (Louette and Smale, 2000; Soleri *et al.*, 2000).

These approaches to understanding FK have made valuable contributions to achieving more effective crop improvement for farmers' conditions. However, the possible conceptual basis of FK has not usually been fully or even partially considered, and rigorous comparisons with SK have not been carried out; 'opportunities rarely develop for interaction between breeders and farmers beyond the survey', with the discussion 'driven by the breeders' concepts of the present situation, making it difficult for farmers to express their views in the context of their reality' (Weltzien *et al.*, 2000: 51). It may also be difficult for farmers to communicate to outsiders their knowledge that goes beyond description or discrimination. Thus, still lacking is an overarching approach to FK and SK that is broadly applicable and has the objective of facilitating understanding and interactions between the two. Below ('A holistic model of knowledge') we suggest an alternative perspective on farmer and scientist collaboration based on a new definition of knowledge.

## Defining participation

Participatory research to include farmers has been an important movement in agricultural development, with the goal of making formal science more useful to farmers and more efficient from the scientist's viewpoint (Chambers *et al.*, 1989). The explicit application of participatory research to plant breeding is relatively new, and there is a wide range of understandings of what it entails, and a wide range of activities present in PPB projects (Friis-Hansen and Sthapit, 2000). The relative contribution and control by farmers or plant breeders in PPB varies substantially and is one of its most important and discussed aspects.

The lack of rigour in using the term and the lack of scientific evaluation of participatory development are seen as threats to the continued use of participation in agricultural development (Ashby, 1997).

One approach to classifying participation is to distinguish between farmers' participation in formal scientists' research ('formal-led' PPB) on the one hand, and formal researchers' participation in farmers' research ('farmer-led' PPB) on the other hand (McGuire *et al.*, 1999; Weltzien *et al.*, 2000). The definition of PPB, however, often implies that farmers 'participate' in scientists' breeding, for example that PPB 'denotes a range of approaches that involve *users* (emphasis added) more closely in crop development or seed supply' (McGuire *et al.*, 1999: 7). A review of 11 examples of farmer-led plant breeding, including PPB projects, shows that the emphasis in scientist involvement has been on transferring SK and scientific practices directly to farmers, and secondarily from farmer to farmer, with almost no acknowledgement of the possibility of theoretical FK, or of transferring FK to scientists, or of farmers leading the research (although they may be involved in defining goals) (McGuire *et al.*, 1999). A companion review of 49 formal-led PPB projects found that researchers usually use 'participation' to refer to the stage in the breeding cycle where farmers are involved and the degree or amount of their involvement at that stage, but that what farmers actually do or how they affect the breeding process is 'usually left analytically vague' (Weltzien *et al.*, 2000: 59).

Multi-level or multi-stage taxonomies are also common in participatory research (see Joshi *et al.*, Chapter 10, this volume, for a detailed discussion of this), and tend to emphasize the social and institutional participation of farmers and scientists. The implicit assumptions are that FK can be complementary to SK, and that SK can strengthen farmer knowledge and practice. For example, Biggs has suggested a typology of four modes of 'farmer participation' (contractual, consultative, collaborative and collegial) (1989) that has been successfully applied in PPB (e.g. see Joshi *et al.*, Chapter 10, this volume). As with many other approaches to participatory research, Biggs' typology is meant to help research managers 'increase the cost effectiveness of research and . . . keep research priorities focused on the clients [resource poor farmers]' (1989: 1). In terms of FK, scientists recognize that FK and SK are complementary, and that FK is useful to them, and collect and use it. In the collegial mode scientists 'work to strengthen farmers' informal research and their ability to request information and services' (Biggs, 1989: 3, 8).

Another frequently used distinction in the discussion of PPB is a dichotomy between 'functional' (biological, product-oriented) benefits and 'empowering' (social, process-oriented) benefits of PPB (Ashby, 1997, for discussion see Weltzien *et al.*, 2000: 5–6). The modes described above are frequently interpreted and used as if the increasing social and particularly physical involvement of farmers at successive levels (e.g. from contractual to collegial) is synonymous with increasing equity and empowerment, and that empowerment and biological effectiveness are not necessarily related and may actually be in conflict (e.g. Bellon, 2000). On the other hand, some researchers see empowerment and biological effectiveness as causally related and synergistic (Ceccarelli *et al.*, 2000; Ceccarelli and Grando, Chapter 12, this volume).

## Theory for Understanding Farmers' Knowledge

### A holistic model of knowledge and collaboration

We suggest what we call a *holistic* model of knowledge that has as its goal minimizing deductive assumptions about the nature of FK (and SK), and that inductively tests ideas based on the possibility that FK may be both socially constructed and a verifiable description of objective reality, and consists of intuition, skill, empirical data and theory. The goal is to support collaboration between farmers and scientists.

We use the word *theory* to mean knowledge of the way things (namely, plant genotypes and growing environments) in the world relate to each other, including causal relationships, on which predictions and action can be based, and which are generalizable (but not necessarily universal) (cf. Hull, 1988: 485; Medin and Atran, 1999: 9). Our use of the word theory thus includes two important aspects that are often not differentiated in discussions of FK and SK.

**1.** Consciously developed theory intended to be universally applicable is often associated with modern science (but may also be carried out by farmers). The fact that farmers do not have access to the same information that scientists have (e.g. of microorganisms) is not an adequate basis for saying that it is not generalizable, because all theory is partial, and leaves things out; it could not function unless it did (Hull, 1988: 485). Theory in this sense, including modern scientific theory, is also influenced by personal psychology, historical contingencies and the social context of its production (Giere, 1999).
**2.** Unconscious 'heuristics' build on experiences with particular genotypes and environments and are often associated with non-scientific thinking, but may also be an important part of modern science (see Duvick, Chapter 8, this volume), such as the ecologically rational 'simple heuristics' discussed by Gigerenzer and Todd (1999). In this sense, theory pervades all human observation to some degree; according to some philosophers of science 'Theory-free observation, languages and classifications are impossible' (Hull, 1988: 485).

We originally suggested the term collaborative plant breeding (CPB) as an alternative to PPB, to 'remind ourselves that this effort should not privilege either men's or women's, or farmers' or formal plant breeders' approaches, values, etc., but aim for a true collaboration based on mutual respect, regardless of the proportional contribution of each' (Cleveland and Soleri, 1997). We think that this goal can be served by using a holistic model of knowledge that supports communication across the cultural and disciplinary divides that may separate farmers, plant breeders and social scientists. The relative merits of different knowledge in terms of contribution to CPB need to be empirically assessed in each situation, and successful collaboration requires mutual respect that is based on an understanding of differences, similarities and objectives. No a priori judgements need to be made about the relative quantitative contribution of farmers and plant breeders to collaboration. In this chapter we use the term PPB to talk about farmers and plant breeders working together in a general way that includes, for example, the four stages defined by Biggs. We use the term CPB to refer to situations in which the whole range of FK is considered, including theoretical, and in which SK is not privileged. ('Collaborative' in CPB is not, therefore, the equivalent of Biggs's 'collaborative' in his modes of participation.)

## A biological model

As a framework for evaluating farmer breeding we use the elementary biological model on which plant breeding is based, as it is presented in standard texts (e.g. Simmonds, 1979; Falconer and Mackay, 1996). First, variation in population phenotype ($V_P$) on which choice and selection are based is determined by genetic variation ($V_G$), environmental variation ($V_E$), and variation in genotype × environment (G × E) interaction ($V_{G \times E}$), thus $V_P = V_G + V_E + V_{G \times E}$. Broad sense heritability (H) is the proportion of $V_P$ due to genetic variance ($H = V_G/V_P$), while narrow sense heritability ($h^2$) is the proportion of $V_P$ due to additive genetic variance ($h^2 = V_A/V_P$), that is, the proportion of $V_G$ considered directly transmissible from parents to progeny, and therefore of primary interest to breeders.

Second, response to selection (R) is the difference, for the traits measured, between the mean of the whole population from which the parents were selected and the mean of the next generation that is produced by planting those selected seeds under the same conditions. R is the product of two different factors, $h^2$ and S ($R = h^2S$), where S is the selection differential, the difference between the mean of the selected group and the mean of the whole original population selected from. Expression of S in standard deviation units (the standardized selection differential or selection intensity; Falconer and Mackay, 1996: 189) permits comparison of selections among populations with different amounts or types of variation. The results of selecting for a given trait improve as the proportion of $V_P$ contributed by $V_A$ increases.

The biological relationships described in these simple equations underlie plant breeders' understanding of even the most complex phenomena that they encounter (Cooper and Hammer, 1996; DeLacy et al., 1996). For example, two widely respected English language plant breeding texts state that the relationship between genotype and phenotype is 'perhaps the most basic concept of genetics and plant breeding' (Allard, 1999: 48), and of $R = h^2S$, that 'If there were such a thing as a fundamental equation in plant breeding this would be it' (Simmonds, 1979: 100).

We use the biological model to understand farmers' perceptions in fundamental terms in order to facilitate collaboration, including increasing farmers' status in plant breeders' eyes, and increasing farmers' ability to use their own knowledge of their FVs and growing conditions. Potentially our research could also enable farmers to compare plant breeders' theories with their own. We are aware of the 'intimate links between knowledge and power' that have been ignored by many indigenous knowledge advocates who, perhaps unconsciously, privilege scientific knowledge while simultaneously lauding indigenous knowledge (Agrawal, 1995: 430). We seek methods to identify similarities and differences but do not assume that when there are differences between farmers and breeders, that the farmer is always 'wrong', nor, on the other hand, do we assume that outsiders have been negligent in understanding farmer knowledge and practice in their own terms (see Scoones and Thompson, 1993). Indeed, it is these similarities and differences that we have found the most challenging and stimulating to our own understanding and thinking.

It is important to note that while the emphasis here is on exploring FK, an understanding of SK may be equally critical to the long-term success of CPB, and we use the model in other parts of our research to understand plant breeders' knowledge and *differences among* plant breeders (Cleveland, 2001; Soleri and Cleveland, 2001; Cleveland and Soleri, 2002).

## Methods

### Interview scenarios

The method we use in framing questions to farmers is based on hypothetical scenarios that build on the key concepts of the biological model: genotype × environment interaction, heritability and genetic gain from selection. The method was developed in Oaxaca, Mexico, and is being modified and applied in a research project with collaborating scientists and farmers in Syria, Cuba and Nepal. The range of sociocultural and biological variables (including farmers, scientists, crops and environments) requires adapting scenarios to each of the different sites by: (i) referring to the biological model; (ii) including both components that are familiar and those that are unfamiliar to farmers; (iii) referring to crop-specific reproductive systems and local propagation methods; and (iv) addressing issues and practices central to scientists' and farmers' approaches. An example of a novel component in the scenarios is an *optimal* field that is uniform and in no way limits plant growth, in contrast to a farmer's *typical* field that is relatively variable and often has high levels of biotic and abiotic stresses. The optimal field facilitates an understanding of farmers' knowledge in terms of the biological model because in the optimal field the source of any variation among plants will be primarily genetic, not environmental.

Translating the 'optimal field' into terms familiar to farmers was a major challenge and took many iterations. During the first year in Mexico we defined the optimal field as one in which nothing was lacking for plant growth, there was plenty of water, the soil was good (i.e. there was plenty of manure, compost or chemical fertilizer), and there were no insects or other pests and diseases, or any abiotic stresses, such as flooding or high winds. We included uniform planting of seed, i.e. equal spacing. Farmers thought of many ways in which the planting or the plants resulting might not be equal, such as a farmer dropping many more seeds in one planting hole by accident, or a maize plant producing two rather than the usual one ear, and we gradually developed the scenarios to account for all of these factors.

We used lots of visual aids in the scenarios (Figs 2.1 and 2.2) and these proved to be very valuable both for ourselves and for the farmers we interviewed. For example, in Mexico we used maize ears from local farmers' fields when talking with farmers about ear length and photographs of different tassels from local fields when talking about tassel colour. Beginning in Syria, we used ten each of small rocks or crumpled paper balls of three sizes to represent years with good, 'normal' and poor rainfall; farmers selected from these to create a distribution of years defined by rainfall during a typical 10-year period (Fig. 2.3).

## Other materials and methods

In Syria, Cuba and Nepal the work involved collaboration with plant breeders and/or social scientists working with farmers on PPB or related work. In all sites we interviewed farmers from two communities with contrasting biophysical environments, one typically characterized as favourable and the other as difficult for the cultivation of the crop of interest. These contrasts were often based on different average precipitation, but also on soil quality and availability of agricultural inputs

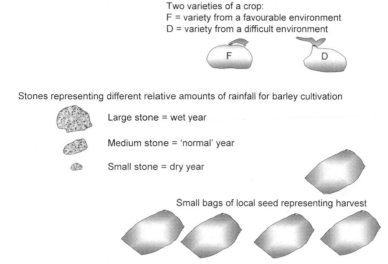

**Fig. 2.1.**  Visual aids used for G × E scenarios.

**Fig. 2.2.**  Nicasio Hernandez Sanchez and Daniela Soleri discuss the expression of traits with high and low average heritabilities in different environments described in scenarios used in Oaxaca, Mexico. Pictured are colour photos of tassels and ears of maize. (Photo by D.A. Cleveland, used with permission of subjects.)

**Fig. 2.3.** Farmer Juri Aboud discusses local rainfall distribution, risk and its affect on her choice of a barley variety in a scenario described to her in Mardabsi, northern Syria. She is selecting stones of three different sizes to represent the average distribution of dry, normal and wet years in a 10-year period. (Photo by D. Soleri, used with permission of subject.)

(irrigation, agrochemicals, machinery) (Table 2.1). The crop species we focused on differed by site, but in each case was the major local crop: maize (> 95% outcrossing; Craig, 1977) in Mexico and Cuba, barley (0.6–3.8% outcrossing; Allard, 1999: 41) in Syria (the major crop for small ruminant feed, the primary source of livelihood) and rice (0–3.0% outcrossing; Poehlman and Sleper, 1995) in Nepal.

In the following three sections we present results from our ongoing research investigating FK related to genotype × environment interaction, heritability and genetic gain. We include specific methods and findings in each section.

## Farmers' Knowledge of Genotype × Environment Interaction

The environmental scale for which crop varieties should be developed is an important decision for both farmers and plant breeders, and is directly related to interactions between variations in plant genotypes and those growing environments, $V_{G \times E}$. Environmental variation can be partitioned into several components: $V_E = V_L + V_T + V_M$ ($V_L =$ variance due to *location*, e.g. soil and climatic variables; $V_T =$ variance due to *time*, e.g. season or year; and $V_M =$ variance due to human *management*). $V_{G \times E}$ represents the degree to which genotypes behave

**Table 2.1.**   Descriptions of the study sites.

| Site and community[a] (no. of farmers) | Crop and % outcrossing | Elevation (masl) | Average annual precipitation (mm) | Community population size | Average field size (ha) | Average yield (t ha$^{-1}$) |
|---|---|---|---|---|---|---|
| Mexico (13) | Maize (*Zea mays mays*), > 95% | | | | | |
| D (5) | | 1780 | 468 | 2533 | 0.7 | 0.5 |
| F (8) | | 1490 | 685 | 2800 | 0.4 | 0.8 |
| Cuba (31) | Maize (*Zea mays mays*), > 95% | | | | | |
| D (20) | | 80 | 1350 | 204 | 0.5 | 1.5 |
| F (11) | | 15 | 1320 | 8000[b] | 27.9 | 1.5[c] |
| Syria (40) | Barley (*Hordeum vulgare* L.), 0.6–3.8% | | | | | |
| D (20) | | 495 | 300 | 1450 | 7.3 | 0.9 |
| F (20) | | 360 | 340 | 6000 | 2.0 | 3.0 |
| Nepal (10) | Rice (*Oryza sativa* L.) 0–3.0% | | | | | |
| F | | 660–1200 | 3979 | 5458 | 0.7 | 2.4 |

[a]D, community in relatively more difficult growing environment; F, community in relatively more favourable growing environment.
[b]Population size is for town, however, only members of three production cooperatives (total population approximately 1220) within that town area were interviewed.
[c]Yields at the favourable location have dropped significantly since the economic crisis of the 'special period' began in 1989, see Ríos Labrada *et al.*, Chapter 9, this volume.

consistently across a number of environments. Low quantitative $G \times E$ means little relative change in performance over environments (Fig. 2.4a). High quantitative $G \times E$ is characterized by marked changes in performance with changes in environmental factors (Fig. 2.4b) and is associated with reduced stability of performance (defined as variance across environments) of an individual genotype. Qualitative $G \times E$ between two or more varieties means that they change rank across environments (Fig. 2.4c), and is often referred to as a 'crossover' because the regression lines for yield (or other traits) cross over at some point (see Bänziger and De Meyer, Chapter 11; Ceccarelli and Grando, Chapter 12; and Ríos Labrada *et al.*, Chapter 9, this volume).

In the presence of qualitative $G \times E$, distinguishing between forms of $V_E$ based on their repeatability and predictability is part of an approach proposed by some plant breeders to develop crop varieties appropriate for difficult environments (Cooper, 1999; Ceccarelli and Grando, Chapter 12, this volume). A narrow range of repeatable and predictable $V_E$ would be addressed through selection for *specific* adaptation, while *broad* adaptation would be sought to less predictable or

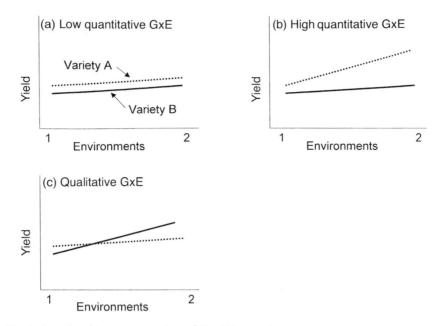

**Fig. 2.4.**    Graphic representation of G × E interactions.

non-repeatable $V_E$. That is, seeking broad adaptation means identifying the variety with the best performance over the range of environments (best overall average), while specific adaptation is achieved by identifying the variety with the best performance (absolute value) in a single environmental type in the range. Temporal $V_E$ ($V_T$ e.g. within and between season precipitation) and variance due to location ($V_L$ e.g. durable soil characteristics, slope and orientation), also referred to as spatial $V_E$, differ in that the relative unpredictability of $V_T$ makes it particularly challenging. For this reason specific adaptation to spatial $V_E$ and broad adaptation to $V_T$ may be sought under this approach.

Farmers' perceptions of different forms of qualitative G × E across different spatial and temporal environments were elicited through scenarios based on the biological model, $V_P = V_G + V_E + V_{G \times E}$. We asked farmers: if the same two varieties were compared in different growing environments, would their relative phenotypes for yield be the same or different? In other words, does a change in environments result in a *rank* change in terms of yield (Figs 2.5 and 2.6)? A key assumption in the scenarios was that local crop populations originating in contrasting growing environments would represent the possibility of such interactions that farmers may have observed or at least the components of the scenarios (some of the genotypes and environments) would be familiar to them. Because of the substantial evidence that a desire for

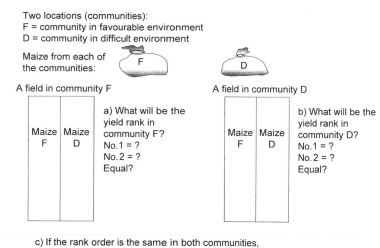

Two locations (communities):
F = community in favourable environment
D = community in difficult environment

Maize from each of the communities:

A field in community F

| Maize F | Maize D |
|---|---|

a) What will be the yield rank in community F?
No.1 = ?
No.2 = ?
Equal?

A field in community D

| Maize F | Maize D |
|---|---|

b) What will be the yield rank in community D?
No.1 = ?
No.2 = ?
Equal?

c) If the rank order is the same in both communities, will these always be their ranks wherever they are compared?

**Fig. 2.5.** Qualitative spatial G × E: interlocation scenario.

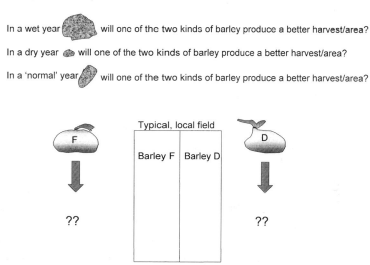

In a wet year will one of the two kinds of barley produce a better harvest/area?

In a dry year will one of the two kinds of barley produce a better harvest/area?

In a 'normal' year will one of the two kinds of barley produce a better harvest/area?

Typical, local field

F

D

| Barley F | Barley D |
|---|---|

??                                ??

**Fig. 2.6.** Qualitative temporal G × E scenarios.

diverse postharvest or other qualities is often the reason for maintaining multiple varieties of one species (e.g. see Joshi *et al.*, Chapter 10; McGuire, Chapter 5; Smale, Chapter 3; and Zimmerer, Chapter 4, this volume), the scenarios described varieties of a crop that originated in different and distant environments, but were exactly the same in all other ways. These scenarios were not used in Mexico as they were developed after that work was completed.

## Spatial G × E

Farmers often have to choose varieties for different locations (spatial environments) such as a range of fields for a given planting season (i.e. time is held constant). If farmers do not perceive qualitative G × L (crossovers) in varietal performance between fields, then there may be no agronomic reason for them to grow different varieties. When they do perceive crossovers between varieties for two locations, then they may have to decide whether to grow one variety in both, or if the extra yield obtained by growing two different varieties in the two locations, compared with the extra effort required, will produce a net benefit.

In our research we asked 'Is spatial $V_E$ important to farmers and, if so, at what scale?' (e.g. see Fig. 2.5). Our scenarios regarding qualitative G × E in response to spatial $V_E$ looked at three levels of spatial variation: (i) *between locations* (typically represented by distinct communities) with contrasting growing conditions (e.g. relatively favourable and difficult for the crop considered), one of them being the farmers' own location; (ii) *among fields* within the farmers' own location; and (iii) *within one* typical *field* in the farmers' own location. The null hypothesis in each case was that farmers would not be aware of such interactions.

The proportions of positive responses concerning the presence of qualitative G × E at some scales in Syria and Nepal represented significant deviations from the null hypothesis (Table 2.2). In addition, the site-based findings suggest a few overall trends. First, between location G × E is the most frequently recognized at all sites but the percentage of farmers noting this varied substantially from 25% in Cuba to 100% in Nepal. Second, for this same scenario in the two sites where farmers from communities in contrasting growing environments were interviewed (Cuba and Syria), more farmers in the favourable growing environment noted the potential for G × E, while those in the difficult environment more frequently gave different interpretations of the interactions they predicted between genotypes and environments (Figs 2.7 and 2.8). Third, at all sites, the percentage of farmers stating that G × E may occur between locations, as compared to within locations, and then compared to within a field in one location decreased, although at different rates in different sites.

The trends noted in the first and third points can be understood to a large extent through consideration of the context of this knowledge. The majority of responses appear to reflect differences in crop mating systems and their impact on the capacity of selection to eliminate intrapopulation diversity in favour of locally beneficial alleles (Allard, 1988). Environmental heterogeneity is seen as differentiating between varietal performances at all spatial levels most frequently among farmers working with crop species with low rates of outcrossing and least frequently among those cultivating a highly outcrossing species.

Combined with crop mating system, extent and scale of environmental heterogeneity as well as scale and type of management may account for these response trends as well. For example, in comparing the communities with difficult environments included in this study, Syrian fields were almost 15 times the size of Cuban and 10 times the size of Nepalese fields (see Table 2.1), and the production in the former is entirely mechanized while today in Cuba it rarely is and in Nepal all fields are hand worked. The scale and type of management in Nepal permits identification and use of rice varieties that may be more specifically responsive to particular sub-locations between and within

**Table 2.2.** Summary of farmers' perceptions of qualitative spatial and temporal $G \times E$ interactions for their primary crop in their environments; Cuba, Syria and Nepal.

| Site and community[a] (no. of farmers) | Percentage of farmers responding spatial $V_E$ as potential source of qualitative spatial $G \times E$ at the level of | | | Percentage of farmers responding temporal $V_E$ as potential source of qualitative temporal $G \times E$[b] |
|---|---|---|---|---|
| | Between locations | Between fields in one location | Within fields in one location | Between years with contrasting precipitation in one location |
| Cuba total (31) | 25 | 4 | 0 | 13 |
| D (20) | 6 | 0 | 0 | 20 |
| F (11) | 55 | 13 | 0 | 0 |
| Syria total (36) | 67* | 22 | 16 | 39* |
| D (19) | 63* | 21 | 11 | 32 |
| F (18) | 71* | 22 | 22 | 47* |
| Nepal[c] (10) | | | | |
| F | 100[d]* | 100[d]* | 30[d] | 60 |

[a]D: community in relatively more difficult growing environment, F: community in relatively more favourable growing environment.
[b]Responses to scenarios regarding two varieties originating in two contrasting locations, one relatively wet, the other relatively dry, due to either precipitation or reliable irrigation.
[c]The work in Nepal is just starting, only ten households in one community had been interviewed at the time of writing.
[d]Responses to scenarios regarding two varieties originating in contrasting, farmer-defined sub-locations occurring in one location.
*Significant $\chi^2$ test for goodness of fit to the null hypothesis that farmers would not perceive qualitative $G \times E$ interactions, $P \le 0.05$.

fields that may not be possible or necessary with the scale of Syrian barley cultivation, or with levels of gene flow due to cross-pollination by wind in maize in large, heterogeneous, areas of cultivation in small fields in Cuba.

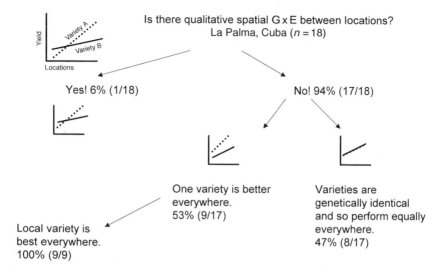

**Fig. 2.7.** Farmers' characterization of the relationship of varieties' performance across environments in spatial G × E scenario in a Cuban community with a relatively difficult growing environment.

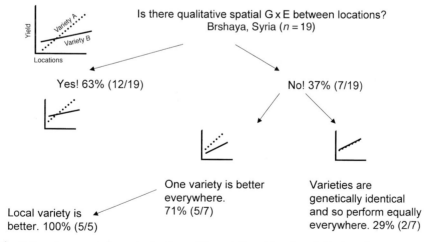

**Fig. 2.8.** Farmers' characterization of the relationship of varieties' performance across environments in spatial G × E scenario in a Syrian community with a relatively difficult growing environment.

Our results suggest that in some cases farmers are very aware of qualitative spatial $G \times E$. Some explicitly stated that they maintain or seek multiple, distinct varieties because of this, as for rice in Nepal within specified sub-locations or fields within those sub-locations, or in Syria for particular soil types within a location or even within fields at a location, while in other spatial environments they were not concerned by it. In those cases the alternative interpretations offered by farmers directly paralleled interpretations of $G \times E$ interactions observed and reported by some plant breeders: for example, describing the presence of high quantitative $G \times E$ (see Fig. 2.4b) with one variety always better than another variety but in some places this superiority being greater; or no $G \times E$ because $V_G = 0$, thus all 'varieties' are actually the same genetic population that has been given different names in the locations where it is being grown. In practice, both of these interpretations imply that there is no need for farmers to maintain or seek distinct (named) varieties for their growing environments.

## Temporal variation and risk perception

An important factor affecting a farmer's choice between varieties to grow in a given location is how she perceives variation in yield (and income) over time for that location (temporal $G \times E$), as well as average yield. Important but unpredictable components of the growing environment like precipitation can result in yield variation among varieties over time that could be expressed as quantitative or/and qualitative $G \times E$ (Cooper, 1999; Cooper *et al.*, 1999). If variety A has a higher average yield and lower yield variance than variety B through time in a given location, then the choice would be A. However, if variety A has a higher average yield but also a larger yield variance, then the choice between the two varieties will depend on her attitudes towards risk and on her ability to manage it; she may or may not be willing to sacrifice average yield in order have a more stable yield, or a 'smoother income stream' through time (Walker, 1989).

In the growing environments included in this study, precipitation is considered either the most, or one of the most important constraints to agricultural production and the main cropping season depends on rainfall for growth, not only stored soil moisture. We asked farmers about the potential for qualitative $G \times E$ in response to annual variation in precipitation defined broadly by them as 'wet' vs. 'dry' rainfall years for crop yield, not addressing timing or distribution of rainfall. They were asked to make this comparison between a local variety originating and maintained in a relatively wet environment, and one from a dry environment, as defined by average rainfall. The same biological model as described above for spatial G × E applies here, the only difference being that the environments referred to are temporal and are characterized by precipitation, as opposed to being spatially delineated.

The findings regarding farmers' perceptions of qualitative temporal G × E as defined by year-to-year variation in rainfall (Table 2.2) show the same trend as shown for perceptions of spatial G × E. Increasing proportions of farmers interviewed in Cuba, then Syria and finally Nepal foresaw possible qualitative interactions among those two varieties over time. Again, the different mating systems of their crop species may contribute to these differences. If that is the case, we could interpret the responses as indicating that in highly selfing species specific adaptation to a particular level of average rainfall was foreseen by a greater percentage of farmers than was the case for those growing more highly outcrossing species.

To understand farmers' attitudes to the risks posed by qualitative temporal G × E we asked them to choose which of two varieties would be best for them, one with high average yields and high yield variance over precipitation regimes (a highly responsive variety, HRV), the other with relatively low average yields but with higher yield stability (a stable variety) (Fig. 2.9). Because farmers' knowledge of future rainfall is uncertain or imperfect, the HRV represents a greater risk if farmers have few resources to rely on in cases of a poor harvest.

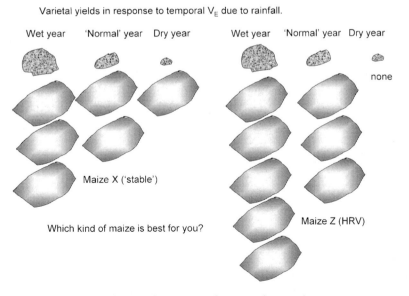

**Fig. 2.9.** Scenario for eliciting farmers' preference of varietal response to qualitative temporal G × E.

At each site farmers' responses were nearly evenly divided between the HRV and stable variety (Table 2.3). However, in Cuba and Syria, where farmers from communities with relatively difficult and favourable growing environments were interviewed, a significantly higher percentage of those in difficult environments preferred the stable variety. To interpret these results it seems useful to note that a higher proportion of Syrian as compared to Cuban farmers had experienced crop failure, and within each site a far larger percentage of farmers in difficult environments had experienced crop failure (Table 2.3). Although farmers at both sites and locations within these described a similar proportion of wet and dry years, expectations for qualitative temporal $G \times E$ between two varieties in response to rainfall variation were lower among Cuban farmers at all levels than among Syrian farmers. Once again, mating system and its impact on selection and adaptation seems likely to contribute to the differences between Cuban and Syrian responses. Still, biological factors are not necessarily complete explanations of farmers' choices. Other potentially important

**Table 2.3.** Farmers' estimations of rainfall distribution, experience of crop failure and choices between varieties that are highly responsive (HRV) vs. stable under temporal $V_E$, characterized by annual variation in precipitation useful for agriculture.

| Location and community[a] (no. of farmers) | Percentage of farmers choosing a | | Percentage of farmers choosing stable variety in communities with D vs. F[a] environments | Farmers' estimations of rainfall distribution over time (% years typically wet- 'normal'-dry) | Percentage of farmers reporting having experienced crop failure |
|---|---|---|---|---|---|
| | HRV | Stable variety | | | |
| Cuba total (29) | 41 | 59 | * | 30-40-30 | 28 |
| D (18) | 28 | 72 | 76 | 30-40-30 | 33 |
| F (11) | 64 | 36 | 24 | 30-50-20 | 14 |
| Syria total (40) | 48 | 52 | * | 30-40-30 | 51 |
| D (20) | 25 | 75 | 71 | 20-50-30 | 89 |
| F (20) | 70 | 30 | 29 | 30-40-30 | 15 |
| Nepal | – | – | – | – | – |
| F (10) | 50 | 50 | – | 30-50-20 | 0 |

[a]D: community in relatively more difficult growing environment, F: community in relatively more favourable growing environment.
*Significant $\chi^2$ test for goodness of fit for distribution of respondents choosing a stable variety across communities with difficult and favourable environments within a study site, $P \le 0.05$.

factors in farmers' choice between HRV and stable varieties not considered in this research include household economy and social structure,
community support networks, cultural values concerning risk, storage
capability and markets.

These findings concerning farmers' genetic perceptions of qualitative spatial and temporal $G \times E$ can have practical significance for CPB.
First, farmers' locally specific insights into scales at which qualitative
$G \times E$ may be present can inform and alter experimental design (cf.
Zimmerer, Chapter 4, this volume). This has happened in Syria through
close, long-term interactions between scientists and farmers; whether
these changes would have been accomplished more rapidly had farmers' conceptual knowledge been included earlier in that work remains
to be tested elsewhere. Second, spatial scales considered important by
farmers for varietal discrimination may be indicative of their receptiv-
ity to new material and the extent to which it may be used across a
range of local growing environments. Third, concern for qualitative
temporal $G \times E$ and the risk that may accompany it may not always be
the same across or within communities. If other factors remain the
same, changes may be required in breeding strategies, and explicit
evidence of this, to address unpredictable sources of crossover interactions and the consequent risk to some households.

## Farmers' Knowledge of Heritability and Implications for Selection

The heritability scenarios were designed to improve understanding of
how farmers perceive the influence of $V_G$ and $V_E$ on expression of a
particular trait and implications of this for selection. Building on the
biological model, a simple interpretation of realized $h^2$, an estimate of
$h^2$ based on a population's response to selection (Falconer and Mackay,
1996: 197ff.), was used in these scenarios. Based on this, the greater
the similarity of the progeny population phenotype to the parental
phenotype across different environments, the greater the $h^2$.

In the scenarios the contribution of $V_A$ was represented by the
relationship between phenotypes of maternal and progeny generations
(Fig. 2.10). The contribution of $V_E$ was represented by the contrasting
growing environments described; a typical, variable field vs. an optimal, uniform field. The null hypothesis was that farmers see a relatively small contribution by $V_A$ to total $V_P$ (low $h^2$) saying that seeds
from plants with a given trait would produce a progeny population
with diverse phenotypes of that trait when planted in a typical field,
and mostly progeny with the same phenotype as the parents when
planted in an optimal field, attributing $V_P$ predominantly to $V_E$ and

**Fig. 2.10.**    Photographs of different coloured tassels and maize ears to represent the seed planted from plants with different coloured tassels, used in scenarios in Oaxaca, Mexico (photo by D.A. Cleveland).

$V_{G \times E}$. The alternative hypothesis was that farmers see the trait as primarily determined by $V_A$, thus the progeny plants' phenotypes would be the same as the parents' regardless of the field environment. Our hypotheses did not include xenia (the effects of the pollen parent) or of segregation in the formation of progeny phenotypes, although some farmers did mention this. In these scenarios we compared traits known in the literature to have low (Fig. 2.11) and high (Fig. 2.12) average heritabilities relative to each other, and were also traits noted by farmers.

To interpret the findings of the $h^2$ scenarios requires two nested comparisons at each site that refer to the null hypothesis. First, considering each trait separately, in a typical vs. optimal field, 'Does the *proportion* of responses predicting progeny phenotypes being the same as parental ones differ according to the environment?' The proportion of responses differed significantly from that expected under the null hypothesis for high $h^2$ traits of tassel (Mexico), husk (Cuba) and seed colour (Syria and Nepal) in a typical environment (Table 2.4). That is, the number of responses foreseeing progeny phenotypes the same as parental ones did not change across environments for those traits, leading to rejection of the null hypothesis, and implying that farmers considered these traits to have high realized $h^2$. The null hypothesis was, however, accepted based on responses for traits of low average $h^2$.

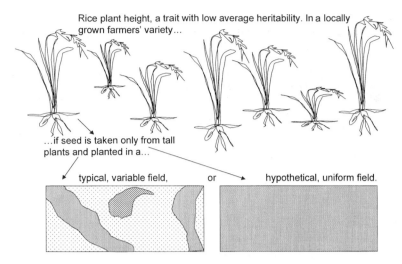

Rice plant height, a trait with low average heritability. In a locally grown farmers' variety…

…if seed is taken only from tall plants and planted in a…

typical, variable field,        or        hypothetical, uniform field.

Will the height of the progeny population be the same as the parents?

**Fig. 2.11.** Heritability scenario for trait with low average heritability (rice plant height, from Nepal scenario).

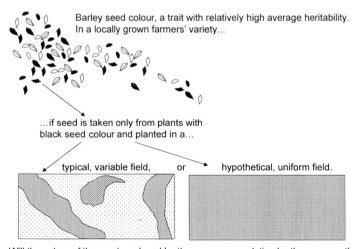

Barley seed colour, a trait with relatively high average heritability. In a locally grown farmers' variety…

…if seed is taken only from plants with black seed colour and planted in a…

typical, variable field,        or        hypothetical, uniform field.

Will the colour of the seed produced by the progeny population be the same as the parents?

**Fig. 2.12.** Heritability scenario for trait with high average heritability (rice grain colour, from Nepal scenario).

Across sites most farmers saw $V_E$ as influencing progeny phenotypes for these traits, while a small proportion of responses anticipated a replication of parental phenotypes for traits with low $h^2$ when sown in a variable environment. A much higher proportion stated that the

**Table 2.4.** Testing the $H_0$: Farmers do not recognize the contribution of $V_A$ to $V_P$. Farmers' perceptions of heritabilities for traits of high and low average $h^2$ in Mexico, Cuba, Syria and Nepal.

| Site, crop (no. of farmers) | Percentage of farmers stating that progeny will have same phenotype as parents when grown from seed sown in a/n | | | |
|---|---|---|---|---|
| | Trait with low average $h^2$ | | Trait with high average $h^2$ | |
| | Typical, variable field | Optimal, uniform field | Typical, variable field | Optimal, uniform field |
| Mexico, maize* | Ear length (< 0.50[a]) | | Tassel colour** ([b]) | |
| (13) | 0 | 92 | 70 | 70 |
| Cuba, maize* | Ear length (< 0.50[a]) | | Husk colour** ([b]) | |
| (31) | 3 | 97 | 77 | 71 |
| Syria, barley* | Plant height (0.62[c]) | | Seed colour** (0.90[c]) | |
| (36) | 8 | 86 | 69 | 69 |
| Nepal, rice* | Plant height (0.85[d]) | | Seed colour** ([e]) | |
| (10) | 0 | 100 | 100* | 100 |

[a]Hallauer and Miranda (1988).
[b]Published $h^2$ values not available, many pigmentation traits of maize have been used as easy to observe genetic markers, and as such are considered to have relatively high $h^2$ (Coe and Neuffer, 1988: 135).
[c]Rasmusson (1985).
[d]Calculated from Sthapit (1994).
[e]Not available.
*Fisher's exact test of independence of farmers' response distributions for high vs. low average $h^2$ traits in a typical field, $P \leq 0.001$.
**Fisher's exact test of independence of farmers' response distributions for similarity between parental and progeny phenotypes for the same trait in typical vs. optimal fields, $P \leq 0.001$.

parental population's phenotype for the same trait *would* be replicated in a uniform environment, that is $V_E$ and perhaps $V_{G \times E}$ make a larger contribution to $V_P$ than does $V_A$.

The second comparison was between responses for the two traits (high vs. low $h^2$) at one site, asking, 'Does the *pattern* of response differ?' The answer here was clearly 'yes'; the pattern of responses regarding those two types of traits was significantly different at all sites (Table 2.4).

While the $h^2$ scenarios are based on a number of assumptions that we are continuing to clarify, the results have yielded some useful insights. In all locations the traits of high average $h^2$ are ones that at least some farmers have manipulated via selection, implying an awareness of an aspect of $V_P$ similar to what $V_G$ represents in SK. The low $h^2$ traits are all ones farmers stated as criteria they use to define a desirable crop; for example, long maize ears in Mexico and Cuba, moderate (neither tall nor short) plant height in Syria and Nepal. However, despite using ear length as a criterion in their selection, farmers in Mexico (Soleri and Cleveland, 2001) and Cuba did not expect a response to this selection in the form of an increase in the frequency or magnitude of that trait. In Mexico this may be explained by farmers selecting large ears to ensure the quality of their planting material (large, heavy kernels), not for cumulative genetic improvement of their crop population.

Simple and self-evident as this may seem, it took us 3 years of interviews, selection exercises, participant observation and field trials to reach this understanding! We had initially shared the assumption underlying most plant breeding, that conscious human selection for yield-related traits has directional population change as its primary goal (Simmonds, 1979). However, the evidence that ear length is a primary selection criterion (from interviews, participant observation and selection exercises) was inconsistent with the results anticipated by farmers (from interviews and participant observation) and confirmed in our field trial (Soleri et al., 2000). Responses to the $h^2$ scenarios and other questions offered a different perspective (Soleri and Cleveland, 2001). Responses to the $h^2$ scenario for ear length in Mexico suggested to us that farmers saw that trait to be overwhelmingly the result of the growing environment. If they recognize a genetic component for ear length, it was not evident in the interviews, and farmers do not believe that they can make lasting changes in their varieties by selecting seed based on ear length. Instead, farmers appear to value ear size because of its correlation with seed size (e.g. as 100 grain weight, $r = 0.32$; Soleri, 1999) and their perception (conscious or unconscious) that large seed size is in turn positively correlated with such traits as seed quality and seedling vigour (Fig. 2.13).

One reason for the difficulty we and others may experience in understanding farmers' selection could be the result of our assumptions about farmers' and plant breeders' seed systems. A fundamental difference between them is that in farmers' systems genetic resources conservation, crop improvement and seed production functions are all accomplished primarily within the same seed lots or local populations, whereas in plant breeders' systems these functions are spatially and temporally distinct (Soleri and Smith, 1995; Smale et al., 1998). It seems likely that in farmers' systems some functions take precedence over others at different times depending on both socio-economic and biophysical circumstances. Outsiders operating under the assumptions

**Fig. 2.13.**    Delfino Jesus Llanez Lopez performing selection exercise on a sample of 100 ears from a study plot in one of his fields, Oaxaca, Mexico. (Photo by D.A. Cleveland, used with permission of subject.)

of formal plant breeding systems may have difficulty seeing the possibility of alternative functions for practices such as selection.

## Farmers' Expectations for Response to Selection

Over time, selection of plants from a heterogeneous population to obtain planting material for the next generation can affect allelic frequencies, and thus response to selection, R. Mass selection appears to be the most common form of selection used by farmers, involving the identification of superior individuals in the form of plants, and/or propagules, from a population and the bulking of seed or other planting material to form the planting stock for the next generation. This approach requires only a single season, and relatively little effort compared with other selection methods (Simmonds, 1979; Weyhrich *et al.*, 1998). If practised season after season with the same seed stock, mass selection has the potential to maintain or even improve a crop population, depending upon the mating system, trait heritability, trait $G \times E$, the selection intensity and gene flow in the form of pollen or seeds into the population.

Surprisingly little research has been done on farmers' selection goals considering their importance for the selection process, especially for farmers in marginal environments (Weltzien *et al.*, 1998). The implicit assumption has often been that farmers must be attempting directional selection for quantitative, relatively low $h^2$ traits like yield, the main goal of plant breeders. However, there appear to be relatively few data demonstrating that farmers have directional selection for quantitative traits as a conscious goal, in contrast with data on farmers' conscious choice of new varieties.

Though not necessarily constant in its response from one generation to the next, and with a diminishing rate of change over generations due to reductions in both $V_P$ and $h^2$ (Falconer and Mackay, 1996: 201ff.), in plant breeding directional selection is conceived and practised as a cumulative process. This research sought to directly test the null hypothesis that farmers' seed selection was intended to produce cumulative, directional change in their crop population, as assumed for directional selection in formal plant breeding. The alternative hypothesis was that farmers have other reasons for discriminating within their harvest to identify planting material.

Based on results of the $h^2$ scenarios in Mexico, we created scenarios for work at the other sites to gain greater understanding of farmers' expectations for selection. Beginning in Cuba, a scenario was created presenting a comparison between the results within a subpopulation (identified as S) of typical, farmers' selection using specified selection criteria, and results within a subpopulation (identified as R) of random selection (i.e. a relaxation of all artificial selection pressures). The scenarios specified that R and S were subunits of the same original population in the same growing environment, and that selection occurred in both over the same time period (Fig. 2.14). Farmers were asked 'After 10 years of selection, how would subpopulations R and S compare for yield?' The scenario then described an additional cycle in which artificial selection was used in both subpopulations; S for the 11th year ($S_{11}$) and R for the first time after 10 years of random selection ($R_{10 + 1}$). Again, farmers were asked to compare yields between the progeny subpopulations grown from the seed selected according to this scenario (Fig. 2.15).

The same scenario was adjusted for interviews in Nepal and Syria because at those sites the most common means of identifying planting material from within the entire population is by choosing between field plots planted to the same variety in the former, and by identifying an area within one field from which seeds are obtained in the latter. In the scenarios, typical local selection was compared with a random choice of field plot or area within a field.

Almost all farmers at all sites saw their selection as providing benefits over random selection (Table 2.5, $R_{10}$ vs. $S_{10}$). However, based on responses to the comparison of $S_{11}$ vs. $R_{10 + 1}$, many of them also felt

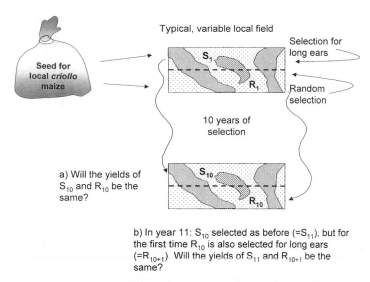

**Fig. 2.14.**   Scenario to understand farmers' expectations for response to selection over time.

that the benefits of selection are not cumulative, since they can be attained in one cycle. Interaction with plant breeders can influence farmers' expectations. For example, in Cuba, 80% of those who did see cumulative benefits from selection were participants in the beginnings of an on-farm PPB project that included presentation of basic plant breeding concepts. However, there are other variables such as age, education, experience with other plant or animal selection that were not considered in this research that may account for the difference in responses to this scenario.

The basis of the responses foreseeing no cumulative change from selection may well differ between and within sites and could include low $h^2$ due to large $V_E$ and/or small $V_G$, or lack of effective methods. Identifying reasons for this response seems important both for understanding farmers' practices and goals, but also for detecting impediments to gains from selection, or even why such 'gains' may not be beneficial. For example, in Syria scientists did not have evidence that farmers were aware of intrapopulation selection and its potential benefits (Ceccarelli and Grando, 2000, ICARDA, personal communication). However, discussing a scenario in which they were asked what they would do with a highly variable new barley population including some of the best and worst plants they had ever seen, 43% of those farmers ($n = 40$) stated that they would select the best individuals for planting the following season. Despite the farmers' awareness of intrapopulation selection, based on past experience this is often abandoned after two

**Fig. 2.15.** Farmer Loreto Mederos explains his perception of scenario outcomes to Humberto Ríos Labrada in La Palma, Cuba. (Photo by D. Soleri, used with permission of subjects.)

**Table 2.5.** Farmers' expectations for response to selection.

| Location, crop (no. of farmers) | Percentage of farmers stating that the yield of $R_{10} > S_{10}$ | Percentage of farmers stating that the yield of $R_{10+1} \geq S_{11}$ |
|---|---|---|
| Cuba, maize (29) | 3 | 59 |
| Syria, barley (20) | 0 | 45 |
| Nepal, rice (10) | 0 | 90 |

$R_{10}$ = population randomly selected in the local growing environment for 10 consecutive years.

$S_{10}$ = population intentionally selected in the traditional, local manner and environment for 10 consecutive years.

$R_{10+1}$ = population randomly selected in the local growing environment for 10 consecutive years and then intentionally selected in the traditional, local manner and environment for 1 year.

$S_{11}$ = population intentionally selected in the traditional, local manner and environment for 11 consecutive years.

cycles because progeny are not true to type. If presented with more advanced segregating populations these farmers might be motivated to make greater use of their knowledge of the benefits of individual plant selection.

Farmers' perceptions of the potential to improve their populations via selection – and thus their selection expectations and goals – will probably be influenced not only by their understanding of genetic variation in the population and $h^2$ for traits of interest, but also of alternative uses of their time and labour. If they do not believe population improvement to be possible or cost-effective, one alternative may be to choose different varieties or populations or infuse their own varieties or populations with new genetic variation as has been documented for maize farmers in Mexico (Louette *et al.*, 1997).

# Conclusion

## Approach to knowledge in CPB

Do farmers have a conceptual knowledge of critical issues concerning the relationships between their crop genotypes and growing environments and, if so, is that congruent with scientists' knowledge of those same issues? Testing this hypothesis is central to our research and exploration of knowledge in relation to application in CPB. There are two major points suggested by the findings to date.

First: Do farmers have *conceptual* knowledge and is it *congruent* with SK? Although these findings do not conclusively affirm our central hypothesis, they are in no way inconsistent with it. They suggest the presence of a conceptual component in farmers' knowledge of relationships between their crop genotypes and growing environments. That this component is in some ways congruent with scientists' knowledge is supported by their ability to understand and answer scenario questions about abstract ideas such as $G \times E$, $h^2$ and R. It is also supported by the explanations presented by some for their responses that corresponded in many ways to plant breeders' expectations regarding varietal performance across environments (e.g. Figs 2.7 and 2.8). Indeed, interpretation of these findings – though ongoing – is facilitated by some plant breeders' analyses of their own knowledge, in terms of theory, interpretation, intuition and practice (Bänziger and Cooper, 2001; Duvick, Chapter 8; Ceccarelli and Grando, Chapter 12, this volume).

Second: Why are there *differences* and *similarities among* farmers and *between* farmers and scientists? These findings indicate that farmers' conceptual knowledge is partially defined by context. As with formally trained researchers, it appears that most farmers base their understanding of $V_G$ and $h^2$ on their own experiences. That is, differences in knowledge may be more the result of differences in crops, environments and sociocultural variables, rather than in knowledge of basic biological relationships. For example, farmers' perceptions of the potential for qualitative $G \times E$ (Table 2.2) may be at least partially attributed to the mating system of the crop and the growing environments they are working in. Farmers' responses to $h^2$ scenarios (Table 2.4) may not always deny the presence of $V_G$ in their populations for traits of low average $h^2$, but reflect their unfamiliarity with optimal growing environments and indicate the overwhelming influence of $V_E$ in local fields, obscuring $V_G$ in low $h^2$ traits (Fig. 2.16a). Similarly, it has been suggested that the interpretations of theory underlying some plant breeders' practices reflects their experiences (Ceccarelli, 1989, 1996; Cleveland, 2001). For example, contrasting assumptions among plant breeders regarding appropriate selection environments for highly stress-prone target environments have been attributed to contrasting experiences with range and type of $V_E$, affecting the likelihood of anticipating the $G \times E$ interactions that might occur in the marginal fields of many farmers (Fig. 2.16b).

Thus, FK can be congruent with the basic biological model, and thus generalizable, and at the same time be in conflict with the basic

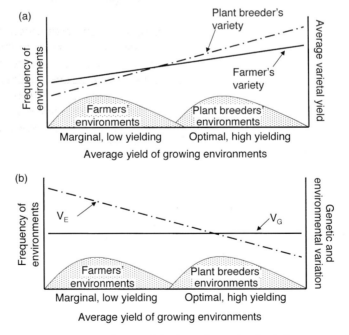

**Fig. 2.16.** Graphic representation of the hypothesis that both farmers' and plant breeders' knowledge may be contextual. (a) Range of $V_E$ experienced limiting plant breeders' observation of $G \times E$ and affecting their interpretation of theory. (Partially based on Ceccarelli, 1989.) (b) $V_E$ in their fields limiting farmers' perception of $V_G$ for low $h^2$ traits.

biological model, and be locally specific. This is similar to the situation with scientific plant breeding (Cleveland, 2001; Cleveland and Soleri, 2002). What is often not appreciated is the extent to which scientists think that their practice is the result of applying grand theory, when in fact practice may be based on heuristics developed for their particular circumstances and may not be generalizable. Very few plant breeders have specifically questioned the extent to which accepted conventional practice, generally believed to be based on grand theory and thus generalizable to farmers' contexts, is actually based on context-specific heuristics, thus not necessarily generalizable to all contexts (Ceccarelli and Grando, Chapter 12, this volume).

## Methods

This discussion of our ongoing research demonstrates the development and use of a method for understanding farmers' conceptual plant breeding knowledge. While much remains to be investigated, we feel that the results to date show that the scenario methodology, even when applied to a small sample of farmers, can provide information concerning FK that is relevant and useful for CPB. The fundamental elements of this methodology, the holistic model of knowledge and the basic biological model, each contribute to its utility. The holistic model of knowledge proposes a perspective focused on the components of knowledge and their similarities and differences. In so doing it avoids essentialisms, instead supporting inductive investigation of those components. The basic biological model provides a valid comparator for exploring knowledge of farmers and plant breeders about empirical relationships between their crop populations and their environments. As such, it creates a common ground on which issues fundamental to plant breeding can be discussed and locally relevant solutions sought.

## Implications for collaboration

This research suggests three ways that understanding farmers' conceptual knowledge can be useful for CPB. It can:

1. Demonstrate to outside researchers that farmers do indeed have a conceptual basis for their choice and selection practices and that, therefore, their role in CPB can be more substantive and central than had been previously thought, regardless of the actual strategies that are chosen.

**2.** Alert outside researchers to issues they may be unaware of, assumptions that may be incorrect about environments, genotypes or farmers' interests, e.g. scales of $V_E$ influencing varietal choice, desirability of HRV vs. stable varieties, knowledge of and experience with within-population plant selection.

**3.** Inform researchers of perceptions of farmers that underlie practices; e.g. why certain things are done (as with selection in Mexico, Cuba and Syria), and so offer opportunities to discuss, investigate and improve those practices. Similarly, this approach may be useful in other fields such as soil management and conservation biology.

In addition, the application of the methodology we have outlined can provide new insights into the differences that exist among plant breeders, often resulting in very different interpretations and applications of plant breeding theory. In so doing this approach may help to elucidate problems in applying SK to CPB.

What do points 1–3 imply for how farmers and scientists can work together? The distinction between 'functional' and 'empowering' participation may be less relevant when using a holistic model of knowledge, since social and biological benefits of collaboration would be expected to be more synergetic (see Ceccarelli *et al.*, 2000; Ceccarelli and Grando, Chapter 12, this volume).

This approach suggests an alternative to common uses and measures of FK in participation, a change from a quantitative to a qualitative emphasis. A *qualitative* approach means that the source, amount and other aspects of ideas and effort are not the defining characteristics of participation. This relationship would be present regardless of the specific strategy or level of physical involvement of either farmers or breeders. A qualitative approach might foster a relationship between farmers and plant breeders characterized by ongoing substantive interaction including discussion of the conceptual basis of plant breeding practice, mutual respect and the common goal of meeting local needs. Achieving such a relationship will probably require deeper understanding of farmers' and plant breeders' knowledge and the similarities and differences between them. Testing ideas on the *conceptual* content of FK, and on similarities and differences between the knowledge of farmers and scientists is a new area of research opening up new possibilities for PPB and other activities.

There is a small but increasing number of crop improvement projects worldwide in which perceptive breeders have included farmers' participation to attain positive results and in some cases unprecedented successes (e.g. Bänziger and de Meyer, Chapter 11; Ceccarelli and Grando, Chapter 12; Joshi *et al.*, Chapter 10, this volume). The research reported here in no way detracts from those successes, and indeed has had the benefit of many insights provided by them. The empirical results of these projects demonstrate that significant results can be obtained in PPB that are not based on results of research on the theoretical component of FK.

However, it may be that the results of the projects just mentioned depend in an important, but undocumented way, on the personalities of the scientists involved (Ceccarelli *et al.*, 2000; Friis-Hansen and Sthapit, 2000), which has contributed to their willingness to try new approaches, including listening to farmers, and rethinking the application of basic plant breeding theory in terms of farmers' circumstances. Indeed, they may have already, at an informal or even unconscious level, incorporated many insights into farmers' theoretical knowledge into their work. Therefore, if the methodology we are testing is adequately robust and adaptable, it could provide a means to enable more scientists to approach their plant breeding challenges with the perceptiveness of those already successful practitioners.

We are suggesting the exploration of a new approach to participation that appears to have potential, but which has not been applied systematically over the long term; this is the next step. The results reported in this chapter are based on small samples and further testing and methodological development is required and will undoubtedly provide new or different insights. Still, these findings suggest the value of expanding this applied research within CPB projects to facilitate practical results. An approach to collaboration that includes conceptual contributions and interactions from both farmers and scientists appears to have the potential of increasing both biological and social effects, and thus, hopefully, realizing some of the real benefits collaboration has to offer.

## Acknowledgements

For the research in Mexico we thank the collaborating farm households in Oaxaca who taught us much and brought great patience and good humour to our work together; the municipal authorities of the study communities for permission to conduct this work; M. Smale and M.E. Smith for discussion and advice. That research was supported in part by grants to DS from the American Association for University Women, Association for Women in Science; Research Training Grant for the Study of Biological Diversity (University of Arizona), Sigma Xi, the US-Mexico Fulbright Commission, and the USAID-CGIAR Linkage Program; grants to DAC from the Committee on Research, Faculty Senate, UC Santa Barbara; and the financial and logistic support of

CIMMYT's Economics Program through M. Smale, and Maize Program through S. Taba.

For research in Syria, Cuba and Nepal, we thank the farming households we are working with and our collaborating institutions (ICARDA, INCA and LI-BIRD, respectively) and scientists in each location; the Institute for Social, Behavioural and Economic Research, UC Santa Barbara, and the National Science Foundation (SES-9977996) for financial support.

Any inaccuracies or misunderstandings are the sole responsibility of the first three authors.

# References

Agrawal, A. (1995) Dismantling the divide between indigenous and scientific knowledge. *Development and Change* 26, 413–439.

Allard, R.W. (1988) Genetic changes associated with the evolution of adaptedness in cultivated plants and their wild progenitors. *Journal of Heredity* 79, 225–238.

Allard, R.W. (1999) *Principles of Plant Breeding*, 2nd edn. John Wiley & Sons, New York.

Ashby, J.A. (1997) What do we mean by participatory research in agriculture? In: CGIAR Systemwide Project (ed.) *New Frontiers in Participatory Research and Gender Analysis: Proceedings of the International Seminar on Participatory Research and Gender Analysis for Technology Development*. CGIAR Systemwide Project, Cali, Colombia, pp. 15–22.

Bänziger, M. and Cooper, M. (2001) Breeding for low-input conditions and consequences for participatory plant breeding: examples from tropical maize and wheat. *Euphytica* 122, 503–519.

Bellon, M.R. (2000) Of participation in participatory plant breeding: an analysis of two common assumptions. In: GRCP (ed.) *Scientific Basis of Participatory Plant Breeding and Conservation of Genetic Resources, Oaxtapec, Morelos, Mexico, 8–14 October 2000. Abstracts.* University of California Division of Agriculture and Natural Resources, Genetic Resources Conservation Program, Davis, California.

Berlin, B. (1992) *Ethnobiological Classification: Principles of Categorization of Plants and Animals in Traditional Societies.* Princeton University Press, Princeton, New Jersey.

Biggs, S.D. (1989) *Resource-Poor Farmer Participation in Research: a Synthesis of Experiences from Nine National Agricultural Research Systems.* OFCOR Comparative Study Paper 3. International Service for National Agricultural Research (ISNAR), The Hague.

Brown, A.H.D. (1999) The genetic structure of crop landraces and the challenge to conserve them *in situ* on farms. In: Brush, S.B. (ed.) *Genes in the Field: On-farm Conservation of Crop Diversity.* Lewis Publishers, Boca Raton, Florida; IPGRI, Rome; IDRC, Ottawa, pp. 29–48.

Ceccarelli, S. (1989) Wide adaptation: how wide? *Euphytica* 40, 197–205.

Ceccarelli, S. (1996) Positive interpretation of genotype by environment interactions in relation to sustainability and biodiversity. In: Cooper, M. and Hammer, G.L. (eds) *Plant Adaptation and Crop Improvement*. CAB International in association with IRRI and ICRISAT, Wallingford, UK, pp. 467–486.

Ceccarelli, S., Grando, S., Tutwiler, R., Bahar, J., Martini, A.M., Salahieh, H., Goodchild, A. and Michael, M. (2000) A methodological study on participatory barley breeding. I. Selection phase. *Euphytica* 111, 91–104.

Chambers, R., Pacey, A. and Thrupp, L.-A. (1989) *Farmer First: Farmer Innovation and Agricultural Research*. Intermediate Technology Publications, London.

Cleveland, D.A. (2001) Is plant breeding science objective truth or social construction? The case of yield stability. *Agriculture and Human Values* 18(3), 251–270.

Cleveland, D.A. and Soleri, D. (1997) Posting to the Farmer-breeding list serve <farmer-breeding-l-postmaster@cgnet.com> sponsored by the CGIAR's Systemwide Program on Participatory Research and Gender Analysis, http://www.prgaprogram.org/prga/, 27 September 1997.

Cleveland, D.A. and Soleri, D. (2002) Indigenous and scientific knowledge of plant breeding: similarities, differences, and implications for collaboration. In: Sillitoe, P., Bicker, A.J. and Pottier, J. (eds) *'Participating in Development': Approaches to Indigenous Knowledge*. Routledge, London.

Cleveland, D.A., Soleri, D. and Smith, S.E. (2000) A biological framework for understanding farmers' plant breeding. *Economic Botany* 54, 377–394.

Coe, E.H., Jr and Neuffer, M.G. (1988) The genetics of corn. In: Sprague, G.F. and Dudley, J.W. (eds) *Corn and Corn Improvement*, 3rd edn. ASA, CSSA, SSSA, Madison, Wisconsin, pp. 111–223.

Cooper, M. (1999) Concepts and strategies for plant adaptation research in rainfed lowland rice. *Field Crops Research* 64, 13–34.

Cooper, M. and Hammer, G.L. (1996) Synthesis of strategies for crop improvement. In: Cooper, M. and Hammer, G.L. (eds) *Plant Adaptation and Crop Improvement*. CAB International in association with IRRI and ICRISAT, Wallingford, UK, pp. 591–623.

Cooper, M., Rajatasereekul, S., Immark, S., Fukai, S. and Basnayake, J. (1999) Rainfed lowland rice breeding strategies for Northeast Thailand. I. Genotype variation and genotype x environment interactions for grain yield. *Field Crops Research* 64, 131–151.

Craig, W.F. (1977) Production of hybrid corn seed. In: Sprague, G.F. (ed.) *Corn and Corn Improvement*. American Society of Agronomy, Madison, Wisconsin, pp. 671–719.

DeLacy, I.H., Basford, K.E., Cooper, M., Bull, J.K. and McLaren, C.G. (1996) Analysis of multi-environment trials – an historical perspective. In: Cooper, M. and Hammer, G.L. (eds) *Plant Adaptation and Crop Improvement*. CAB International in association with IRRI and ICRISAT, Wallingford, UK, pp. 39–124.

Duvick, D.N. (1996) Plant breeding, an evolutionary concept. *Crop Science* 36, 539–548.

Ellen, R. and Harris, H. (2000) Introduction. In: Ellen, R., Parkes, P. and Bicker, A. (eds) *Indigenous Environmental Knowledge and Its Transformations: Critical Anthropological Perspectives.* Harwood Academic Publishers, Amsterdam, pp. 1–33.

Falconer, D.S. and Mackay, T.F. (1996) *Introduction to Quantitative Genetics.* Prentice Hall/Pearson Education, Edinburgh.

Friis-Hansen, E. and Sthapit, B. (2000) Concepts and rationale of participatory approaches to conservation and use of plant genetic resources. In: Friis-Hansen, E. and Sthapit, B. (eds) *Participatory Approaches to the Conservation and Use of Plant Genetic Resources.* International Plant Genetic Resources Institute, Rome, pp. 16–19.

Giere, R.N. (1999) *Science Without Laws.* University of Chicago Press, Chicago.

Gigerenzer, G. and Todd, P.M. (1999) Fast and frugal heuristics: the adaptive toolbox. In: Gigerenzer, G., Todd, P.M. and ABC Research Group (eds) *Simple Heuristics That Make Us Smart.* Oxford University Press, Oxford, pp. 3–34.

Hallauer, A.R. and Miranda, J.B. (1988) *Quantitative Genetics in Maize Breeding,* 2nd edn. Iowa State University Press, Ames, Iowa.

Hull, D.L. (1988) *Science As a Process: an Evolutionary Account of the Social and Conceptual Development of Science.* The University of Chicago Press, Chicago.

Joshi, A. and Witcombe, J.R. (1996) Farmer participatory crop improvement. II. Participatory varietal selection, a case study in India. *Experimental Agriculture* 32, 461–467.

Louette, D. and Smale, M. (2000) Farmers' seed selection practices and maize variety characteristics in a traditional Mexican community. *Euphytica* 113, 25–41.

Louette, D., Charrier, A. and Berthaud, J. (1997) In situ conservation of maize in Mexico: genetic diversity and maize seed management in a traditional community. *Economic Botany* 51, 20–38.

McGuire, S., Manicad, G. and Sperling, L. (1999) *Technical and Institutional Issues in Participatory Plant Breeding – Done from the Perspective of Farmer Plant Breeding: a Global Analysis of Issues and of Current Experience.* Working Document No. 2. CGIAR Systemwide Program on Participatory Research and Gender Analysis for Technology Development and Institutional Innovation. Cali, Colombia.

Medin, D.L. and Atran, S. (1999) Introduction. In: Medin, D.L. and Atran, S. (eds) *Folkbiology.* MIT Press, Cambridge, Massachusetts, pp. 1–15.

Poehlman, J.M. and Sleper, D.A. (1995) *Breeding Field Crops,* 4th edn. Iowa State University Press, Ames, Iowa.

Rasmusson, D.C. (1985) *Barley.* ASA, CSSA, SSSA, Madison, Wisconsin.

Scoones, I. and Thompson, J. (1993) *Challenging the Populist Perspective: Rural People's Knowledge, Agricultural Research and Extension Practice.* Discussion Paper 332. Institute of Development Studies, University of Sussex, Brighton, UK.

Selener, D. (1997) *Participatory Action Research and Social Change.* The Cornell Participatory Action Research Network, Cornell University, Ithaca, New York.

Sillitoe, P. (1998) The development of indigenous knowledge: a new applied anthropology. *Current Anthropology* 39, 223–252.

Simmonds, N.W. (1979) *Principles of Crop Improvement.* Longman Group, London.

Smale, M., Soleri, D., Cleveland, D.A., Louette, D., Rice, E., Blanco, J.L. and Aguirre, A. (1998) Collaborative plant breeding as an incentive for on-farm conservation of genetic resources: economic issues from studies in Mexico. In: Smale, M. (ed.) *Farmers, Gene Banks, and Crop Breeding: Economic Analyses of Diversity in Wheat, Maize, and Rice.* Kluwer Academic, Norwell, Massachusetts, pp. 239–257.

Soleri, D. (1999) Developing methodologies to understand farmer-managed maize folk varieties and farmer seed selection in the Central Valleys of Oaxaca, Mexico. PhD dissertation, University of Arizona, Tucson, Arizona.

Soleri, D. and Cleveland, D.A. (2001) Farmers' genetic perceptions regarding their crop populations: an example with maize in the Central Valleys of Oaxaca, Mexico. *Economic Botany* 55, 106–128.

Soleri, D. and Smith, S.E. (1995) Morphological and phenological comparisons of two Hopi maize varieties conserved in situ and ex situ. *Economic Botany* 49, 56–77.

Soleri, D., Smith, S.E. and Cleveland, D.A. (2000) Evaluating the potential for farmer and plant breeder collaboration: a case study of farmer maize selection in Oaxaca, Mexico. *Euphytica* 116, 41–57.

Srivastava, J.P. and Jaffee, S. (1993) *Best Practices for Moving Seed Technology: New Approaches to Doing Business.* World Bank Technical Paper Number 213. The World Bank, Washington, DC.

Sthapit, B.R. (1994) Genetics and physiology of chilling tolerance in Nepalese Rice. PhD thesis, University of Wales, Bangor.

Sthapit, B.R., Joshi, K.D. and Witcombe, J.R. (1996) Farmer participatory crop improvement. III. Participatory plant breeding, a case for rice in Nepal. *Experimental Agriculture* 32, 479–496.

Walker, T.S. (1989) Yield and household income variability in India's semi-arid tropics. In: Anderson, J.R. and Hazell, P.B.R. (eds) *Variability in Grain Yields: Implications for Agricultural Research and Policy in Developing Countries.* Johns Hopkins University Press, Baltimore, Maryland, pp. 309–319.

Wallace, D.H. and Yan, W. (1998) *Plant Breeding and Whole-System Crop Physiology: Improving Crop Maturity, Adaptation and Yield.* CAB International, Wallingford, UK.

Weltzien, E., Smith, M.E., Meitzner, L.S. and Sperling, L. (2000) *Technical and Institutional Issues in Participatory Plant Breeding – From the Perspective of Formal Plant Breeding. A Global Analysis of Issues, Results, and Current Experience.* Working Document No. 3. CGIAR Systemwide Program on Participatory Research and Gender Analysis for Technology Development and Institutional Innovation, Cali, Colombia.

Weltzien, R.E., Whitaker, M.L., Rattunde, H.F.W., Dhamotharan, M. and Anders, M.M. (1998) Participatory approaches in pearl millet breeding. In: Witcombe, J., Virk, D. and Farrington, J. (eds) *Seeds of Choice.* Intermediate Technology Publications, London, pp. 143–170.

Weyhrich, R.A., Lamkey, K.R. and Hallauer, A.R. (1998) Responses to seven
    methods of recurrent selection in the BS11 maize population. *Crop Science*
    38, 308–321.

# Economics Perspectives on Collaborative Plant Breeding for Conservation of Genetic Diversity On Farm

**3**

MELINDA SMALE

*International Plant Genetics Resources Institute (IPGRI), via dei Tre Dinari 472/a, 00057 Maccarese (Fiumicino), Italy, and International Food Policy Research Institute (IFPRI), 2033 K Street, NW, Washington, DC 20006, USA*

## Abstract

One of the theoretical benefits of CPB is that it can support on-farm conservation of genetic diversity by enhancing the value to farmers of some of the varieties they are growing, thereby generating 'incentives' for farmers to continue growing varieties which benefit society at large by conserving crop genetic diversity. In contrast to subsidies or cash payments, such incentives are 'in kind', yet generating them requires sound understanding of the genetics and biological implications of farmers' practices. It also requires in-depth social and economic assessments, much of which are location-specific and may be costly. Costs are borne in large part by the farmers' themselves. Will farmers perceive that the benefits of CPB that conserves genetic diversity are worth the additional investment of their labour and management? This chapter summarizes some perspectives that economic theory and methods can provide for those engaged in the applied research and practice of CPB.

## Introduction

This chapter summarizes some perspectives that economic theory and methods can provide for those engaged in the applied research and practice of farmer plant breeding, especially in the context of conserving crop genetic resources (CGRs). For this purpose, farmer plant breeding is broadly defined to include farmers' choice of varieties to grow and in which relative amounts, as well as selection of plants within populations, and the relationship of these decisions and other farmer management practices to crop genetic diversity (see Chapter 1, this volume).

The basic economic question involved in conserving cultivated plant species differs fundamentally from that related to preserving wild species in protected areas because crops and their genetic diversity are shaped by human management practices as well as natural selection. While the cost of species preservation in protected areas consists primarily of the opportunity costs of land use and enforcement, the cost of conserving crop genetic resources on farms is the opportunity cost to farmers who grow them, which may be relatively great or small, according to who they are and the economic conditions they face.

Since this chapter is geared primarily to the non-economist reader, it begins with a review of some basic definitions and relevant economics concepts. In particular, the concept of economic 'incentives' is discussed. Next, to illustrate these concepts, summaries of related research are provided with some empirical examples. The process of applying economics methods to farmer plant breeding and on-farm conservation is in its infancy, few published case studies exist, and many problems still need to be resolved.

## Economics Definitions and Concepts

Economics is the study of the decisions people make regarding the allocation and use of resources, based on how they as individuals and as a society value these resources. Resources include human, natural and human-made capital. Economics is a utilitarian discipline; it focuses on human society rather than a biological system, and its purpose is to choose the best means of achieving a predetermined social goal (Randall, 1986). 'Best' is defined according to the incidence of 'costs' and 'benefits' among social groups and the trade-offs that inevitably occur because not all groups can 'win' at the same time. The economic value of crop genetic resources derives from human use, though human use can refer not only to the consumption or commercial sale of a harvested crop, but also to aesthetics, ecosystem and social support functions (Brown, 1990). Economic value can also derive from the recognition that genetic resources are stocks on which the future of humankind depends; concepts of this type are difficult to quantify compared with the value of a traded commodity on a well-functioning commercial market.

The survival of the crop varieties and the genetic diversity of which they are composed *on farms* depends critically on the decisions of the farmers who grow them. In interaction with the elements of the natural environment, farmers' decisions about which crop varieties to grow and in which amounts, combined with the practices they use to select or procure seed and manage their fields, shape the genetic diversity of their crops and therefore the stock of genetic resources that plant breeders and CGR scientists seek to conserve.

As the role of a crop in its agroecosystems is important for us to understand in studying genetic diversity, so is the farmer's place within a social and economic system. This is because the benefits of crop genetic resources, like other natural resources such as air and water, are shared with other people. The seed a farmer plants each year is two things at once: a physical input to crop production and genetic information that is unique to each variety. Therefore it generates two types of value at the same time. The first, 'private' value is the harvest the farmer enjoys either directly as food or feed or indirectly, through the cash obtained by selling it and purchasing other items. The second, 'public' value is the genetic information evolving today in the fields of farmers, from which modern commercial agriculture and future generations of farmers and consumers will also benefit (Smale *et al.*, 2001).

As they are defined in each research context, private and public values of crop varieties are important for understanding farmers' 'incentives' to keep growing certain varieties. When we refer to farmers' 'incentives' to grow crop varieties, we mean the extent to which these varieties have attributes that satisfy farmers' objectives, as *they* define them. Since most small farmers consume a significant proportion of the food crops they produce, such traits often include not only agronomic characteristics such as tolerance of biotic and abiotic stress but also some consumption characteristics such as their suitability for the preparation of special dishes that are of cultural significance in local communities.

Economists generally hypothesize that the changes accompanying modernization reduce the interest farmers have in growing diverse crops and crop varieties. As agriculture becomes more resource intensive and commercialized, farmers tend to specialize in the crops and varieties that they can sell for a profit on the market, directing their choices towards the tastes and preferences of distant urban consumers and buying what they need for their own consumption. Farms grow fewer and larger, and labour moves to urban areas (Pingali, 1997). Sometimes government policies encourage such changes.

It follows that only some farmers will have the economic 'incentives' to maintain diversity as the economic, social and cultural environment in which they grow their varieties changes. In the high potential environments where commercialization occurs, farmers' varieties (FVs, including landraces, modern varieties modified by adaptation to farmers' environments by farmer and natural selection, and progeny from crosses between landraces and modern varieties) may be grown only when unique end-use characteristics for which specialized markets exist cannot be transferred efficiently through advanced breeding techniques (biotechnology) to modern varieties (MVs). In the worst growing environments, crop production may be abandoned entirely, as has already occurred in parts of Asia (Pingali, 1997). When there are limited opportunities for emigration from these environments, farmers may remain on small landholdings growing FVs for subsistence.

Between these two extremes, the prospects for continued landrace cultivation and CPB are more interesting. In more difficult growing environments, FVs may remain the choice of farmers since germplasm that better meets their needs may not be available. Even when an agroclimatic zone is suitable for the production of MVs, the development of commercial seed systems is not sufficient to ensure that modern types will replace FVs because of market imperfections.

As long as farmers themselves find it in their own best interests to grow FVs, both farmers and society as a whole will benefit at no extra cost to anyone. Where genetic diversity is considered to be important, but farmers are revealed to have few social, cultural or market-based incentives to maintain it, then specific publicly funded initiatives may be needed. Public funds are those generated by taxes or donations, transferring income from one segment of the world population to another. Economists generally believe that these forms of interventions are more 'costly' to society than market-based incentives, though the expense may be justified if the magnitude of benefits that accrue to society outweigh the cost (Stiglitz, 1987). Whether this is true or not depends on the society, and the perceptions of social welfare held by those responsible for such decisions, or how decision makers weigh various social goals.

These are hypotheses based on economics principles, assumptions about human behaviour and historical experience. Like any hypotheses, they need to be tested under a wide range of conditions with respect to crop varieties and societies before generalizations can be made. The following sections summarize some approaches that can be used to test them, with the goal of understanding the role of farmers in plant breeding and conserving CGD: how we can assess the current value of varieties to farmers; a summary of a large body of applied economics literature on variety choice with economic change, highlighting only the research that has attempted to link variety choice directly to crop genetic diversity; the limitations of variety choice models for explaining farmer management of crop genetic diversity; and an attempt at explaining the relationship between private and public attributes of the varieties grown by farmers, as these relate to incentives for conservation on farms.

## The Current Value of Crop Varieties to the Farmers who Grow Them

Genetically desirable crop varieties will continue to be cultivated as long as they are valuable to the farmers who grow them. There are at least two features of crop production in many environments where such varieties are still grown that make it difficult to estimate their value to farmers in terms of money or prices.

First, many small farmers in areas of importance for crop genetic diversity are neither fully commercialized growers nor are they strictly consumers of their crop. Often they consume more of their harvest than they sell in one year, and sell more than they buy in another. Under these circumstances the implicit value of their crop output lies somewhere between the producer and consumer prices on the market. The location of the implicit value – or 'price' which governs decision making but cannot be observed – is unique to each farm household and is determined by its socio-economic characteristics.

Second, FVs also differ in their production and consumption attributes. Some yield better in a year with mid-season dry periods, others have a grain type that, when processed on the farm, is particularly well suited for the preparation of a special dish.

The current value of crop varieties accrues to farmers directly as a consequence of their involvement in a local economy and culture, and is derived either from some form of market exchange or their own valuation of the crop, which can be reinforced by the culture of which they are a part. When market valuations are feasible and farmers are commercially oriented, we can assess the relative costs and benefits of growing some varieties using standard economic analysis. When market prices also vary by trait, we can apply hedonic models to estimate the implicit prices consumers assign to particular traits. Hedonic models are consistent with the economic theory of utility and have been applied extensively to study the demand for rice quality in Asia (see Unnevehr *et al.*, 1992).

In the case of cooking quality in Asian rice, consumer preferences are closely linked to observable features of shelled grain, and markets are well developed. This may not be the case for other crops and locations, however. In many cases, market valuations are infeasible or may carry limited information. Then, utility rankings may provide an alternative means of estimating the relative value-in-use of varieties grown by farmers.

'Utility' refers only to the usefulness or overall satisfaction derived from the choices that people make. Utility is not confined to the consumption of physical commodities; nor does it imply that 'everything has a price' (Arrow, 1997: 759). Preference orderings can encompass altruism, peer pressure and culture. In economic theory, utility is represented, for theoretical convenience, as a continuous mathematical function that embodies preference orderings over goods or states, and rules out certain types of behaviour such as an infinite preference for one thing over another. When the goods and states over which it is defined are measurable and divisible, then marginal variations in their use are comparable and can be equated to changes in purchasing power to derive predictions of behaviour known as demand functions.

Since the obstacles to measuring utility are insurmountable, utility is largely used in applied economics as an analytical device. The axiom of revealed preference can be invoked when working with observed preferences. The axiom says that if an individual with some price and income level can afford two alternatives but chooses one, that choice is revealed as 'preferred' and is consistent with maximizing a utility function. We might say, for example, that a farmer who grows more of one variety than another, though its per unit costs are at least as high, is expressing a revealed preference for that variety.

Sociologists and anthropologists have developed relatively straight-forward methods for eliciting utility rankings (for example Chambers, 1988). When there are no market prices to use in assessing the relative value of varieties to farmers, or market information is imperfect, it is possible to: (i) elicit from farmers the characteristics that matter most to them, including both production and consumption characteristics; and (ii) ask them to assess the extent to which each variety of interest satisfies these characteristics. Production and consumption character-istics might include 'costs' of production in terms of hours of crop management or on-farm processing.

Table 3.1 provides an example of such analyses. Both men and women in 240 survey households were asked to rate a set of 25 maize characteristics in terms of their importance (1 = very important; 2 = more or less (neither very important nor unimportant); 3 = not important). This table shows the percentage of men and women respondents who rated each maize characteristic as 'very important'. For all respondents, the most important five characteristics were: (i) drought-tolerance; (ii) resistance to insects in storage; (iii) disaster avoidance; (iv) yield by weight; and (v) the taste of tortillas. The relative importance of some characteristics differs between men and women. For example, though grain weight and yield per hectare were of equal importance, producing 'something' even in a bad year, and a number of consumption traits were more important to women than to men. The next step in the analysis was to ask men and women the extent to which different maize types met these needs. The analysis revealed differences in the perceived performance of maize types by characteristic. Modern maize types were particularly weak on

**Table 3.1.** Men and women's perceptions of most important maize variety characteristics (source: Smale *et al.*, 1999).

| Characteristic | | Percentage of decision makers rating characteristic as 'very important' | | |
| --- | --- | --- | --- | --- |
| | | Men | Women | Both |
| Grain weight (kg per volume) | | 76.3 | 76.6 | 76.4 |
| Grain yield (kg ha$^{-1}$) | | 52.8 | 66.1 | 59.4 |
| Length of growing period | | 46.5 | 46.9 | 46.7 |
| Produces 'something' even in bad years | * | 63.8 | 89.8 | 76.8 |
| Drought-tolerant | | 91.1 | 89.9 | 90.5 |
| Weed-tolerant | | 26.7 | 39.8 | 33.2 |
| Disease-resistant | * | 31.5 | 61.4 | 46.4 |
| Resistant to insects in storage | | 79.7 | 75.5 | 77.6 |
| Taste of tortillas | * | 50.8 | 78.4 | 64.6 |
| Good for atole[a] | * | 34.0 | 60.2 | 47.1 |
| Good for tamales[a] | * | 14.9 | 38.4 | 26.6 |
| Good for pozol[a] | * | 8.3 | 25.4 | 16.9 |
| Good for tlayudas[a] | | 27.5 | 50.7 | 39.1 |
| Good for forage | | 30.9 | 51.4 | 41.2 |
| Good for feed | | 37.1 | 50.0 | 43.1 |
| Good for sale | | 32.4 | 53.6 | 43.0 |
| Produced with little labour | | 37.4 | 43.5 | 40.0 |
| Produced with few purchased inputs | | 48.2 | 57.5 | 52.9 |

*Chi-square test of homogeneity shows significant differences between men and women at the 0.01 level of significance.
[a]Local maize foods primarily produced in the household.

consumption traits and superior for fodder and grain yield under favourable growing conditions.

Either hedonic analysis or utility rankings can be used to help identify benefits that might be gained through CPB by indicating which varieties and traits are most important to farmers. Both methods are relatively simple to make operational but have recognizable limitations. First, we often have a long list of characteristics that may make it cumbersome to use the results in other types of statistical analysis. If the purpose of the analysis is to identify the pros and cons of different varieties from the viewpoint of a sample of farmers who both produce and consume the crop, these methods are likely to be adequate.

Second, to be of use for CPB, we need to be able to relate the abstract 'characteristics' that interest farmers to the crop's physiological traits as recognized by plant breeders. Research by Louette and Smale (2000), Soleri *et al.* (2000) and Soleri *et al.* (Chapter 2, this volume) are examples of efforts to systemically relate the selection criteria and genetic perceptions of farmers to the phenological and morphological descriptors, and the genetic characteristics such as genetic variance and heritability used by plant breeders. Finally, it is important to consider the possibility that, if CPB is successful, it may alter the relative rank of varieties and/or change either the private or the public genetic value of a variety, enhancing or diminishing crop genetic diversity.

In general, values assessed in this fashion are current and do not inform us about what happens as economies and cultures change. For this purpose, applied economists use economic theory to design econometric models, testing hypotheses statistically with data from a relatively large sample of farm households in which cross-sectional variation is substituted for temporal variation. These are discussed in the next section.

## Economic Change and Opportunity Costs

In much of the literature about crop genetic diversity, the overriding concern has been whether FVs will continue to be grown at all, rather than whether those that are grown are more genetically desirable or diverse than others. This concern has reflected the beliefs that: (i) economic change inevitably leads to the substitution of MVs for FVs; and (ii) MVs are less diverse. As a consequence, on-farm conservation of crop genetic diversity – continued farmer management of FVs – is necessarily associated in this literature with major economic costs in terms of development opportunities forgone (Wright, 1997).

Most of the research by applied economists has therefore focused on the determinants of the probability that farmers will grow MVs or FVs, even though farmers often choose to grow both! Viewed as partial adoption, this observed pattern is not usually explained by conventional neoclassical economics. In a neoclassical economic model of decision making, an unconstrained, risk-neutral farmer who maximizes profits would choose to grow only the variety with the highest profits per hectare. Understanding why a farmer grows more than one variety has motivated economists to propose a number of competing explanations that are extensions of the neoclassical model. Broadly categorized, alternative explanations include farmer attitudes towards yield, price or consumption risk (Just and Zilberman, 1983; Hammer, 1986; Carter and Wiebe, 1990; Fafchamps, 1990), experimentation and learning under uncertainty (Hiebert, 1974; Tsur *et al.*, 1990), missing (de Janvry *et al.*, 1991) or imperfect (McGuirk and Mundlak, 1991) markets for fertilizer, or a jointly produced output, such as fodder (Renkow and Traxler, 1994).

In this literature, growing MVs and FVs simultaneously is referred to as partial adoption and is considered transitional. We might argue instead that the coexistence of MVs and FVs represents an equilibrium if one or several of the factors mentioned above persists despite economic change. Meng *et al.* (1998) concluded that multiple factors, including missing markets, yield risk, grain quality and agroclimatic constraints influence the probability that a Turkish household will grow a wheat landrace; a change in any single economic factor is unlikely to cause farmers to cease growing it. Case studies demonstrate that in many of the regions of the developing world where FVs are still grown, either markets for commercially produced seed or markets for the attributes demanded by farmers are incomplete (Brush *et al.*, 1992; Brush and Meng, 1998; Smale *et al.*, 2001). In such cases, farmers must rely on their own or neighbours' production for their seed supply.

Though environments may be favourable for the production of MVs, economic growth occurs, and commercialization is complete, the coexistence of MVs and FVs may persist through the development of niche markets or markets for specialized attributes. In advanced agricultural economies, demand grows for arrays of increasingly specialized goods and services (Antle, 1999), which may support limited markets for 'heirloom' varieties.

In summary, concerns for the high costs of on-farm conservation follow from the assumption that MVs will inevitably replace FVs as economies develop. While this may be true in many parts of the world, some evidence suggests that there will be some areas where: (i) MVs will not be grown at all; (ii) MVs will be cultivated along with FVs because it is economically optimal for farmers; or (iii) FVs will continue to be grown even in fully commercialized agriculture as 'heirlooms' that have value in some market niches. Therefore, the opportunity costs of growing FVs are not uniformly high. Moreover, crop genetic diversity is not simply a question of whether modern or traditional types are grown. In other words, there is a role for CPB.

## Variety Choice and Crop Genetic Diversity

Clearly, the probability that farmers will grow crop varieties is positively related to their market value or to the utility rankings they ascribe to them. But the models of variety choice developed so far have limitations when used to address plant breeding and on-farm conservation issues.

First, such models have not incorporated environmental variation. Typically, differences in growing environment have been treated indirectly with statistical distributions or in fairly simplistic ways with proxy variables. In most cases, economic data were not collected in conjunction with detailed measurements on physical and biological variables. Bellon and Taylor's 1993 article is a rare example in which soil quality is treated explicitly. Research in Kenya, summarized by Hassan (1998), has demonstrated the usefulness to plant breeding of nesting social and economic data collection in a geographically referenced database.

Second, we need to be able to separate the determinants of diversity that lie outside farmers' control or influence from those that are a direct consequence of their choices and actions. For example, the extent to which crop genetic diversity is determined by variety choice depends very much on the breeding system of the crop, the crop's natural environment, and a number of farmer practices other than the choice of which varieties to plant each season. A species' breeding system affects the exchange of genetic material within and between varieties. There are great differences between cross-pollinating, self-pollinating and vegetatively propagated crops. Though the plant's breeding system is not within the control of farmers, farmers' decisions concerning plot location, as well as their seed procurement and exchange practices with other farmers, can influence the migration of new genetic material into their crop varieties. When they select the seed for future generations of varieties, they also exert selection pressures on the agro-morphological characteristics that matter to them, including maturity, unique aspects of adaptation, or particular uses or even qualities of its useful parts (Brown, 2000; Jarvis and Hodgkin, 2000).

Third, though we can test hypotheses statistically regarding the economic factors that influence farmers' choices among varieties, the results have no meaning to plant breeders or conservationists unless they are linked to knowledge of their genetic structure. So far, published attempts to relate microeconomic models of variety choice to genetic analyses are few. At the same time, formal studies that have related farmer selection criteria to genetic structure have not been linked to microeconomic theory.

Brush *et al.* (1992) defined potato diversity as the number of FVs cultivated by a farm household, modelling it simultaneously with the area planted to improved types as a function of household production characteristics and socio-economic status. Using number of FVs as an indicator of diversity assumes that each FV contributes equally to genetic diversity, which is not likely to be the case. In Turkey, Meng (1997) analysed the probability that a household cultivates wheat FVs as a function of household attitudes towards risk, agroecological conditions on the farm and market development. The diversity among FVs, calculated from measurements taken on their morphological characteristics, was specified as an outcome of variety choice. Relating wheat diversity directly to variety choice was difficult in that study because morphological variation was observed and measured only for FVs.

Smale *et al.* (2001) analysed percentage of maize area farmers allocate to named maize FVs (area shares) as a function of: (i) variety traits; (ii) household characteristics; (iii) production environment and market infrastructure of the region; and (iv) the maize diversity farmers observe in their communities.

Table 3.2 shows the results of a two-limit tobit estimation, a procedure commonly used for econometric regressions with censored dependent variables, which relates area shares to variety attributes, household characteristics, maize diversity in the community, and agro-ecological and marketing features of the region in which the communities and their farmers are located. Hypothesis tests were conducted with Z-tests on individual variables and likelihood ratio tests on sets of variables. Results demonstrate the importance of variety attributes such as suitability for preparing tortillas, relative to price and costs or socio-economic characteristics of the household, in determining area shares. This made sense given that costs and returns differed little among the FVs; less so, for example, than would be apparent between FVs and MVs. Regional features, such as infrastructure and productivity potential in interaction with infrastructure, were of importance in explaining the area shares planted to named varieties, as well as the number of soil types on the farm. Maize diversity in the community was measured with: (i) richness and evenness indices adapted from ecological indices of species diversity; and (ii) farmers' perceptions of 'loss' of varieties in the community. Richness and evenness indices were constructed with data on the frequencies of racial types among maize ears sampled from farmers' harvests. Races of maize in Mexico are defined by plant and ear characteristics that are visible to farmers (Wellhausen *et al.*, 1952), providing a convenient overarching taxonomy for varieties that shared the same name but differed morphologically (or vice versa) across study sites. Of the three diversity variables, however, only farmers' perceptions were statistically significant in explaining the areas farmers allocate among maize varieties.

Other hypotheses than those tested in the Smale *et al.* (2001) study are essential to on-farm conservation and CPB efforts, including: (i) To what extent does the variety choice of farmers determine genetic diversity? (ii) Which variables explain greater (lesser) cultivation of the named varieties that contribute most to maize genetic diversity? To formulate and test such hypotheses, more genetic information on the population structure and more interdisciplinary work is essential. It is important to consider the possibility that if CPB is successful, it may alter the relative rank of varieties and/or change both the private and

**Table 3.2.** Estimated variety area share equation (source: Smale *et al.*, 2001).

| Explanatory variable | Marginal effects | SE |
|---|---|---|
| Constant | 0.619++ | 0.354 |
| Productivity potential of region | 0.0742 | 0.0508 |
| Infrastructure development of region | 0.274+ | 0.0877 |
| Productivity–infrastructure interaction | −0.233+ | 0.0839 |
| Variety best for sale = 1, 0 otherwise | 0.137 | 0.127 |
| Variety best for tortillas, 0 otherwise | 0.286+ | 0.113 |
| Variety least cost in inputs, 0 otherwise | 0.115 | 0.115 |
| Variety best for preparing a special dish = 1, 0 otherwise | −0.213+ | 0.112 |
| Variety best for avoiding disastrous harvests = 1, 0 otherwise | 0.119 | 0.117 |
| Variety best for livestock feed or forage = 1, 0 otherwise | 0.072 | 0.155 |
| Household total maize area preceding year | −0.00114 | 0.00395 |
| Household farm has irrigated land = 1, 0 otherwise | 0.0532 | 0.0676 |
| Household farm has only rainfed land = 1, 0 otherwise | −0.0237 | 0.0572 |
| Number of soil types on household farm | −0.0178** | 0.138 |
| Percentage of maize output sold, preceding year | −0.000207 | 0.000493 |
| Household received remittances in preceding year | 0.0259 | 0.0421 |
| Maize percentage of farm cultivated area, preceding year | −0.0578 | 0.584 |
| Household tractors owned = 1, 0 otherwise | −0.0544 | 0.0626 |
| Margalef richness index for landraces in community | 0.00622 | 0.0334 |
| Evenness richness index for landraces in community | −0.212 | 0.381 |
| Farmer perceived losses of varieties in communities | −0.132+ | 0.0406 |

*Test statistics for likelihood ratio tests (d.o.f., level of significance) of joint hypothesis*

| | | |
|---|---|---|
| Equation | $\lambda(20,0.05)$ | 119.43 |
| Maize diversity in community | $\lambda(3,0.05)$ | 11.27 |
| Agroecology/infrastructure | $\lambda(6,0.05)$ | 17.70 |
| Household characteristics | $\lambda(8,0.05)$ | 6.35 |
| Variety attributes | $\lambda(6,0.05)$ | 107.96 |

$n = 319$; dependent variable is area share planted by household to each farmer-named variety; marginal effects are partial derivatives of expected value, computed as means of variables.
+ significant at 0.05 with two-tailed Z test; ++ significant at 0.10 with two-tailed Z test;
**significant at 0.10 with one-tailed Z test.

public genetic value of a variety, either enhancing or diminishing crop genetic diversity.

Another limitation of much of the variety choice literature is the assumed negative relationship between MVs and crop genetic diversity. In some cases the presence of MVs among FVs expands the breadth of the traits available to farmers (Dennis, 1987). When MVs serve for generating cash, they can support the production of FVs which satisfy other consumption needs in farm households (Zimmerer, 1996). However, dividing a finite planting area between a few local populations with rare alleles and MVs with common alleles may not lead to much of a net gain in terms of the public value of conservation. Testing the relationship between MVs and crop genetic diversity with a good research design would advance scientific knowledge. This will be especially important in CPB efforts, which may include the introduction of exotic materials or the improvement of certain varieties within a larger population structure.

## The Public and Private Characteristics of Seed

As stated above, seed has both private and public characteristics. These characteristics have important implications for the way farmers maintain genetic diversity in their fields (Heisey *et al.*, 1997; Smale *et al.*, 2001) and the way seed industries are organized (Morris *et al.*, 1998).

In economic theory, the extent to which any economic good is classified as public or private along a continuum depends on two essential criteria: subtractability and excludability. Because a handful of seed is both a physical input to crop production and a genetic resource, the extent to which it is considered to be public or private depends on whether we are considering it as a farm production input or as a genetic input to crop improvement by breeders.

*Subtractability* refers to the degree to which use of a good or service by one person precludes its use by another. Seed as a seasonal input to crop production has high subtractability because no two farmers can plant the same bag of seed. Seed as a genetic resource has low subtractability since many farmers can grow the same genetic material (e.g. variety or hybrid) simultaneously (Morris *et al.*, 1998).

*Excludability* refers to the extent to which a seller of a good or service can deny access to other users. Seed as a physical input has high excludability since anyone who possesses the seed, including farmers or companies, can deny others access by refusing to give it to them or sell it to them below a certain price. The degree of excludability differs for different methods of seed production, however. Seed of either open-pollinated or self-pollinating crops has low excludability since it is easily reproduced by obtaining a few seeds. Hybrid seed on the other hand has high excludability, because it can only be reproduced by those who know the pedigree or have access to the parent lines (Morris *et al.*, 1998).

In addition, both seed as a physical input and seed as a genetic resource score low on a third criterion of economic goods: *transparency*. Potential users cannot easily observe whether it is good quality simply by looking at it in a bag. For that reason, farmers prefer to see a new variety grown out in the fields of other farmers before they buy the seed themselves, in order to combine observations on its performance with evidence of its form, colour and shape as a physical unit of seed. Similarly plant breeders wish to test and evaluate new materials before using them in breeding programmes.

These three attributes of the seed of crop varieties have important implications for CPB. First, lack of transparency creates an asymmetry between sellers who know the seed and buyers who cannot easily ascertain its value. This leads to such institutions as government regulation of seed quality and the practice of purchasing from companies or farmers whose seed is known to have been good in the past (brand or farmer loyalty). As most plant breeders have experienced first-hand, farmers will not 'adopt' either the seed produced by the efforts of other farmers or the seed imported into their area through a formal seed production system until they are convinced, by observed trial and error, that planting it may be worth the gamble.

Second, the fact that subtractability and excludability of seed as a genetic resource varies by the seed production system of the crop means that institutional arrangements needed to effectively develop, produce and distribute improved seed will tend to differ (Morris *et al.*, 1998). The combination of high subtractability and high excludability, as in hybrid seed for either cross- or self-pollinating species, is associated with products that can be handled effectively through commercial markets since buyers have incentives to pay for it and sellers will realize a return on their investment. The combination of low subtractability and low excludability, as in improved varieties of either self- or open-pollinated species, but in a far more pronounced way for open-pollinated, is associated with products than cannot be handled well through commercial markets. The farmer is less willing to pay and the seller is not able to gain a return on research investments. For example, this explains why, even if modern seed types could be produced that would generate advantages for them over FVs, there will continue to be many farmers 'left out' of commercial maize production in the developing world. Meeting their needs requires alternative institutional arrangements, such as CPB combined with less formal seed systems or small-scale seed multiplication by farmers.

These attributes of seed also have implications for managing crop genetic diversity. For example, in a system of FVs, one farmer's enjoyment of a diverse local genepool does not necessarily diminish another's. It is hard to exclude another farmer from using any variety belonging to a local genepool since it can be reproduced with a few seeds, although there are numerous examples of farmers' desire and attempts, sometimes successful, to do so (Soleri *et al.*, 1994; Cleveland and Murray, 1997). In any local system, farmers may be aware of changes in the supply of distinct varieties or genetic materials in their communities, although the extent to which they can, as individuals, exert control over it is likely to be slight as, for example, the flow of genes in the pollen of cross-pollinating crops (Bellon and Brush, 1994; Louette, 1994).

Individually, farmers can influence the genetic structure of the group of varieties they grow through allocating more land to one or the other. In south-east Guanajuato, Mexico, there was a high correlation between farmers' recognition of loss of distinct materials in communities and the numbers of varieties they grew (Aguirre, 1999). Farmers' deliberate decisions to save seed from the harvest or mix it with seed procured from other sources also affect the genetic diversity of the varieties they grow. In some cases, they may even be able to influence the variety choices of their neighbours. With respect to the disappearance of some materials when farmers cease to cultivate them, farmers can organize socially or be predisposed culturally to consider the effects of their actions on others and vice versa. Then the 'externality', whether it is positive or negative, can be 'internalized', meaning that what happens to the crop genepool in a local community matters to individual farmers and figures in their variety choice.

At a more aggregate level, the fact that improved wheat varieties are low in subtractability, excludability and transparency can explain why farmers as a group are not likely to choose the portfolio of varieties that provides the best genetic resistance to disease (Heisey *et al.*, 1997). First, they may grow varieties that are higher yielding though susceptible to disease; planting them in any given year is to gamble on whether the disease will occur and, if so, whether losses will be great enough to offset the relative gain in yield potential. Second, since the genes that confer resistance are invisible to them, many farmers may choose to grow varieties that happen to have similar genetic bases of resistance. When many farmers choose to grow a variety or a set of varieties with the same genetic mechanism for resistance, the probability of crop disease epidemics rises. If epidemics occur, both individual farmers and society in general may suffer, particularly if the costs of disease control are high and the crop losses are large. The genetic mechanism for resistance bestows a public benefit through forgoing losses when it is effective, and a public cost when it is overcome. In cropping systems dominated by MVs, expected disease losses can be reduced by ensuring a more diversified genetic background in released varieties, promoting a greater evenness in the spatial distribution of area among varieties carrying different types of resistance, or by enforcing a temporally changing list of varieties known to be resistant. All of these measures require public investments in terms of government regulations, however. In the empirical application shown in Heisey *et al.* (1997), which is based on variety area shares sown to wheat varieties in the Punjab of Pakistan over time, a more genealogically diverse area-weighted mix of modern wheat varieties would, in fact, have been associated with lower farm-level production, since in the time period covered there was not a high incidence of rust disease. Many farmers chose to grow genealogically similar varieties with higher yields in the absence of rust, even though they may have been more susceptible to that disease. Genealogical similarity was measured in terms of coefficient of parentage.

Empirical results of such studies clearly depend on the setting and diversity indicator employed (as with the studies discussed in the preceding section). Heisey *et al.* (1997) employed coefficients of parentage as an indicator of genealogical diversity in wheat MVs. Use of another indicator might have generated different results. Coefficients of parentage have been used as indicators of latent genetic diversity but their limitations are well-known. In a system dominated by traditional FVs, coefficients of parentage could not be used because parentage is unknown. Measurement problems aside, a trade-off between crop yields and a more diverse bundle of varieties grown by farmers in a region over a series of individual seasons is in no way inconsistent with a longer-term positive association between advances in yield potential and the incorporation in breeding stocks of novel sources of genetic diversity. In fact, there are theoretical and empirical bases for expecting that over the longer term genetic diversity and increased yield could be positively correlated as a result of successful plant breeding programmes involving farmers and scientists (Ceccarelli and Grando, Chapter 12, this volume; Joshi *et al.*, Chapter 10, this volume; Ríos Labrada *et al.*, Chapter 9, this volume).

## Summary

Economic analysis can offer an important perspective on farmers' knowledge of their crop varieties, including their management decisions, and thus on their incentives to collaborate with plant breeders. Since the choices of farmers figure so heavily in whether varieties will be grown tomorrow, much of the work of applied economists has focused on explaining variety choice. This work has been hampered by an obvious bias: the belief that all farmers' varieties will eventually be replaced by MVs as the inevitable outcome of economic development and cultural change. More recent work has questioned this assumption. This chapter has argued that there may indeed be instances in which FVs, and FVs grown simultaneously with MVs, are the economically optimal choices of farmers. Understanding when this is true requires tests of economics hypotheses. Finally, whether this is positive in terms of the conservation of genetic variation will depend on additional variables including actual size of the area sown, effective population size and mating system.

Information from economic analyses is of little relevance to the design of agricultural innovations or to the identification of policies to support on-farm conservation unless it is linked to the analyses of geneticists, plant breeders and other social scientists. Farmers work with varieties and their observable traits rather than the unobservable genes they embody, and crop genetic diversity cannot be fully predicted from farmers' varietal choice unless farmers' named varieties are quantitatively defined in terms of genetic diversity. Unless we can associate data on farmers' valuation of varieties with data on their potential as genetic resources, we cannot speak concretely about the varieties whose cultivation we seek to support.

A critical economic issue amplified by the discussion of CPB for CGR conservation is that varieties have both private value as the seed used by farmers to produce a crop and public value as germplasm. For economists, the fact that varieties have both signals an important divergence that may occur between what farmers choose to do and what society in general would choose when it takes the needs of future generations into account. Whether or not such a divergence occurs in practice, and its extent, depends on the crop, community, agroecological and economic environment in which farmers make choices.

To encourage a farmer to grow a crop population that is genetically desirable in genetic terms for society in general, but is not recognized by farmers or their communities as valuable, would require other public-funded initiatives, such as subsidies, taxes or promotion, to redirect (or in economic terms to 'distort') farmers' choices. Such actions are costly since they require public funds, and typically involve some general losses to society due to a reallocation of resources from which no one in particular gains. However, economic theorists have advanced a number of arguments to justify such government interventions, which include public goods and externalities similar to those described above for germplasm. What is justifiable and under which circumstances will depend on the society and decision makers in question.

CPB offers an important potential alternative for increasing both the private and the public value (utility) of individual varieties and of varietal mixes grown by farmers. CPB can do this either through increasing the production value of FVs so that they maintain much of their CGD, or through the development and adoption of exotic varieties, including in some instances MVs, that complement FVs either in genetic or in economic terms and so contribute to CGD on farms.

## Acknowledgements

An earlier version of sections 1–3 was developed for *A Training Guide to In Situ Conservation On-Farm*, published by the International Plant Genetic Resources Institute, Rome, Italy. The ideas presented draw principally from the author's work with scientists and economists at CIMMYT and IPGRI. In particular, the author would like to acknowledge the work of Alfonso Aguirre, Mauricio Bellon, Pablo Eyzaguirre, Devra Jarvis, Paul Heisey, Toby Hodgkin, Erika Meng and Michael Morris, as well as that of the editors of this volume, Daniela Soleri and David Cleveland.

## References

Aguirre, J.A. (1999) Análisis regional de la diversidad del maíz en el Sureste de Guanajuato. Tésis de doctorado, Universidad Nacional Autónoma de México, Facultad de Ciencias, México, D.F.

Antle, J.M. (1999) The new economics of agriculture. Presidential Address prepared for the Annual Meetings of the American Agricultural Economics Association, Nashville, August, 1999.

Arrow, K.J. (1997) Invaluable goods. *Journal of Economic Literature* 35, 757–765.

Bellon, M.R. and Brush, S.B. (1994) Keepers of maize in Chiapas, Mexico. *Economic Botany* 48, 196–209.

Bellon, M. and Taylor, J.E. (1993) 'Folk' soil taxonomy and the partial adoption of new seed varieties. *Economic Development and Cultural Change* 41, 763–786.

Brown, A.H.D. (2000) The genetic structure of crop 'landraces' and the challenge to conserve them in situ on farms. In: Brush, S.B. (ed.) *Genes in the Field: On Farm Conservation of Crop Diversity*. IPGRI, Rome; IDRC, Ottawa; Lewis Publishers, Boca Raton, Florida.

Brown, G.M. (1990) Valuation of genetic resources. In: Orians, G.H., Brown, G.M. Jr, Kunin, W.E. and Swierbinski, J.E. (eds) *The Preservation and Valuation of Biological Resources*. University of Washington Press, Seattle.

Brush, S.B. and Meng, E.C.H. (1998) Farmers' valuation and conservation of crop genetic resources. *Genetic Resources and Crop Evolution* 45, 139–150.

Brush, S.B., Taylor, J.E. and Bellon, M.R. (1992) Biological diversity and technology adoption in Andean potato agriculture. *Journal of Development Economics* 38, 365–387.

Carter, M.R. and Wiebe, K.D. (1990) Access to capital and its impact on agrarian structure and productivity in Kenya. *American Journal of Agricultural Economics* 72, 1146–1150.

Chambers, R. (1988) *An Interim Note on Ranking Methods*. Institute of Development Studies, University of Sussex, Brighton, UK.

Cleveland, D.A. and Murray, S.C. (1997) The world's crop genetic resources and the rights of indigenous farmers. *Current Anthropology* 38, 477–515.

Dennis, J.V. (1987) Farmer management of rice variety diversity in Northern Thailand. PhD thesis, Cornell University, Ithaca, New York.

Fafchamps, M. (1990) Cash crop production, food price volatility, and rural market integration in the Third World. *American Journal of Agricultural Economics* 74, 90–99.

Hammer, J.S. (1986) Subsistence first: farm allocation decisions in Senegal. *Journal of Development Economics* 23, 357–369.

Hassan, R.M. (ed.) (1998) *Maize Technology Development and Transfer: a GIS Application for Research Planning in Kenya*. International Maize and Wheat Improvement Center (CIMMYT), Mexico D.F.; Kenya Agricultural Research Institute (KARI), Nairobi; CAB International, Wallingford, UK.

Heisey, P.W., Smale, M., Byerlee, D. and Souza, E. (1997) Wheat rusts and the costs of genetic diversity in the Punjab of Pakistan. *American Journal of Agricultural Economics* 79, 726–737.

Hiebert, D. (1974) Risk, learning, and the adoption of fertilizer responsive varieties. *American Journal of Agricultural Economics* 56, 764–768.

de Janvry, A., Fafchamps, M. and Sadoulet, E. (1991) Peasant household behaviour with missing markets: some paradoxes explained. *Economic Journal* 101, 1400–1417.

Jarvis, D. and Hodgkin, T. (2000) Farmer decision making and genetic diversity: linking multidisciplinary research to implementation on-farm. In: Brush, S.B. (ed.) *Genes in the Field: On Farm Conservation of Crop Diversity*. IPGRI, Rome; IDRC, Ottawa; Lewis Publishers, Boca Raton, Florida.

Just, R.E. and Zilberman, D. (1983) Stochastic structure, farm size and technology adoption in developing agriculture. *Oxford Economic Papers* 35, 28–37.

Louette, D. (1994) Gestion Traditionnelle de Variétés de Maïs dans la Réserve de la Biosphère Sierra de Manantlán (RBSM, états de Jalisco et Colima, Méxique) et Conservation In Situ des Ressources Génétiques de Plantes Cultivées. Thèse de doctorat, Ecole Nationale Supérieure Agronomique de Montpellier, Montpellier.

Louette, D. and Smale, M. (2000) Farmers' seed selection practices and maize variety characteristics in a traditional Mexican community. *Euphytica* 113, 25–41.

McGuirk, A.M. and Mundlak, Y. (1991) *Incentives and Constraints in the Transformation of Punjab Agriculture*. IFPRI Research Report 87. International Food Policy Research Institute (IFPRI), Washington, DC.

Meng, E.C.H. (1997) Land allocation decisions and *in situ* conservation of crop genetic resources: the case of wheat landraces in Turkey. PhD thesis, University of California, Davis, California.

Meng, E.C.H, Taylor, J.E. and Brush, S.B. (1998) Implications for the conservation of wheat landraces in Turkey from a household model of varietal choice. In: Smale, M. (ed.) *Farmers, Gene Banks, and Crop Breeding: Economic Analyses of Diversity in Wheat, Maize, and Rice*. Kluwer Academic Publishers, Boston and CIMMYT Mexico, D.F.

Morris, M., Rusike, J. and Smale, M. (1998) Maize seed industries: a conceptual framework. In: Morris, M. (ed.) *Maize Seed Industries in Developing Countries*. Lynne Rienner, Boulder, Colorado.

Pingali, P. (1997) From subsistence to commercial production systems. *American Journal of Agricultural Economics* 79, 628–634.

Randall, A. (1986) Human preferences, economics, and the preservation of species. In: Norton, B.G. (ed.) *The Preservation of Species: the Value of Biological Diversity*. Princeton University Press, Princeton, New Jersey.

Renkow, M. and Traxler, G. (1994) Incomplete adoption of modern cereal varieties: the role of grain-fodder tradeoffs. Selected paper, *American Agricultural Economics Association Annual Meetings, San Diego, 7–10 August 1994*.

Smale, M., Aguirre, A., Bellon, M., Mendoza, J. and Rosas, I.M. (1999) Farmer management of maize diversity in the Central Valleys of Oaxaca, Mexico: CIMMYT/INIFAP 1998 Baseline Socioeconomic Survey. CIMMYT Economics Working Paper 99–09. CIMMYT, Mexico, D.F.

Smale, M., Bellon, M. and Aguirre, A. (2001) Maize diversity, variety attributes, and farmers' choices in southeastern Guanajuato, Mexico. *Economic Development and Cultural Change* 50(1), 201–225.

Soleri, D., Cleveland, D.A., Eriacho, D., Bowannie, F. Jr, Laahty, A. and Zuni Community Members (1994) Gifts from the Creator: intellectual property rights and folk crop varieties. In: Greaves, T. (ed.) *IPR for Indigenous Peoples: a Sourcebook*. Society for Applied Anthropology, Oklahoma City, Oklahoma, pp. 21–40.

Soleri, D., Smith, S.E. and Cleveland, D.A. (2000) Evaluating the potential for farmer and plant breeder collaboration: a case study of farmer maize selection in Oaxaca, Mexico. *Euphytica* 116, 41–57.

Stiglitz, J. (1987) Some theoretical aspects of agricultural policies. *The World Bank Research Observer* 2, 43–60.

Tsur, Y., Sternberg, M. and Hochman, E. (1990) Dynamic modelling of innovation process adoption and learning. *Oxford Economic Papers* 42, 336–355.

Unnevehr, L.J., Duff, B. and Juliano, B.O. (eds) (1992) *Consumer Demand for Rice Grain Quality*. IRRI, Los Baños, the Philippines; IDRC, Ottawa.

Wellhausen, E., Roberts, J., Roberts, L.M. and Hernández, X.E. (1952) *Races of Maize in Mexico: Their Origin, Characteristics, and Distribution*. The Bussey Institution, Harvard University, Cambridge, Massachusetts.

Wright, B.D. (1997) Crop genetic resource policy: the role of ex situ genebanks. *Australian Journal of Agricultural and Resource Economics* 41(1), 81–115.
Zimmerer, K.S. (1996) *Changing Fortunes: Biodiversity and Peasant Livelihood in the Peruvian Andes.* University of California Press, Berkeley, California.

# Social and Agroecological Variability of Seed Production and the Potential Collaborative Breeding of Potatoes in the Andean Countries

**4**

KARL ZIMMERER

*Department of Geography, University of Wisconsin, 550 N Park St, Madison, WI 53706, USA*

## Abstract

Small-scale Andean farmers of Peru manage highly diverse repertoires of potatoes of both traditional farmer varieties (FVs) and improved, modern varieties (MVs). Much potential exists for the combination of local and scientific knowledge in collaborative plant breeding (CPB) for these farmers and their crops. This chapter is based on long-term fieldwork with Quechua farmers in the area of Eastern Cuzco. Its focus is the role of socioenvironmental and agroecological variability in seed management. Farmer knowledge and selection practices concerning the size of seed potatoes are different according to socio-economic level and gender. The majority of poorer farmers prefer seed tubers of the medium and small-medium categories. The intra-varietal knowledge that supports their preference is an important complement to variety-level preferences and know-how. Agroecological variability is important in the movement of seed potatoes across the landscape. These seed flows demonstrate that seed production and exchange networks are oriented towards 'farm spaces' defined by farmers in terms of both management and environmental variables, and not to narrow tiers of agroenvironments based solely on elevation. The results suggest that successful CPB will have to incorporate understanding of farmers' intra-variety preferences and the agroecology of farmers' varieties that are conditioned through existing management practices.

## Introduction

The majority of the production of potatoes, the world's premier tuber crop, is based on modern varieties (MVs) of *Solanum tuberosum* subsp. *tuberosum*, the so-called Irish potato. The MVs account for nearly 100% of the potato crop of the US, Canada and Europe, which together amount to about 22% of world production. The MVs are common, though less predominant, in other world regions, where farmer varieties (FVs) are utilized to varying extents. The FVs, which include landraces or traditional varieties, are potatoes produced among the cultivators of farm communities who often maintain and innovate with them for many generations. The FVs also include those varieties that were bred approximately 30 years ago or longer with the use of modern agronomic science and that have since been modified and maintained by farmers (Salaman, 1985; Horton, 1987; Douches *et al.*, 1996; Almekinders and Louwaars, 1999).

The use of FVs, along with MVs, is a cornerstone of potato production in the Andean countries of Peru, Bolivia, Ecuador, Colombia, Venezuela and Chile. Approximately 8–10 million small-scale farmers – about half of whom are Quechua and Aymara Indian peasants – rely on both FVs and MVs of the potato crop as staple foodstuffs as well as for seed production and market sales (Mayer, 1980; Horton, 1986; van der Ploeg, 1990; Zimmerer, 1991a, 1996; Brush, 1992, 1995; Franco and Godoy, 1993). These farmers make use of modest-sized fields, whose area typically sums to less than 2 ha per farmer. Their agriculture as well as their non-farm work activities is integrated tightly into markets, while some production is typically slated for local consumption. Their agriculture tends to be highly diversified, risk averting and often part-time. Both their FVs and MVs belong to an ever-evolving complex of eight species of domesticated potatoes (Hawkes, 1990; Zimmerer, 1998a). The Andean potato or *S. tuberosum* subsp. *andigenum* is especially diverse and contains approximately four-fifths of the 2000 potato varieties that are estimated to be in current cultivation (CIP, 1997). Geographic ranges and diversity of the cultivated potatoes display an area of maximal convergence in the Andean highlands between central Peru and central Bolivia (Hawkes, 1990).

Locally relevant plant breeding and improvements in seed production are needed to aid the Andean small farmers who incorporate the FVs into contemporary farm strategies (Brush, 1991; Eyzaguirre and Iwanaga, 1996; Tripp, 1996; Thiele *et al.*, 1997; Almekinders and Louwaars, 1999; McGuire *et al.*, 1999; Thiele, 1999; Friis-Hansen and Sthapit, 2000; CIP and PNUMA, 2000; Huamán, 2000). Many of these growers, the majority in many places within Peru, Bolivia and Ecuador, combine the cultivation of both a considerable diversity of FVs along with the new MVs (Zimmerer, 1991a, 1992, 1996; Brush, 1992, 1995; Wood and Lenné, 1997; Bianco and Sachs, 1998). The FVs are still preferred locally for their hardiness and capacity to produce in marginal growing environments, for their culinary traits and for other cultural and social reasons. None the less a growing number of small farmers no longer cultivate FVs, producing only MVs (Zimmerer, 1991a, 1992, 1996; Brush, 1992; Brush and Taylor, 1992; Dueñas et al., 1992; Ortega, 1997; Mayer and Glave, 1999). Increased potential for higher yields and for maintenance of preferred characteristics would make the FVs more competitive when farmers compare them with MVs. Research towards these goals in Peru and Bolivia is introduced below (Methods and Materials) and is then set in the context of potato breeding and seed production. The remainder of this chapter focuses on two major factors that strongly affect Andean farmers' procurement and selection of potato seed: social variability and environmental variability in farmlands (agroecological). Understanding the role of these factors has important implications for designing collaborative plant breeding (CPB) programmes.

Social variability is a distinguishing feature of seed production among the small-scale farmers of low-income countries, who are a high priority of 'participatory, client-driven research and technology development' (Ashby and Sperling, 1995). Social variability of various sorts is recognized as important. Gender and resource level (wealth) are usually considered among the most important features of local social variability. These differences have been correlated with varietal preferences of several crops; such preferences have been expressed both within and between types of FVs and MVs (Ashby et al., 1989; Thiele et al., 1997; Almekinders and Louwaars, 1999; Cleveland et al., 2000; Soleri et al., 2000). This chapter focuses on resource-level differences among farm families that influence their seed management of potato FVs. Such local differences in resource level are common within complex rural societies in developing countries (see McGuire, Chapter 5, this volume). The examples presented below demonstrate how social variability among local small farmers is related to one of the most important steps in the seed production of potato FVs, namely the selection of seed tubers (also referred to as clonal seed) on the basis of size (see also Ortega, 1997). The perspective in this chapter draws attention to the intra-varietal level of seed management (and its relations to varietal preferences), which is needed as a complement to the current emphasis on variety-related seed and selection dynamics.

Variability of growing environments is also a central concern of collaborative plant breeding (Almekinders et al., 1994; Almekinders and Louwaars, 1999; McGuire et al., 1999; Cleveland et al., 2000; Soleri et al., Chapter 2, this volume). Variation of farm environments in the

Andean countries has helped to generate a series of concerted efforts to involve farmers in agricultural research and development (Fernández and Salvatierra, 1989; Rhodes, 1989; Altieri, 1996; Thiele, 1999). Such efforts are premised on the awareness that diverse Andean agro-environments are often subject to environmental stress (e.g. drought and frost). This means that many Andean agroenvironments are unsuited or poorly matched to the potatoes that are produced through conventional breeding and formal seed outlets. The productive and sustainable use of this broad range of agroenvironments should be a goal of collaborative breeding and seed production (Cleveland *et al.*, 1994; Witcombe *et al.*, 1998; Zimmerer, 1998a, 1999, 2000, 2001b; Almekinders and Louwaars, 1999; McGuire *et al.*, 1999). The studies presented in this chapter demonstrate how farm-level variability of agroecological conditions relates to the selection of seed tubers and their allocation to growing environments.

## Methods and Materials: Field Studies in Eastern Cuzco

The design of this research is a case study. It involves field investigations working with a local farm organization and individual farmers in the Paucartambo Andes region of Eastern Cuzco. The region's population is roughly 20,000 rural inhabitants. Most farmers belong to one of the approximately 100 or so communities or rural hamlets, while approximately 3000 townspeople reside in three villages (Zimmerer, 1996). The region's farmers are bilingual Quechua–Spanish speakers who identify themselves as both Quechua Indians and peasant agri-culturalists. Potatoes, including a vast range of diverse FVs, are their staple crop, followed by barley and maize. The field studies analysed in this chapter were conducted in Eastern Cuzco in 1985–1987, 1990 and 1998. Related field studies, which were conducted in Bolivia (1991, 1992, 1993, 1995, 1996, 1997, 2001), are part of work in progress (Zimmerer, 2001a,b).

Methods included ethnographic participation followed by a field interview on farm-level resources and the seed knowledge and manage-ment of farmers who grow the FVs of potatoes. Methods also included the genetic analysis of a sample of potato FVs. The interview was designed with the input of farmers who were asked about how FV production could be evaluated as one aspect of present-day farm livelihoods that also involve the growing of MVs. The interview on farm-level resources was conducted with 33 small-scale farmers who are active growers of mixed potato FVs as well as MVs and a variety of other crops. These farmers were chosen from communities that belong to three sub-regions of Eastern Cuzco: the upper Paucartambo valley

(12 farmers), the lower Paucartambo valley (8 farmers) and the Paucartambo interior (13 farmers). The triad of sub-regions was chosen to represent the range of major environmental and social-cultural differences that are characteristic of the region. The upper valley is an area of sub-humid growing environments and populous ex-haciendas; the lower valley consists of humid climate nearing that of cloud forest, and more sparsely populated former haciendas (with many recent immigrants); and the interior is made up of long-time indigenous communities and near-semiarid environments (for details see Zimmerer, 1996).

Field interviews were conducted with farmers who belong to a range of resource levels. The 33 persons were categorized, by themselves and by one another, as 'poor' (13 farmers), 'middle' (17 farmers) or 'rich' (3 farmers) (note that all belong to the category of small-scale farmers as defined above; also discussed further below). The sample included 16 men and 17 women. The study included a structured interview about seed production practices and preferences that are related to the size of seed tubers of FVs. The farmers were interviewed in or near their fields and homes. They were asked to inspect a representative sample of 15 tubers of *qompis*, a common farmer potato variety cultivated throughout the region. The name of this FV is also spelled *ccompis*, for example to the south near Lake Titicaca (Valdivia *et al.*, 1996; see also Ortega, 1997; Zimmerer, 2001a). The example of *qompis* in Eastern Cuzco is discussed in this chapter; research on the farmer potato variety known as *mariva* and a farmer variety of the ulluco crop are analysed elsewhere (Zimmerer, 2001a). The sample of *qompis* used in the research consisted of tubers that ranged widely in the primary characteristics of size, colour and shape. Sizes of tubers (in grams) were: 15, 22, 28, 29, 33, 34, 41, 44, 52, 61, 75, 76, 130. Farmers were asked to identify the types that they would typically plant and explain their reasons for these preferences. (Further detail on the methods and materials used in the seed preference interviews are given below in the section, Social Variability in Potato Breeding and Seed Production.)

Methods for the analysis of agroecological variability consisted of the systematic observation of crop field types and measurements of elevation, slope and soil moisture. These data were collected at 100 m elevation intervals along a set of four transects (for details see Zimmerer, 1999: 146), with measurements made at approximately 85 points. Participating farmers discussed their own classifications during the transect studies. In addition detailed field-by-field interviews were conducted with 15 families with multiple fields located across a wide range of agroenvironments. These interviews led to a sample of 946 fields that were categorized by local farmers in terms of their own units of landscape organization (for details see Zimmerer, 1999: 143–145). Estimates of seed flow within the locally recognized farm spaces – the 'hill' space is chosen as the example – were based on interviews with 90 farmers (for details see Zimmerer, 1998a: 449).

## Modern Potato Breeding and Seed Systems in the Andean Countries

Modern varieties of potatoes have been produced in substantial number since the mid-1800s in the UK and Western Europe, Russia and the former Soviet Union, and the USA and Canada (USDA, 1961; Salaman, 1985; Horton, 1987; Hawkes, 1990; Harris, 1992; Douches *et al.*, 1996). In the Andean countries, and more recently other developing countries (e.g. China, India), the scientific breeding of MVs of potatoes grew rapidly in the post-Second World War period. Not until the 1970s, however, were the MVs widely incorporated into the agriculture of small-scale farmers in the Andes. National potato programmes and the Lima-based International Potato Centre (CIP, *Centro Internacional de la Papa*, founded and funded by the Consultative Group of International Agricultural Research or CGIAR) have produced most of the MVs grown in the Andean countries. The MVs in these countries are taxonomic members of either the Irish potato subspecies (*S. tuberosum* subsp. *tuberosum*), which is grown mostly in coastal valleys, or the Andean potato (*S. tuberosum* subsp. *andigenum*), which is produced mainly in the mountains above 2000 m (CIP, 1984; Hawkes, 1990).

Breeding for increased yield and traits to increase market value has been a main emphasis of the production of MVs (Howard, 1970; Jellis and Richardson, 1987; Harris, 1992; Bradshaw and Mackay, 1994; Douches *et al.*, 1996). Increased yields of the MVs are highly correlated with increase in tuber number, weight and size (Howard, 1970), which are a result of both improvement of growing environments and the genetic changes due to breeding (David Douches, personal communication, Michigan State University, 2001; Henry de Jong, personal communication, Agriculture and Agri-Food Canada, 2000). Selection for yield increase has focused on resistance to pests and diseases, response to nitrogen fertilizers with ample water and earliness. To date in the Andean countries the breeding of potato MVs has focused more on resistance to disease (late blight) and pests (nematodes) than on size *per se* (CIP, 1984). Still, the size of tubers – with a preference for large tubers – has been among the most important general concerns of potato producers and industrial and urban consumers (Horton, 1987: 46; Thiele *et al.*, 1997: 284). Tuber size may become a focus of future breeding in the Andean countries as it has in the USA and Europe. It is likely, moreover, that biotechnology advances will greatly enhance the identification and manipulation of genetic controls of yield and its components (such as tuber size) that were previously unexploited.

Tuber traits affected by breeding include a number of properties that are related to seed quality in addition to production and consumption parameters (compare Soleri *et al.*, Chapter 2, this volume). Since small-scale farmers of the Andean countries base nearly all potato production on seed tubers from their own harvests, the qualities of yield must also be seen from the perspective of seed needs (Brush *et al.*, 1981; Monares, 1981; Zimmerer, 1991b, 1996; Thiele *et al.*, 1997; Thiele, 1999). For these Andean farmers the customary use of their own seed tubers reduces the costs that are associated with potato planting, especially of the FVs (Mayer and Glave, 1999; Zimmerer, 2001a). Use of their own seed tubers also provides the advantage of maintaining varietal distinctness or purity. Reliance on this clonal propagation means that tubers themselves are the targets of intense scrutiny and selection pressure by farmers. The farmers' criteria for potato seed include tuber 'health' (evident lack of disease or stress symptoms), tuber colour, shape, indicators of seed dormancy (evidence of sprouting), and, not least, size (Brush *et al.*, 1981; Zimmerer, 1991b, 1996; Brush, 1992; Thiele *et al.*, 1997; Thiele, 1999). Small tubers contain more eyes (anatomical buds that produce sprouts) per unit of surface area and weight of the seed tuber (Allen *et al.*, 1992; Allen and Wurr, 1992). None the less, large seed tubers have the advantage of better withstanding physiological stress, especially moisture deficit or drought. A large seed tuber helps to ensure a healthy seedling by reducing the risk of germination or seedling failure since it contains a greater volume of stored energy reserves per individual sprout. Overall the large tubers contain more eyes (per seed tuber), which in turn produce more sprouts and thus the potential for more tuber offspring and greater total yield per seed tuber (Allen *et al.*, 1992; Allen and Wurr, 1992).

Potatoes can also be propagated with botanical or sexual seed, also known as true potato seed (TPS). Botanical potato seed is incorporated both intentionally and inadvertently by Andean farmers (Fernández and Salvatierra, 1989; Zimmerer and Douches, 1991), although it can also be produced and distributed by seed programmes and development institutions (CIP, 1984; Horton, 1987). Purposeful production of TPS by either farmers or formal seed programmes may hold promise as a source of lightweight, disease-free, low-cost planting material (CIP, 1984; Horton, 1987). Farmers' incorporation of TPS into their planting material takes place principally in fields of mixed FVs (Fernández and Salvatierra, 1989). Growers of these fields are most likely to notice and take interest in multiple varieties and off-types (Zimmerer and Douches, 1991). The periodic incorporation of TPS is a major contributor to the typically high level of genetic variation in FVs of Andean potatoes, estimated as the heterozygosity (multiple alleles at single genetic loci) of a group of common FVs sampled in Eastern Cuzco (Table 4.1). The laboratory results of allozyme analysis are based on starch-gel electrophoresis of 139 tubers, and are estimated statistically using Nei's measure of total allelic diversity (Zimmerer and Douches, 1991: 182). These findings establish the relatively minor genetic differences among the six FVs (the genetic diversity measures off-diagonal) and the high level of heterozygosity within them (on-diagonal). In effect each FV is a mixture or population of clones, rather than a single clonal type (Zimmerer and Douches, 1991: 185). The high level of heterozygosity found in the FVs adds to a greater overall cost of formal potato breeding, which has frequently started with FVs or incorporated them into screening and improvement programmes (CIP, 1984; Horton, 1987). This heterozygosity thereby contributes to the incentive for collaborative breeding that offers the potential for lower-cost techniques.

In the Andean countries, the importance of the formal seed system is largely for the initial introduction of MVs, including the release of new varieties through programmes for certified seed. Certified seed refers to seed tubers that are inspected, usually by a public sector agency, in order to ensure conformity with varietal types and the absence or low incidence of disease. The contribution of certified seed to annual planting is estimated at 1–5% in Bolivia, Peru and Ecuador (Monares, 1981; Crissman and Uquillas and PROSEMPA in Thiele, 1999). Reasons for the minor importance of the formal seed system in the Andean countries include low demand for this seed, which is due in large part to its high cost relative to producers' incomes, and an insufficient supply of seed (Thiele, 1999). There is also a problem with the inadequate adaptation of varieties released through the formal system to the marginal growing environments of many farmers, whose cultivation sites are often subject to environmental stresses such as

**Table 4.1.** Genetic diversity measures of six farmer varieties (FVs) in terms of average within-variety heterozygosity (on diagonal) and genetic differences between varieties (off diagonal) (from Zimmerer and Douches, 1991).

|  | 1. Kusi | 2. Pitikina | 3. Puka mama | 4. Qompis | 5. Suyt'u | 6. Wakoth'u |
|---|---|---|---|---|---|---|
| 1. Kusi | 0.303 | 0.024 | 0.084 | 0.022 | 0.031 | 0.023 |
| 2. Pitikina |  | 0.188 | 0.092 | 0.027 | 0.017 | 0.022 |
| 3. Puka mama |  |  | 0.297 | 0.06 | 0.085 | 0.046 |
| 4. Qompis |  |  |  | 0.294 | 0.042 | 0.017 |
| 5. Suyt'u |  |  |  |  | 0.267 | 0.026 |
| 6. Wakoth'u |  |  |  |  |  | 0.268 |

drought and frost (Fernández and Salvatierra, 1989; Rhodes, 1989; Altieri, 1996; Thiele, 1999).

Farmers of the Andean countries rely heavily on informal systems for seed tubers. Informal seed systems, like their formal counterparts, integrate breeding, management, replacement and distribution of seed. The most important sources of informal-system seed tubers are farmers' own production, farmer–farmer exchange, and farm markets and development institutions (Mayer, 1980; Brush *et al.*, 1981; Monares, 1981; Horton, 1987; Dueñas *et al.*, 1992; Fano and Benavides, 1992; Zimmerer, 1996, 1998a, 2001b; Thiele, 1999). This rank order is similar for both FVs and MVs, though market acquisition and involvement of development institutions are relatively more common in the case of MVs. Replacement of seed tubers is essential since potato seed 'degenerates' due to worsening viral infection. In general many small-scale farmers utilize their own production as replacement seed, thus lowering the costs and reducing the risks of market-based exchanges (Mayer and Glave, 1999). Replacement seed is usually first multiplied in fields in the cooler and more disease-free sites (Almekinders and Louwaars, 1999: 31). This general sort of site is favoured for the multiplication of seed of both FVs and MVs, with the latter typically located at somewhat lower elevations. Replacement rates vary greatly, though they are generally higher in the commercial production of MVs. Replacement rates are also higher when plantings suffer widespread damage as a result of frost, hail or drought. Mixed-purpose production of FVs is typically dependent on seed replacement every 3–10 years (Brush *et al.*, 1981; Horton, 1987; Zimmerer, 1991c, 1996, 1998a; Fano and Benavides, 1992; Thiele, 1999).

Linkages between informal seed tuber production and formal production promise several benefits that include the farmer-based selection of seed traits and more effective distribution of seeds (Almekinders *et al.*, 1994; Almekinders and Louwaars, 1999; Thiele, 1999). Investigation of the potential of these benefits requires the evaluation of social and environmental variability. Examples of the importance of this variability are discussed in the next two sections.

## Social Variability in Potato Breeding and Seed Production

Social variability exerts strong influences, in general, on the seed knowledge and practices of Andean farmers. Household resource level and gender are two common types of prominent variation that have been associated with differences in seed management. Resource level varies considerably among small-scale farmers and in many Andean regions they distinguish one another as 'poor peasants', 'middle peasants' and 'rich peasants' (Zimmerer, 1991a, 1996; Dueñas *et al.*, 1992; Bianco and Sachs, 1998; Mayer and Glave, 1999). Resources include land, livestock, labour and monetary wealth. In Eastern Cuzco the range of total field holdings varies between approximately five fields ('poor peasants') and 45 fields ('rich peasants') (Zimmerer, 2001a). Between roughly one-half and one-third of these fields are typically cultivated each year, since a number of fields are usually fallowed, partly as a means to manage soil fertility and pests (Zimmerer, 2001b). About one-half of the cultivated fields are usually sown with potatoes, ranging from an average of 2.9 actively planted potato fields per household to 7.7. This range of field holdings traverses the spectrum of 'poor peasants', who may cultivate no fields at all, to locally 'rich peasants', who actively cultivate upwards of 25 fields (Zimmerer, 2001a).

One general influence of social variability on seed management is related to the overall strategies of many small-scale farmers, who grow both MVs (often grown as varietal monocultures) as well as highly diverse mixtures of FVs (varietal polycultures) (Zimmerer, 1991b, 1996; Brush, 1992). Of their several potato fields these farmers typically grow 1–2 fields of such FV polycultures. These fields of FV polycultures consist of highly diverse mixtures that average more than 20 varieties per 125 plants (Zimmerer, 1991b,c). Some fields contain upwards of 30 varieties, although the survey of on-farm storage has resulted in estimates as low as 6–10 varieties per family (Brush *et al.*, 1981; Brush, 1992; Dueñas *et al.*, 1992; Tapia and Rosas, 1993). Fields of mixed FVs are often grown in the most marginal growing environments. Currently large numbers of farmers belonging to the 'poor', 'middle' and 'rich' resource levels continue to grow the highly diverse mixtures of FVs. Trends indicate those small-scale farmers in the 'middle peasant' group may be most likely to curtail their production of diverse FVs. They are the most pressed and unable, in many cases, to supply sufficient resources (fields, labour) for both FVs and the production of MVs (Zimmerer, 1991c, 1996; Bianco and Sachs, 1998).

One example documenting this trend is based on a farm survey of agrodiversity changes in the northern Paucartambo valley (Zimmerer, 1991a). In this case the FVs of the early or chaucha potato (*Solanum phureja*), which are extremely demanding of farm-level resources (especially labour-time and planting sites), were abandoned by 44 out of 50 grower households during the 1970–1991 period. The locally better off farmers were disproportionately represented in the ranks of those who continued growing the chaucha potatoes, since they were most able to hire labourers and control access to sufficient farmland.

This section describes social variability with regard to the resource level of farm households, with a particular emphasis on the local group of cultivators referred to locally as 'poor peasants'. It focuses on the association of farmer resource level with seed knowledge and management, especially concerning the size of tuber seed. A review of *in situ* conservation among Peruvian farmers suggests that the selection of small tubers is a distinctive feature of their management of diverse potato types (Ortega, 1997). Selection practices among these potato farmers is likely to be influenced by a pair of factors that set the stage for seed management and the preferences for different types of tubers (Brush, 1992; Fano and Benavides, 1992; Zimmerer, 2001a). First, the costs of potato seed are calculated to account for 10–30%, on average, of the direct costs of potato production in Eastern Cuzco. Seed expenditures are a considerable expense to all growers and to poorer farmers in particular due to their severely limited resources. Second, small-scale farmers often save seed from their own harvest, while at the same time they depend heavily on the harvest for other usage as well, principally for income from marketing and for food. Balancing the different end-uses of their production is a major challenge. This balancing consists of trade-offs in the allocation of a household's harvest. Such trade-offs impact the poorer farmers, especially, since they typically must make do with the most limited amounts of harvest, which both influences and constrains their decisions about seed selection.

In the preference evaluation for *qompis* the farmers were asked to choose the tuber or tubers of this FV that would be most suitable for seed and to explain their rationale and selection criteria. The results in Table 4.2 show that the selection criteria of these Andean farmers are weighted more heavily to the size of the potato tuber than to its shape or colour, which are also important factors. Size of the seed tuber was rated most important as a criterion for selection among both the 'poor' farmers (13 persons) and the 'middle farmers' (17 persons). (Seed tuber sizes contained in the trial are listed in the Methods section above.) Only the locally better off farmers tended to focus more attention on other selection criteria (the shape and colour of the tuber). These specific results are in agreement with the observations on seed

**Table 4.2.** Farm resource level and frequency of selection criteria for seed tubers of *qompis* potatoes, a FV.

| Socio-economic category | Number of farmers interviewed | Size as first criterion | Other criteria first (shape, colour) | Small-medium size as first choice | Medium or large size as first choice |
|---|---|---|---|---|---|
| 'Poor' | 13 | 0.77 | 0.23 | 0.85 | 0.15 |
| 'Middle' | 17 | 0.65 | 0.35 | 0.53 | 0.47 |
| 'Rich' | 3 | 0.33 | 0.67 | 0.00 | 1.00 |

selection related to agrodiversity conservation offered by Ortega (1997). Generally, the findings resemble a number of other studies that have shown that small-scale farmers make use of seed size as an important criterion for the choice of planting propagules from unsorted mixtures, and also when they undertake the final choice of seed (Zimmerer, 1991b, 1996, 2001a; see also Ashby *et al.*, 1989 regarding common beans in Colombia and Soleri *et al.*, 2000 on maize in Mexico).

The actual size of the customarily preferred seed tubers is in the range of medium or small-medium relative to other tubers that could possibly be chosen. This general preference is demonstrated in selection preferences for *qompis* seed (Table 4.2). The preferred tuber weighs 28 g, the second most preferred 52 g, and the overall range of tuber weights in the trial is 15–130 g. Although these evaluations were not controlled experiments, since sample tubers displayed differences in more than one parameter, the salience of size as a chief selection criterion is supported strongly by farmers' comments. A total of 28 of the 33 farmers who participated in the interviews offered detailed descriptions of how the intermediate and smaller sizes of seed make the most efficient allocation among end uses that is possible of their limited harvest. The evidence here consists of the five or six common phrases that farmers frequently relied on and emphasized to me in discussing their preference for these sizes. Their phrases about smaller seed, voiced in a typical mix of Quechua and Spanish, were as follows: 'it walks far' (*karuman purin*), 'it walks high' (*hatunman purin*); 'it walks well' (*allinta purin*), 'it advances well' (*avance bien*); 'it covers the mountain' (*cubre el orqo*); or 'enters a large area' (*hatunman haycun* or *hatunman tarpusqa*) (Zimmerer, 2001a). Definite preference for this general category of seed size – small-medium – is also applied to planting tubers of other potato FVs and to the tuberous ulluco crop, which were also investigated (Zimmerer, 2001a).

Preference for seed tuber size is related to the resource level ranking of farmers according to the results in Table 4.2. As mentioned 'poor peasant' and 'middle peasant' growers were the most likely to rank size criterion first in the evaluation of their customary selection practices. Their customary preference is for seed tubers of a small-medium size (Fig. 4.1). In contrast, two of the three better off farmers did not identify seed size as their first criterion, and all of them preferred medium or large size tubers. These relatively 'rich' farmers may be better able to benefit from the higher yields that are correlated with large seed tubers. In contrast, poorer farmers prefer small tubers in order to provide more seed than would large tubers. The poorer cultivators appear to balance the short-term need for seed with the desire for increased consumption and income that the greater yield of larger tubers provides. In sum, the need to balance the multiple uses of their own harvest is associated

**Fig. 4.1.** The planting of seed tubers of the potato crop in Eastern Cuzco. Note that the choice of small tubers for sowing is usually associated with the placement of 2–3 tubers per planting hole, as seen in this photo (photograph by Karl Zimmerer).

with farmers' strong preferences for seed tubers that are small-medium in size.

The effects of social variability on seed selection practices can be used to consider recommendations for seed improvement and the collaborative plant breeding of potatoes among Andean farmers. These recommendations are discussed in the conclusion.

## Agroecological Variability

This section provides an example of how agroecological variation is used in the informal seed system of potato growers in the Andes. Andean farmers make use of growing environments that are highly varied in space and time (Mayer, 1980, 1985; Tapia, 1996). In Eastern Cuzco they use a locally recognized system of land use terminology and classification to identify four principal farm spaces (Zimmerer, 1996, 1999, 2001b). The four spaces are known locally as 'hill' (*loma*), 'ox area' (*yunlla*), 'early season' (*maway*) and 'valley' (*qheshwar*). Each farm space consists of a distinctive set of productive resources including crop, growing techniques, infrastructure technology such as irrigation, and social rules such as property rights (see also Mayer, 1985; Tapia, 1996).

Seed production of potatoes is closely integrated with the main spaces of farming. The example given below shows how the framework

of production spaces of small-scale farmers functions as a set of
guidelines for their management of FVs of potatoes. Seed production
and flows, which include seed sourcing and provisioning, tend to
be circumscribed by the four major farm spaces. As detailed below,
the agroecological range and variability of the farm spaces offers
an important framework for understanding how farmers' existing
management influences the distributions and interactions of a crop.
It is suggested here that the incorporation of this sort of analysis
of farmers' own spatial frameworks of farm management should be a
foundation for collaborative plant breeding and seed production (see
also Zimmerer, 1996, 1998a, 1999, 2001b).

The 'hill' space is the most extensive farm space of Eastern Cuzco,
and it can thus serve as an important example of farmers' spatial
frameworks (Table 4.3). 'Hill' farming reaches to the upper limits of
cultivation at nearly 4100 m, though much 'hill' production occurs
closer to the lower limit at about 3700 m (Table 4.3). The growing

**Table 4.3.** Agroecological variation and farmer variety (FV) potatoes in the 'hill'
space of Eastern Cuzco.

| General parameter | Specific trait |
| --- | --- |
| Farm space, local name | 'Hill' (*loma*) |
| Major agroenvironments within space | Plateaux (0–5 slope)<br>  Uplands (rainfed)<br>  Topographic depressions<br>Slope land (> 5 slope)<br>  Uplands (rainfed)<br>  Topographic depressions |
| Primary elevation range (masl) | 3700–4100 |
| Importance of potato (percentage of total plantings) | 80–95 |
| Potato type(s) grown | Farmer varieties (FVs), mostly in varietal polycultures |
| Main uses | Consumption, seed, commerce |
| Common and scientific names | Andean (*S. tuberosum* subsp. *andigenum*)<br>  Cut-leaf (*S. stenotomum*)<br>  Chaucha (*S. chaucha*)<br>  Bitter or juzepczuk's potato (*S. juzepczukii*)<br>  Bitter or short-lobe potato (*S. x curtilobum*) |
| Sources of potato seed *for* production | 'Hill' space (within and outside community) |
| Destination of potato seed *from* production (estimated proportions) | 'Hill' space (60%)<br>'Ox area' space (35%)<br>'Early planting' space (5%) |

of potato FVs, including seed production, is predominant in 'hill' agriculture (potato farming of different types is also located in other farm spaces; see Zimmerer, 1999, 2001b). Farmers plant the diverse FVs of five distinct species: *S. tuberosum* subsp. *andigenum*, *Solanum stenotomum*, *Solanum chaucha*, *Solanum juzepczukii* and *Solanum ×curtilobum*. The main sources of potato seed for 'hill' agriculture consist of fields that belong to this same space (Table 4.3). The main destinations of the seeds produced in 'hill' agriculture are to other parcels of this space, in addition to fields of other spaces (Table 4.3). 'Hill' fields at higher elevations are favoured for the production of potato seed since they are less beset by the pests and disease that seriously afflict the fields at lower and intermediate elevations (Brush *et al.*, 1981; Horton, 1987; Zimmerer, 1996, 2001b). Nematode and virus infestations, as well as other pest and disease problems that may be potentially devastating at the lower elevations, are noticeably less common in the cool climates located above approximately 3800 m (Zimmerer, 1998b). Although the potato pests and diseases can be controlled through the use of insecticides, pesticides and fungicides, most small-scale farmers in the Andes consider these inputs too expensive for extensive use, especially on their fields of FVs.

The ample flow of potato seed to and from 'hill' fields is controlled largely within the resource stocks of the individual farm family (Zimmerer, 1996). Farmers use the production of at least one 'hill' field site to seed other 'hill' parcels during the next growing season. At times they also use this production for the seeding of fields in other farm units such as the 'ox area' and 'early season' (Zimmerer, 2001b). A survey of 45 farmers in Eastern Cuzco found that most potato seed sown in 1996–1997 was provided by a source that is a different field within the hill unit of the same community (93%; Zimmerer, 2001b). On occasion a family that is planning for the production of potatoes in a 'hill' field will acquire the needed seed through trade, bartering or purchase. Acquisition of the potato seed for 'hill' fields usually occurs through contacts with other farmers in the same community or in the multi-community clusters that centre on extensive areas of diverse potato production referred to as 'cultivar regions' or 'landrace areas' (Zimmerer, 1991a: 175, 1998a). The sites of seed-producing fields in these clusters are concentrated principally in the 'hill' space of local farmers and their communities. The frequency of these acquisitions varies from 5 to 10 years, with an average of 5.8 years. In the case of their diverse FV potatoes, therefore, the small-scale farmers rely on seed that has been grown in 'hill' sites. As a result, the production and acquisition of seed tubers correspond closely to the coverage of this farm space and to the range of major agroenvironments encompassed within it.

Analysis of seed production and flows in the 'hill' space shows that the actual distribution of FVs leads them to be exposed to the substantial range of agroecological variation that is associated with this space. In other words, the actual distributions of crops and crop types in the agriculture of small farmers are shaped through the range of characteristics that are associated with the production spaces of the local farmers and their communities (Fig. 4.2).

In the above example we saw that the seed tubers of the FV potatoes sown in 'hill' agriculture are primarily grown in, sourced from and dispersed to, the same farm space. In the case of the FVs of potatoes in 'hill' agriculture it can be concluded that the networks of seed flow do not correspond to narrow tiers of agroenvironments. Instead, the production of seed and the networks of seed flow are shown to conform most closely to landscape-based farm spaces, which include other important variables in addition to agroenvironments (Zimmerer, 1996, 1999, 2001b). Farm practices related to seed thus help to produce FVs of the Andean potatoes that show a moderate-to-high degree of ecological versatility, as opposed to the highly specialized mode of adaptation that is often assumed (discussed in Zimmerer, 1998a). This pronounced degree of ecological versatility is also evident in such detailed properties of the FVs of Andean potatoes as: widespread or cosmopolitan distributions at the region- and multi-region scales (Zimmerer, 1991c; Zimmerer and Douches, 1991); field experiments that show a general relation of the yield of many FVs to

**Fig. 4.2.**    Potato fields of farmer varieties (FVs) sown in the 'hill' area, an important landscape-based farm unit in Eastern Cuzco. Note the seed flows traverse the variety of growing habitats displayed here, which is only a fraction of the overall range of local 'hill' agroenvironments.

elevation-related environments (Zimmerer, 1991c, 1996, 1998a; see
also Wood and Lenné, 1997); and farmers' customary use of varied
planting sites within highly diverse local environments (for example
within mountain fields; Zimmerer, 1991b, 1998a).


## Conclusions and Recommendations

Social variability must be carefully incorporated into programmes and
policies aimed at the collaborative breeding and seed production of
FVs for the agriculture of small-scale farmers in developing countries.
The examples and analysis of this chapter show that resource levels of
farmers have a close relationship with their preferences for the size of
seed tubers. Poorer farmers tend most strongly to prefer the seed tubers
of small-medium size as they stretch their limited resources as best
they can. These preferences are one of the main foundations of local
seed selection and production. Tubers of the small-medium size category are preferred also for use-related reasons such as quicker cooking
time, saving fuel (Zimmerer, 1992, 1996), and serving many portions
(Ashby et al., 1989). The poorer potato farmers of developing countries
– such as the 'poor peasant' and 'middle peasant' farmers of Eastern
Cuzco – would be well served by the seed production and breeding of
FVs that provide a higher yield of more numerous tubers of moderate
size. Incorporating this strategy would help to address their current
practices of seed tuber use that stem from the real constraints and
immediate concerns of existing social variability.

   As discussed above, the size of potato tubers is a product of both
genotype and environment. Information about the relative influence of
these factors is very incomplete for the FVs of potatoes that belong to
S. tuberosum subsp. andigenum. Still, the findings on social variability
and tuber size suggest a couple of recommendations for seed tuber production and collaborative plant breeding. Seed programmes linking the
informal and formal systems, as are currently proposed (Almekinders
et al., 1994; Tripp, 1996; Wood and Lenné, 1997; Iriarte et al., 1998;
Almekinders and Louwaars, 1999; Iriarte et al., 1999; Thiele, 1999;
Friis-Hansen and Sthapit, 2000; Huamán, 2000), should consider providing healthy seed tubers that are small-medium and medium in size.
Possible benefits of a strategy focused on smaller seed tubers would
include access to more planting material. At the same time this strategy
would help to meet, as efficiently as possible given current conditions,
the existing needs of poorer farmers for food security and marketable
commodities. Possible disadvantages of the use of smaller seed would
include a reduction of yield. The details of this trade-off require further
research. If the trade-offs associated with smaller seed are seen as

untenable by policy makers and seed production programmes, then their attention would be needed to address the incompatibility between officially recommended programme goals (large size of seed tubers) and the existing practices of a majority of Andean farmers (small-medium size of seed tubers).

New seed programmes and projects that occur within the context of existing knowledge and preferences must be geared to very real limits, even as these new initiatives may seek to alleviate such difficulties (Richards, 1995, 1996). For this reason it is important for the institutional and policy support of hybrid-style seed systems to recognize the viability and existing use of seed tubers that are small-medium in size. Such seed tubers are widely relied upon and known about, especially among poorer farmers. Large-size seed tubers, in contrast, are utilized primarily by the locally better off. Programmes that adopted the production solely of large-size planting materials, perhaps even unknowingly, would probably involve and benefit primarily those better-off growers able to produce and afford such seed tubers. Even the use of community gardens for the production of seed tubers, which may often count as a viable proposal (Eyzaguirre and Iwanaga, 1996; Friis-Hansen and Sthapit, 2000), is in need of further analysis, both as a general idea and as a model for specific projects. Since community plots are frequently located in sites near settlements that tend to be the more environmentally favourable or economically valued locales, such planting areas could resemble most closely the farmlands and seed-production strategies of better-off farmers. Their poorer neighbours, in contrast, might not be able to afford or benefit from large-size seed tubers that were produced in such programmes (Zimmerer, 2001b).

Collaborative plant breeding should consider the potential of genetic changes that would support the seed tuber size preferred by the small-scale farmers. It might be that genetic changes could decouple the linkage of yield increase and tuber size increase (Zimmerer, 2001a). For example the yield level of potatoes in the Andean countries could perhaps be improved by concentrating to a greater degree on an increase in the number of tubers per plant. Existing or new breeding techniques might be applied to this goal. Management techniques that are related to plant breeding could also be considered. It is known, for example, that the application of phosphorus fertilizer increases yield through an effect that is concentrated on the number of tubers (rather than tuber size) (Harris, 1992). This sort of yield response might also be amenable to genetic modifications in plant breeding that would make for heavy cropping types with desirable sizes of seed tubers. These sorts of genetic changes and management techniques should be considered as potential steps in collaborative breeding to produce the size category of seed tubers that would fit into the existing requirements of the majority of farmers in the Andean countries.

The role of agroecological variability likewise should be incorporated into potato seed production and farmer breeding. Farmers use the management of their local landscape units as their principal guide to determine the planting locations of different types of potatoes. Seed flows, which comprise both out-flows (seed provisioning) and in-flows (seed acquisition), can be estimated with respect to single farm spaces or, in some cases, to a combination of particular spaces (Dueñas *et al.*, 1992; Zimmerer, 1999, 2001b). As a result, the plants and tubers of FVs appear to become exposed and adapted to certain ranges of characteristic environments (Zimmerer, 1991c, 1998a). With regard to the Andean potato crop, the design and analysis of policies and programmes on seed production, and the environmental component of collaborative plant breeding should become aware of this extremely important role of farm spaces as a guide to farmers' own management of their diverse crop plants. This means that varieties adapted to the management practices and wide range of biophysical variables present in Andean farmers 'spaces' will be more appropriate goals of collaborative plant breeding than varieties adapted to biophysical environments defined by plant breeders.

# Acknowledgements

My deep gratitude is owed to the farmers, communities and farm organizations and NGOs of Cuzco for their cooperation and facilitation of these studies. Agronomy colleagues Martha Rosemeyer, David Spooner and Josh Posner and geneticists David Douches and Henry de Jong helped with advice. People, plant and potato experts Zósimo Huamán, Ramiro Ortega, Judith Carney, Steve Brush, Enrique Mayer, Daniel Gade, Darrell Posey, Fausto Sarmiento and Brent Berlin have commented generously on this chapter. CIP and the University of Wisconsin, Madison helped to facilitate the research. The editors, David Cleveland and Daniela Soleri, provided many useful comments.

# References

Allen, E.J. and Wurr, D.C.E. (1992) Plant density. In: Harris, P.M. (ed.) *The Potato Crop: the Scientific Basis for Improvement*. Chapman and Hall, New York, pp. 292–333.

Allen, E.J., O'Brien, P.J. and Firman, D. (1992) Seed tuber production and management. In: Harris, P.M. (ed.) *The Potato Crop: the Scientific Basis for Improvement*. Chapman and Hall, New York, pp. 247–291.

Almekinders, C. and Louwaars, N.P. (1999) *Farmers' Seed Production: New Approaches and Practices.* Intermediate Technology Publications, London.

Almekinders, C., Louwaars, N.P. and de Bruijn, G.H. (1994) Local seed systems and their importance for an improved seed supply in developing countries. *Euphytica* 78, 207–216.

Altieri, M. (1996) *Enfoque agroecológico para el desarrollo de sistemas de producción sostenibles en los Andes.* CIED, Lima.

Ashby, J.A. and Sperling, L. (1995) Institutionalizing participatory, client-driven research and technology development in agriculture. *Development and Change* 16, 753–770.

Ashby, J.A., Quiros, C.A. and Rivers, T.M. (1989) Farmer participation in technology development: work with crop varieties. In: Chambers, R., Pacey, A. and Thrupp, L.A. (eds) *Farmer First: Farmer Innovation and Agricultural Research.* IT Publications, London, pp. 115–132.

Bianco, M. and Sachs, C. (1998) Growing oca, ulluco, and mashua in the Andes: socioeconomic differences in cropping practices. *Agriculture, Culture, and Human Values* 15, 267–280.

Bradshaw, J.E. and Mackay, G.R. (1994) *Potato Genetics.* CAB International, Wallingford, UK.

Brush, S.B. (1991) A farmer-based approach to conserving crop germplasm. *Economic Botany* 39, 310–325.

Brush, S.B. (1992) Ethnoecology, biodiversity, and modernization in Andean potato agriculture. *Journal of Ethnobiology* 12, 161–185.

Brush, S.B. (1995) In situ conservation of landraces in centers of crop diversity. *Crop Science* 35, 346–354.

Brush, S.B. and Taylor, J.E. (1992) Technology adoption and biological diversity in Andean potato agriculture. *Journal of Development Economics* 39, 365–387.

Brush, S.B., Carney, H.J. and Huamán, Z. (1981) Dynamics of Andean potato agriculture. *Economic Botany* 35, 70–88.

CIP (Centro Internacional de la Papa) (1984) *Potatoes for the Developing World: a Collaborative Experience.* CIP, Lima.

CIP (Centro Internacional de la Papa) (1997) *Diversidad de papas natives en los Andes.* CIP, Lima.

CIP (Centro Internacional de la Papa) and PNUMA (Programa de las Naciones Unidas para el Medio Ambiente) (2000) *Efectividad de las estrategias de conservacion in situ y el conocimiento campesino en el manejo y uso de la biodiversidad.* CIP, Lima.

Cleveland, D.A., Soleri, D. and Smith, S.E. (1994) Do folk crop varieties have a role in sustainable agriculture? *BioScience* 44, 740–751.

Cleveland, D.A., Soleri, D. and Smith, S.E. (2000) A biological framework for understanding farmers' plant breeding. *Economic Botany* 53, 377–394.

Douches, D.S., Maas, D., Jastrzebsi, K. and Chase, R.W. (1996) Assessment of potato breeding progress in the USA over the last century. *Crop Science* 36, 1544–1552.

Dueñas, A., Mendivil, R., Lobaton, G. and Loaiza, A. (1992) Campesinos y papas: a propósito de la variabilidad y erosión genética en comunidades campesinas del Cusco. In: De Gregori, C.I., Escobar, J. and Martitorena, B. (eds) *Perú: El problema agrario en debate, SEPIA IV*, Universidad Nacional de la Amazonia Agraria, Lima, pp. 287–309.

Eyzaguirre, P. and Iwanaga, M. (1996) *Participatory Plant Breeding*. IPGRI, Rome.

Fano, H. and Benavides, M. (1992) *Los Cultivos Andinos en Perspectiva: Produccion y Utilización en el Cusco*. CIP, Lima.

Fernández, M.F. and Salvatierra, H. (1989) Participatory technology validation in highland communities of Peru. In: Chambers, R., Pacey, A. and Thrupp, L.A. (eds) *Farmer First: Farmer Innovation and Agricultural Research*. Intermediate Technology Publications, London, pp. 146–150.

Franco, M. and Godoy, R. (1993) Potato-led growth: the macroeconomic effects of technological innovations in Bolivian agriculture. *Journal of Development Studies* 29, 561–587.

Friis-Hansen, E. and Sthapit, B. (2000) *Participatory Approaches to the Conservation and Use of Plant Genetic Resources*. IPGRI, Rome.

Harris, P.M. (1992) *The Potato Crop: the Scientific Basis for Improvement*. Chapman and Hall, London.

Hawkes, J.G. (1990) *The Potato: Evolution, Biodiversity, and Genetic Resources*. Smithsonian University Press, Washington, DC.

Horton, D. (1986) Farming systems research: twelve lessons from the Mantaro Valley Project. *Agricultural Administration* 23, 93–107.

Horton, D. (1987) *Potatoes: Production, Marketing, and Programs for Developing Countries*. Westview Press, Boulder, Colorado.

Howard, H.W. (1970) *Genetics of the Potato,* Solanum tuberosum. Springer Verlag, New York.

Huamán, Z. (2000) Semilleros comunales de papas nativas del Perú. *Revista AgroNoticias* (Lima) 251, 28–31.

Iriarte, V., Terrazas, F. and Aguirre, G. (1998) *Memoria: Primer Encuentro Taller Sobre el Mantenimiento de la Diversidad de Tuberculos Andinos en sus Zonas de Orígen*. Poligraf, Cochabamba, Bolivia.

Iriarte, V., Lazarte, L., Franco, J. and Fernández, D. (1999) *El Rol del Género en la Conservación, Localización, y Manejo de la Diversidad Genética de Papa, Tarwi, y Maíz*. Lauro, Cochabamba, Bolivia.

Jellis, G.J. and Richardson, D.E. (1987) *The Production of New Potato Varieties: Technological Advances*. Cambridge University Press, Cambridge.

Mayer, E. (1980) *Land Use in the Andes: Ecology and Agriculture in the Mantaro Valley of Peru with Special Reference to Potatoes*. Centro Internacional de la Papa (CIP), Lima.

Mayer, E. (1985) Production zones. In: Masuda, S., Shimada, I. and Morris, C. (eds) *Andean Ecology and Civilization*. University of Tokyo Press, Tokyo, pp. 45–84.

Mayer, E. and Glave, M. (1999) Alguito para ganar (a little something to earn): profits and losses in peasant economies. *American Ethnologist* 26, 344–369.

McGuire, S., Manicad, G. and Sperling, L. (1999) *Technical and Institutional Issues in Participatory Plant Breeding – Done from a Perspective of Farmer Plant Breeding*. CIAT, Cali, Colombia.

Monares, A. (1981) The potato seed system in the Andean region: the case of Peru. PhD dissertation, Cornell University, Ithaca, New York.

Ortega, R. (1997) Peruvian in situ conservation of Andean crops. In: Maxted, N., Ford-Lloyd, B.V. and Hawkes, J.G. (eds) *Plant Genetic Conservation: the In Situ Approach*. Chapman & Hall, London, pp. 303–314.

van der Ploeg, J.D. (1990) *Labour, Markets, and Agricultural Production*. Westview Press, Boulder, Colorado.

Rhodes, R. (1989) The role of farmers in the creation of agricultural technology. In: Chambers, R., Pacey, A. and Thrupp, L.A. (eds) *Farmer First: Farmer Innovation and Agricultural Research*. IT Publications, London, pp. 3–8.

Richards, P. (1995) The versatility of the poor: indigenous wetland management systems in Sierra Leone. *Geo-Journal* 35, 197–203.

Richards, P. (1996) Agrarian creolization: the ethnobiology, history, culture, and politics of West African rice. In: Ellen, R. and Fukui, K. (eds) *Redefining Nature: Ecology, Culture, and Domestication*. Berg, London, pp. 291–318.

Salaman, R.N. (1985) *The History and Social Influence of the Potato*. Cambridge University Press, Cambridge.

Soleri, D., Smith, S.E. and Cleveland, D.A. (2000) Evaluating the potential for farmer and plant breeder collaboration: a case study of farmer maize selection in Oaxaca, Mexico. *Euphytica* 116, 41–57.

Tapia, M.E. (1996) *Ecodesarrollo en los Andes Altos*. Fundación Friedrich Ebert, Lima.

Tapia, M.E. and Rosas, A. (1993) *La Mujer Campesina y las Semillas Andinas*. FAO, Lima.

Thiele, G. (1999) Informal potato seed systems in the Andes: why are they important and what should we do with them? *World Development* 27, 83–99.

Thiele, G., Gardner, G., Torrez, R. and Gabriel, J. (1997) Farmer involvement in selecting new varieties: potatoes in Bolivia. *Experimental Agriculture* 33, 275–290.

Tripp, R. (1996) Biodiversity and modern crop varieties: sharpening the debate. *Agriculture and Human Values* 13, 48–63.

USDA (1961) *Seeds, the Yearbook of Agriculture*. United States Department of Agriculture, Washington, DC.

Valdivia, R., Huallpa, E., Choquehuanca, V. and Holle, M. (1996) Monitoring potato and oxalis varieties in mixtures grown on farm family fields in the Titicaca Lake basin, Peru, 1990–1995. In: Eyzaguirre, P. and Iwanaga, M. (eds) *Participatory Plant Breeding: Proceedings of a Workshop on Participatory Plant Breeding, 26–29 July 1995*. IPGRI, Rome, pp. 144–150.

Witcombe, J., Virk, D. and Farrington, J. (1998) *Seeds of Choice: Making the Most of New Varieties for Small Farmers*. Intermediate Technology Publications, London.

Wood, D. and Lenné, J.M. (1997) The conservation of agrobiodiversity on-farm: questioning the emerging paradigm. *Biodiversity and Conservation* 6, 109–129.

Zimmerer, K.S. (1991a) Labour shortages and crop diversity in the southern Peruvian sierra. *Geographical Review* 81, 414–432.

Zimmerer, K.S. (1991b) Managing diversity in potato and maize fields of the Peruvian Andes. *Journal of Ethnobiology* 11, 23–49.

Zimmerer, K.S. (1991c) The regional biogeography of native potato cultivars in highland Peru. *Journal of Biogeography* 18, 165–178.

Zimmerer, K.S (1992) The loss and maintenance of native crops in mountain agriculture. *Geo-Journal* 27, 61–72.

Zimmerer, K.S. (1996) *Changing Fortunes: Biodiversity and Peasant Livelihood in the Peruvian Andes.* University of California Press, Berkeley and Los Angeles.

Zimmerer, K.S. (1998a) The ecogeography of Andean potatoes: versatility in farm regions and fields can aid sustainable development. *BioScience* 48, 445–454.

Zimmerer, K.S. (1998b) Disturbances and diverse crops in the farm landscapes of highland South America. In: Zimmerer, K.S. and Young, K.R. (eds) *Nature's Geography: New Lessons for Conservation in Developing Countries.* University of Wisconsin Press, Madison, Wisconsin, pp. 262–286.

Zimmerer, K.S. (1999) The overlapping patchworks of mountain agriculture in South and Central America: toward a regional–global landscape model. *Human Ecology* 27, 135–165.

Zimmerer, K.S. (2000) The reworking of conservation geographies: nonequilibrium landscapes and nature–society hybrids. *Annals of the Association of American Geographers* 90, 251–279.

Zimmerer, K.S. (2001a) Just small potatoes (and ulluco)? Seed use strategies among Peruvian farmers as key considerations for agrodiversity conservation. *Agriculture and Human Values* 20(2) (in press).

Zimmerer, K.S. (2001b) Geographies of seed: implications for agrodiversity conservation in the Andean countries. *Society and Natural Resources* 15(1) (in press).

Zimmerer, K.S. and Douches, D.S. (1991) Geographical approaches to crop conservation: the partitioning of genetic diversity in Andean potatoes. *Economic Botany* 45, 176–189.

# Farmers' Views and Management of Sorghum Diversity in Western Harerghe, Ethiopia: Implications for Collaboration with Formal Breeding

**5**

SHAWN J. MCGUIRE

*Technology and Agrarian Development Group, Wageningen University, Hollandseweg 1, 6706 KN Wageningen, The Netherlands*

## Abstract

This chapter examines farmers' goals, concepts and practices in sorghum genetic resource management in highland and lowland ecologies of Western Harerghe, Ethiopia, including how farmers perceive and access diversity, name varieties, make selections and manage their environments. Research results suggest that farmers' perceptions and actions can consist of both more general processes of crop biology or genetics, and more local aspects of environmental conditions, individual actors' goals and sociocultural contexts. The latter are less readily translated into terms familiar to plant breeders, but are none the less important. The differences in the nature and scale of environmental categories of breeders and farmers may be one reason why farmers feel that formally developed varieties rarely perform well in their fields. This analysis, coupled with direct investigation of formal breeding and its institutional context, explores implications for CPB.

## Introduction

Since 1973, the Ethiopian government has prioritized a sorghum breeding programme to improve farmer yields for one of its most important crops (Degu, 1996). A review in 1983 considered it to be 'an exemplary crop improvement program for other national and regional programs to emulate . . . probably one of the very few well-equipped and managed projects in the country' (Yemane and Lee-Smith, 1984: 63–64). Farmers, however, generally appear to have very low adoption rates of the modern varieties (MVs) produced by this programme, preferring their own farmer varieties (FVs) in most instances.

Explanations for this have pointed to poor linkages between research and extension, inadequate seed supply and breeders' unawareness of farmers' goals when developing breeding targets (Yemane and Lee-Smith, 1984; Stroud and Mekuria, 1992; Mulatu, 1996; Ministry of Agriculture, 1998). It is also possible that the differences between breeders' selection conditions and farmers' low-input, variable-stress environments contribute to a poor fit for MVs on farmers' fields (compare Ceccarelli and Grando, Chapter 12, this volume; Bänziger and de Meyer, Chapter 11, this volume).

Greater farmer participation in breeding may be a way to bridge some of these gaps. As farmers actively manage a considerable level of diversity in sorghum, collaborative plant breeding (CPB) should aim to improve understanding of farmers' own systems of crop development and seed exchange (Nyerges, 1997). This may help to identify the nature of gaps in formal work, and point to areas where closer farmer–formal collaboration might be fruitful.

## Methods

This chapter is based on preliminary fieldwork over a 10-month period in 1998/99 in eastern Ethiopia, Western Hararghe Zone, Oromia Region. In both area and production, sorghum (*Sorghum bicolor* [L.] Moench.) is Ethiopia's second crop (after teff (*Eragrostis teff* [Zucc.] Trotter)), and the most important crop in Hararghe, planted to 175,000 ha (CSA, 1995). Though four of the five main racial types of sorghum are found in Ethiopia (Stemler *et al.*, 1977), the Durra type dominates FVs in this area.

Farming systems in Western Hararghe are mainly mixed crop–livestock, with very low use of external inputs and almost no irrigation (CSA, 1997). Fieldwork centred on two adjacent districts (Woredas), Chiro and Miesso, which represent very different agroecologies and populations of sorghum varieties. Chiro, a mountainous, highland district, has longer growing seasons with generally more reliable rainfall onset, though high population densities and soil fertility problems are common. Miesso, a lowland plain adjacent to the mountains, receives less rainfall with variable onset. Farmers in Miesso and Chiro identify moisture stress and soil fertility, respectively, as their main constraint (ICRA, 1996). Table 5.1 lists some characteristics of each district.

Most work concentrated on two individual Farmer Associations (FAs, municipalities clustering several hamlets), one in each district. Funyaandiimo FA (1900–2150 masl) has above-average production for

Chiro, though is 25 km from a significant market centre. Melkaa Horaa FA (1350–1400 masl) is considered to be one of the lower-producing areas of Miesso, and is 5 km from markets and from a lowland research station. Much of my time was spent in the FAs, following the season from May planting until the following February, and I lived in Funyaandiimo itself or in Miesso town, 5 km from Melkaa Horaa. My assistants were the local extension agents who lived full-time in the communities, and I visited farmers daily, following their crops through the season. The study nested sampling at several levels, starting with semi-structured interviews, and using this to select subsets of farmers for closer follow-up, focus group discussions and germplasm collection. A formal survey over a wider region sought to link local findings to a wider context.

Within both FAs, semi-structured interviews asked 141 randomly selected farmers (84 in Funyaandiimo, 57 in Melkaa Horaa) about variety use and seed supply. This helped to inform my choices for focus group discussions in both FAs on soil classification, sorghum varieties, wealth ranking and taxonomy. For the latter half of the season, I selected a subset of 21 farmers to follow more closely (11 in Funyaandiimo, 10 in Melkaa Horaa), seeking diverse soil types, agroecology, sorghum varieties and wealth rank. I made repeated visits to the farms in this subset to observe flowering periods, selection and harvesting practices, to ask more detailed questions, and to carry out a selection simulation (described below). For 15 of these subset farmers (nine in

**Table 5.1.** Some descriptors for Miesso and Chiro Woredas (districts).

| | District | |
| --- | --- | --- |
| Characteristic | Chiro | Miesso |
| Rainfall onset | More certain | Variable |
| Season | > 5 months | 3–5 months |
| Elevation (masl) | 1500–2500 | 1300–1500 |
| Population ha$^{-1}$ | 2.16[a] | 0.47[a] |
| Temperature (°C) | 15–25[a] | 23–31[b] |
| Rainfall (mm) | 650–1000[a] | 420–960[b] |
| Topography | Steep slopes and wide valley bottoms | Flat to undulating |

[a]Cited from ICRA (1996): mean population density calculated from PHC (1989), extrapolated with population estimates in Ministry of Agriculture (1995). Chiro temperature and rainfall ranges from Chiro High School data for 1980s. Note: Chiro High School is lower, and thus warmer and drier than much of the district, including Funyaandiimo.
[b]Miesso: Author's own calculations from Ministry of Agriculture Station data, 1982–1997.

Funyaandiimo, six in Melkaa Horaa), I recorded the dates for 50%
flowering for all varieties in their fields, and in the fields of their
immediate neighbours. Also, with 19 farmers from the same subset (11
in Funyaandiimo, eight in Melkaa Horaa), I ran simulations of selection
and discussed scenarios, to probe farmers' views on heritability.

To gain a better perspective on local sorghum variation, I collected
81 accessions from farmers' fields during harvest, in the FAs and
immediate surroundings (30 in Chiro, 51 in Miesso). Samples from
these accessions, including both common and rare types, were used to
probe farmers' naming systems and definitions of varieties, through
individual interviews and focus group discussions. Also, several
'triads' of three similar grain heads were presented, and farmers asked
to identify the one not belonging and why. They were asked to name
the accessions and to respond to the names given by donors, especially
when this differed from their names.

In September 1998 I administered a survey on variety use, seed
supply, storage and perceptions of seed quality over wider areas of
Chiro, Miesso and adjacent districts, to confirm trends within the
individual FAs. The sample frame involved a rough transect across
a region that brackets the sample districts, interviewing 94 farmers
(54 highland, 40 lowland) across the Miesso and Chiro districts.
Table 5.2 lists the various samples, and Table 5.3 summarizes some
characteristics of the surveyed farms, mainly from farmers' statements.

For simplicity, I will refer to specific data by its source: individual
interviews, focus group discussions or subset (all within target FAs),
collections (within Chiro/Miesso districts) or survey (surrounding
districts). Also, I will use Chiro/Miesso to refer to the FAs and their
surrounding districts in this chapter. Statistical analysis used SPSS,
with $P < 0.05$ as a threshold for significance.

## Diversity at Varietal Level

Ethiopia is a centre of diversity for sorghum, though possibly not its
place of first domestication (Harlan, 1989). In the absence of detailed
genetic characterization of diversity, we need to rely in the first
instance upon farmers' names for varieties, which Teshome et al.
(1997) found related well to formal taxonomies in northern Ethiopia.
Farmers recognize a broad range of types, with little overlap between
districts.

In interviews, when farmers were asked during planting time what
varieties they were sowing, a total of 17 different varieties were named
across the Chiro sample, compared with only seven in Miesso (Table
5.4). Farmers state a wide range of traits that they appreciate in their

varieties, with most frequent mention of quality for food or feed, market value or stress resistance, as well as yield (for more details, see McGuire, 2000). Variation among farmers in their ecological conditions

**Table 5.2.** Summary of different samples of farmers in highland Chiro and lowland Miesso, mostly located in Funyaandiimo and Melkaa Horaa FAs, and overlapping, except for the survey.

| Method | N sampled | | Sampling location | Sampling method |
|---|---|---|---|---|
| | Miesso | Chiro | | |
| Individual interviews on varieties | 84 | 57 | Within focus FAs | Random |
| Subsets for flowering time, selection simulation | 11 | 10 | Within focus FAs | Chosen based on soils, wealth, sorghum types, ecology |
| Observe mid-flowering dates | 9 | 6 | Within focus FAs | Taken from above subset of farmers |
| Selection and heritability questions | 11 | 8 | Within focus FAs | Taken from above subset of farmers |
| Focus group discussions | Varying | Varying | Within focus FAs | Chosen based on knowledge |
| Sorghum collections | 51 | 30 | Across Chiro and Miesso | Random, though deliberately seeking new types |
| Survey | 54 | 40 | Other FAs across Chiro, Miesso and neighbouring districts | Random, stratified by location |

**Table 5.3.** Farm characteristics from survey of 54 highland farmers (Chiro and adjacent districts) and 40 lowland farmers (Miesso and area) in 1998. Means (with standard errors) from farmers' own descriptions, except for presence of MVs, which is author's own assessment.

| Characteristic | Region | |
|---|---|---|
| | Chiro | Miesso |
| Land holdings (ha) | 0.6 (0.3) | 1.28 (0.11) |
| Grain t ha$^{-1}$ 'good year' | 3.23 (0.20) | 2.72 (0.29) |
| Grain t ha$^{-1}$ 'bad year' | 1.03 (0.084) | 0.44 (0.063) |
| % with 0/1/$\geq$ 2 oxen | 22/50/28 | 28/32/40 |
| Presence of MVs | Near nil | Low (near station) |

**Table 5.4.** Levels of on-farm variety diversity in 1998, assessed by different methods: individual semi-structured interviews in FAs during planting season, a survey over wider areas in mid-season, and direct field observations at harvest of sub-sample farmers and some collection points.

| Method | Chiro | | | Miesso | | |
|---|---|---|---|---|---|---|
| | N of farmers | N of different varieties | N of vars per farm[a] | N of farmers | N of different varieties | N of vars per farm[a] |
| Individual interviews | 84 | 17 | 1.57 (0.08)[A] | 57 | 7 | 1.37 (0.08)[A] |
| Survey | 54 | 24 | 1.43 (0.09)[A] | 40 | 13 | 1.40 (0.11)[A] |
| Direct observation at harvest | 15 | 23 | 4.60 (0.46)[B] | 21 | 29 | 4.10 (0.51)[B] |

[a]Means (with standard errors): those followed by a different letter differ significantly from means assessed by other methods.

and needs probably contributes to the overall varietal diversity, similar to the findings of Teshome *et al.* (1999) in northern Ethiopia.

'Wogere Diima' was the most popular variety in Chiro, and 'Masugi Diima' in Miesso, planted on 64% and 88% of farms, respectively. Though the frequency with which certain varieties occur on farms is not directly related to the total area they occupy in the district, my field observations suggest that the most commonly grown variety in each district is also widespread: 'Wogere Diima' and 'Masugi Diima' occupied nearly half the total sorghum area in their respective FAs. In the survey, which covered a wider region, these same two varieties were planted on 24% and 40% of highland and lowland farms, respectively, and more varieties were named in total. Many farmers grow more than one variety. The average number of varieties planted per farm was not significantly different in interviews or the survey. However, at harvest, I discussed and recorded in detail all varieties found on the farms in the subset sample, as well as in other fields where I had collected. On average these farms had more than four varieties, triple the average number mentioned in interviews at the start of the season, a significant increase (F = 130.9, df = 2).

Many of the additional varieties were in small amounts, sometimes a few individuals. While farmers simply may not have mentioned minor varieties in interviews or surveys, discussions revealed that some added new varieties to their fields near the end of planting time, to fill gaps, or to try new material. Also, many of these additional varieties appeared unplanned, and some are unknown to the farmer, possibly a result of hybridization or mechanical mixture (discussed more below). In most of these cases, farmers stated they did not regard these as mere off-types, but considered them as distinct varieties, and many planned to replant them the following year to assess performance. The total number of varieties (by name) in each district would be considerably higher than in Table 5.4 if the list of collections, including all the unknowns, were added.

One conclusion to draw from this is that assessments of varietal diversity depend on the methods used, as well as timing and scale. Farmers' plans at the start of the season may indeed have focused on one or two varieties, but the results suggest that other varieties were added, whether deliberately or otherwise (e.g. as impurities in purchased seed). Also, in assessing diversity of varieties it should be recognized that what counts as a 'variety' may differ among farmers; for instance, what one may name an off-type, another may consider as a separate variety. Naming systems and gene flow are discussed in more detail below.

## Processes Shaping Diversity

Some factors affect farmers' interest in selection within and among crop populations, including economic context, levels of genetic and environmental variations, and crop breeding system (cf. McGuire et al., 1999). Farmers' management interacts with these factors and shapes outcomes through influencing gene flow or by selecting, in one way or another, what makes it to the next generation. They include the naming system, introductions, size of seed lots, degree of isolation of individual varieties and variety populations, selection and environments. Some farmer practices and their attitudes to these are described below.

### Naming system

Sorghum is partially outcrossing, commonly cited at around 5%, though 12% (Ariola and Ellstrand, 1996) or considerably higher rates (e.g. Maunder and Sharp, 1963) have been measured. As Louette (1994) showed for another outcrossing crop, maize, farmers' names can help to guide selection for distinct morphological types, possibly helping to maintain identities under gene flow. Farmers in Western Hararghe refer to particular variety names in introduction, exchange and selection, and refer to individual varieties as components in their mixtures. Thus, farmers' approach to naming may also shape gene flow in sorghum.

Farmers in the study area generally use two, and sometimes three, names to denote their sorghum types. Farmers initially assign primary-level names (described as 'families' by some farmers), such as

Masugi or Muyra, using some features that breeders also use: head morphology, seed shape and size, height, growth-duration, leaf pattern. Secondary names describe variants within primary groups, usually by seed colour (e.g. 'Adii' = white; 'Diima' = red) or head compactness (e.g. 'Bulloo'). Tertiary names are occasionally used to distinguish a sub-type or local population, referring to specific traits, the original source or the introducing person; these are often very local in use.

Farmers associate particular varieties with qualities – generally not sought explicitly by breeders – such as market value, cooking or fodder quality. They place particular importance on stalk sweetness ('Tinkish' in Amharic, 'Ala' in Oromo), linking it to fodder quality and drought tolerance, for example. In 'triad' tests of variety naming, or in discussions, I found that, in the process of determining a variety's name, Western Hararghe farmers sometimes tried to establish stalk sweetness (using leaf midrib colour or grain shape) as one diagnostic character. Indeed, some classifications of sorghum (e.g. Guo *et al.*, 1996; Teshome *et al.*, 1997) make this the key diagnostic trait, treating 'sweet' and 'non-sweet' types as separate groups of varieties. However, individual and focus group discussions on the matter suggest a more nuanced interpretation in Western Hararghe. For a given variety, they assign a particular likelihood of stalk sweetness (i.e. never, sometimes, or often sweet), often linking this to favourable growing conditions. Thus the presence of stalk sweetness may help to rule out some possible names, but does not seem to be sufficient on its own to name a variety. This suggests that farmers consider different varieties have different levels of phenotypic variation (Vp) for sweetness.

Though the commonest primary names are widely known, even beyond their normal ecological zones (e.g. Miesso farmers know of 'Wogere', Chiro farmers of 'Masugi'), there is not always strong consensus on their use. While some accessions inspired general consensus among individual interviews for their naming, others received very different names from individual farmers, and spurred extensive discussion in focus groups. Some primary names, and many tertiary names, seemed to be recognized only at the hamlet or family level. Where some farmers will distinguish a given accession with a particular name, others may 'lump' it under a more familiar type, perhaps as a variant. This seemed to be related in part to the level of awareness or exposure to different types: women (who do most marketing), well-travelled men and older farmers gave more names to the same sample of accessions.

The apparent differences in knowledge and practice in variety naming raise some questions. What levels of distinction in naming (i.e. a primary or tertiary name) lead to distinction in management? How important is variation in naming, where some farmers exchange, select and store an accession separately, while others lump it in with a more common variety, for the genetic structure of these accessions? Given that farmers often confront unfamiliar material through introduction or hybridization (see below), such questions may be relevant. Genetic analysis could shed light on how diversity is structured in these populations, and on how and at what scales farmers' naming systems are linked to this.

## Introduction of new material

Introduction of new material from outside the region plays an important role in current sorghum diversity. Farmers recall the introduction of some of the most established regional varieties, with broad agreement across individuals and communities about times, source areas and even individuals associated with this introduction. For instance, farmers from Funyaandiimo and surrounding areas all agree that their most common variety, 'Wogere', was first introduced from the lower altitudes 50 years ago by their landlord, via a local farmer. Residents in one hamlet may refer to an individual or event responsible for introduction to the immediate area. Finally, there are events, planned or otherwise (such as seed purchase or exchange in response to a crisis), which bring novel material to farmers' fields. These channels of introduction reflect particular economic and social patterns, which are briefly described below.

Patterns of introduction reflect to some extent infrastructure, and historical movements and processes. The pastoral Oromo expanded into Western Harerghe by the mid-16th century (Jalata, 1993; Hassen, 1994). The Oromo adopted crop cultivation in Harerghe as late as the late 19th century; use of the ox-plough – and probably sorghum cultivars – spread through the highlands westward from Eastern Harerghe, where cultivation had long been established around Harer city (McCann, 1995). This reflects farmers' views that the oldest local varieties are 'Muyra' and 'Fandisha', which are dominant in, and quite possibly originate from, the Harer region of Eastern Harerghe.

In the highlands, a minority of Christian Amharas are present, descendants from migrants two or three generations ago, and are generally well-integrated into Oromo social institutions (Tesfaye, 1961; Lewis, 1975). Some varieties common in northern Ethiopia, such as 'Zengada' (Teshome et al., 1997), and 'Qirimindahi', are equally well-integrated into farming systems in Western Harerghe, hundreds of kilometres away.

Though most Miesso farmers have migrated in the last three generations from the adjacent highlands in Chiro, the difference in season length meant that few highland sorghum types are appropriate for lowland conditions. The east–west caravan route passes through this region, from the coast to Ethiopia's centre, strengthened by the Ethio-Djibouti railroad early last century, and by a road in the 1930s (Marcus, 1994). This has helped Miesso farmers in introducing appropriate types to their area, as the nearest lowland sorghum areas to them are separated by wide expanses of uncultivated land to the east and west.

The expansion of Imperial Ethiopia to Harerghe around the start of the 20th century brought farmers under a feudal system (Zewde, 1991). Some landlords were apparently interested in stimulating peasant production, since their tribute was a proportion of the harvest, and since they could extend their power base through patrimonial relationships (Pausewang, 1983). In interviews and focus group discussions, both Chiro and Miesso farmers recall their absentee landlords screening a wide range of germplasm before choosing one type ('Wogere' and 'Abdelota', respectively) and distributing it to all surrounding farmers to encourage (if not necessarily require) its use. Land reform in 1975 in Ethiopia eliminated tribute payments and a wealthy rural class with social and material interests in stimulating peasant production (Rahmato, 1985). Ordinary farmers, however, cannot afford long journeys in search of germplasm, or the resources needed to assess a wide range of material. However, a few relatively prosperous farmers still test and disseminate seed, a notable case being a Sheikh/Member of Parliament near the Chiro site, credited for introducing 'Hadhoo' to the area. Both highland and lowland farmers, however, felt that their access to new seed or other technologies through such introductory channels is less than in the past.

The establishment of a lowland sorghum testing station in Miesso in the early 1980s (ISNAR, 1987; Degu, 1996) has also helped lowland farmers to access MVs, though adoption has been low. Finally, after land reform, producer cooperatives (PCs) were formed in some places, such as Funyaandiimo, collectivizing some farmers, and giving them favourable access to credit, extension and inputs, like new germplasm (Cohen and Isaakson, 1988). These PCs are now disbanded, but were an important channel for MVs to the area.

## Off-farm seed, MVs and seed supply

Farmers in both highland and lowland environments frequently need to seek seed from off-farm, to replenish or replace stocks due to poor harvest, failed germination or pest damage to seedlings. For example, because of unusually late rains during harvest in November 1997 in Chiro, and the failure of the early rains in Miesso in 1998, 40% and 73%, respectively, of those surveyed had germination problems for their March–April 1998 plantings. These were more severe, however, in Miesso, and the majority of farmers were planning to seek off-farm seed (Table 5.5). Environmental stress in the lowlands, more than the highlands, can affect entire communities, so many lowland farmers may need to replenish stocks at the same time, requiring them to source seed from further afield, as revealed in the surveys (data not presented, see McGuire, 2000: 46). Normally, Miesso farmers save significantly more seed, though because they cultivate larger areas, this translates as roughly the same amount per area to sorghum, roughly 30 kg ha$^{-1}$ (Table 5.6). This is double the amount farmers usually sow per hectare (Kefyalew *et al.*, 1996), and 2–4 times the rate that researchers officially recommend (IAR, 1995, n.d.). Thus, saving this amount of seed permits multiple plantings in case of poor emergence. In fact, some farmers set aside 100 kg or more, intending to give unplanted seed to others. However, because of lower and more uncertain yields in Miesso, setting aside sufficient seed even just for oneself may be a problem. The amount of seed Miesso farmers plan to save is a significantly higher proportion of their expected yields for both good and bad years, over

**Table 5.5.** From individual interviews in May–June 1998 (planting time), the number of farmers (with percentages) who reported germination or emergence problems, who had already received off-farm seed for that season, and who stated further plans to seek off-farm seed for planting with the late rains.

| Situation in 1998 | Chiro ($n = 83$) | Miesso ($n = 59$) |
|---|---|---|
| Germination problems | 33 (40.2) | 43 (72.8) |
| Already received off-farm seed | 14 (17.1) | 7 (17.5) |
| Further plans to seek seed | 2 (2.4) | 35 (59.3) |

**Table 5.6.** From surveys, means (with standard errors) for the absolute amount of seed that farmers planned to save after 1998/99 season, this amount relative to sorghum area, and this amount as a percentage of what they consider a good or poor harvest.

| | Chiro ($n = 54$) | Miesso ($n = 40$) |
|---|---|---|
| Total amount of seed planning to save (kg) | 15.6 (2.7)[a] | 28.4 (4.0)[a] |
| Amount of seed planning to save relative to sorghum area (kg ha$^{-1}$) | 29.1 (3.4) | 32.2 (5.5) |
| Seed to be saved as proportion of expected 'good' harvest | 1.25% (0.16)[a] | 2.50% (0.68)[a] |
| Seed to be saved as proportion of expected 'poor' harvest | 5.28% (0.91)[a] | 17.9% (3.66)[a] |

[a]Significant difference between locations.

50% of expected harvest in a bad year for several Miesso farmers. This harvest risk is one of the reasons that seed insecurity is a more common feature of the lowlands.

Formal breeding, starting with the Ethiopian Sorghum Improvement Programme in the early 1970s, has mainly sought to provide smallholders with open-pollinated MVs (Yemane and Lee-Smith, 1984). As the programme developed, it expanded from its exclusive focus on the highlands of the country (Gebrekidian, 1982a), and now structures work along three broad ecological zones (low-, mid- and high-altitude), crossing, selecting progeny and testing separate populations for each zone (Gebrekidian, 1982b). Work currently involves extensive screening, crossing, selection and testing over a range of stations nationally, including one at Miesso. After many years of screening exotic material, only the lowland work now makes significant use of non-Ethiopian germplasm, generally using semi-dwarf, fast-maturing material, often of race Caudatum.

MVs have had low adoption in all environments, in part because of poor seed supply (a tiny fraction of annual planting needs are met through formal multiplication of MVs), but also because farmers prefer FV quality traits or performance under low-input conditions. Lowland farmers prefer their tall, long-maturity FVs to the shorter, faster MVs, because their FVs give more biomass (for fuel, feed and construction) and higher grain yields if the rains are good. Only if the early rains fail do they seek faster material, which can mature after the late (July) rains. For this, they usually seek 'Sarude', a name used in Miesso to refer to any MV (possibly a variant of 'Seredo', a Tanzanian MV released in Ethiopia in the 1980s). The Ethiopian Seed Enterprise does not multiply lowland sorghum seed at all, so the only formal supply of MV seed in Miesso is the local research station, which provides a small amount of MV seed each year to regional government offices and NGOs. Only a fortunate few receive fast-maturing seed this way and on time (Melkaa Horaa, a Farmers' Association with 800 households, received 100 kg in 1998 from the local Woreda office).

Most farmers must find material through informal channels, generally travelling seed merchants, since few farmers save MVs from year to year. Seed merchants bring seed from surplus areas to sell in local markets such as Miesso. This can lead to surprises, occasionally unpleasant. Farmers cannot always identify fast-maturing varieties by seed morphology alone. Either through ignorance or malice, merchants (and sometimes NGOs) supply long-season varieties that fail to mature on time.

Seed obtained through market channels is usually not pure: distinct variants, or entirely different varieties appear when planted. Though some mixtures probably result from hybridization, others may be physical mixtures of different varieties, originating from farmers' fields in other districts. Some farmers I visited could point to scattered individuals, 50 cm taller and a different colour than the 'Sarude' that they bought from the merchant. Even formally supplied seed can be mixed: I visited several of the farmers in Melkaa Horaa who had received 3 kg packets of MV seed from the Woreda office, as mentioned above. Growing among these plants (which I recognized as a released cultivar and they identified as 'Sarude') were at least three distinctly different varieties, completely unfamiliar to them or to me. Since the breeding station in Miesso, as the sole source of MV seed in the region, had supplied the Woreda office with the 100 kg of seed mentioned above, these unknown varieties possibly came from unreleased lines, mixed in from the station threshing ground, where threshing and bagging is done through casual labour. Though occasionally risky, 'surprises' found in off-farm supply are an important means for farmers to acquire new materials. This was more common in Miesso, as many had received exogenous seed. Thus, while the lowland environment appears to offer more seed insecurity and uncertainty, the diversity of channels (and the occasional need for supply from distant regions) provides opportunities for introducing new materials (Fig. 5.1). If this material were seen as a separate type, and especially if it were unfamiliar, most farmers stated they would grow it at least one more season for observations. Farmers generally only recognized any material as MV ('Sarude') if it had the erect, dwarf morphology of lowland released cultivars.

**Fig. 5.1.** An example of a recent variety introduction in Miesso district: the tall plants are readily identified as a variety new to the area, and are locally named 'Turree', after the person who had obtained it from the market 3 years ago. This variety has since appeared in other parts of the district, sometimes as an 'unplanned' introduction, mixed into seed of another variety purchased in local markets (photograph by S. McGuire).

## Seed exchange

New varieties generally spread to other farms via farmer–farmer seed
exchange, often in small gifts of less than 1 kg. Seed exchange is also
the most common way for farmers to replenish lost seed stocks; the role
of local markets or formal supply channels is less important, except
in cases of widespread germination failure, as mentioned above. Farm
visits during harvest, and cross-checking with seed givers, revealed
that those who received seed from neighbours, especially small gifts,
did not always mention this in the interviews in the FAs. Thus, the
proportions of those who admit to receiving seed in individual interviews (Table 5.4), as well as in the survey across the wider region
(Table 5.7), are probably underestimates.

In surveys, the highland farmers in Chiro who did mention receiving seed were not significantly different from others for characteristics
surveyed (Table 5.7). This may be a reflection of the late rains during
the 1997 harvest, which affected farmers across all wealth levels. For
Miesso, however, farmers receiving seed in 1998 had significantly
lower production expectations for poor years. Also, the shorter time
since introduction or last replacement of their variety stocks further
suggests that these may be part of a group of 'chronically seed-insecure'
farmers (Cromwell, 1996). Those possessing their stocks for longer
periods may do so in part because of their greater seed security.
However, except for numbers of oxen, other indicators commonly
associated with production potential show little difference between
those who did and did not receive seed in Miesso in 1998.

Of farmers surveyed in Chiro and Miesso, 35% and 22%, respectively (in interviews, more than a third from both locations), mentioned
giving or selling seed in 1998. Those giving had significantly higher
production expectations, both in good and poor years. While Chiro
seed-givers tended to be older, those in Miesso had significantly more
oxen and land, the two characteristics most commonly mentioned by
farmers in wealth ranking. Even though differences in wealth among
farmers are relatively small, those with slightly more resources play an
important role in supplying neighbours in need. However, in Miesso,
the wealthiest farmers stressed that, in particularly poor years, they
may not always have seed to spare for their neighbours.

## Hybridization

There is considerable opportunity for gene flow between varieties,
populations of the same variety and from weedy relatives. I discussed
seed management with farmers in the subset group, and directly

**Table 5.7.** Means of some characteristics of farmers in Western Hararghe who stated in surveys that they had supplied or had received seed off-farm in 1998.

| | Chiro | | | | Miesso | | | |
| | Received in 1998 | | Gave/sold in 1998 | | Received in 1998 | | Gave/sold in 1998 | |
| Farmer characteristic | Yes | No | Yes | No | Yes | No | Yes | No |
|---|---|---|---|---|---|---|---|---|
| Number responding | 7 | 47 | 19 | 35 | 8 | 32 | 9 | 31 |
| Amount given/received (kg) | 8.8 | – | 46.6 | – | 19.6 | – | 30.4 | – |
| Number given to | – | – | 5.8 | – | – | – | 4.7 | – |
| Age of farmer | 35.7 | 37.6 | 41.7* | 35.0* | 36.6* | 35.3* | 39.3 | 34.5 |
| Family size | 6.0 | 6.7 | 6.9 | 6.5 | 6.5 | 6.5 | 7.3 | 6.3 |
| Number of oxen | 1.0 | 1.4 | 1.4 | 1.3 | 1.9 | 1.2 | 2.3* | 1.0* |
| Area planted to sorghum (ha) | 0.58 | 0.53 | 0.60 | 0.51 | 0.98 | 0.89 | 1.25* | 0.80* |
| Expected production, good year (t) | 1.21 | 1.28 | 1.71* | 1.02* | 1.35 | 1.92 | 2.86* | 1.51* |
| Expected production, bad year (t) | 0.36 | 0.39 | 0.53* | 0.31* | 0.18* | 0.30* | 0.44* | 0.23* |
| Avg no. years current variety stocks on-farm | 6.7 | 13.0 | 14.5 | 10.9 | 4.5* | 11.3* | 15.0 | 8.5 |

*Significant difference between 'yes' and 'no'.

observed their practices, where possible. Farmers growing multiple varieties usually sow them together on the same field; while they may store grain and seed by variety, at planting time, the most common practice is to thresh and bulk the seed, mixing varieties together. Frequent seed exchange also mixes new varieties, or populations of the same variety, into existing populations. When there are multiple sources, farmers manage different populations of the same variety as one. More importantly, fields in both locations are often contiguous with neighbouring holdings, with little or no barriers between adjacent fields. Finally, weedy relatives of sorghum commonly grow in field margins in the lowlands. Recent experiments detected pollen flow between weedy and cultivated sorghums to 100 m (Ariola and Ellstrand, 1996), producing viable hybrids (Ariola and Ellstrand, 1997), suggesting that crop–weed hybrids may occur in this context.

Farmers say they aim for uniform maturity times, in part to minimize risk of bird damage. I recorded 50% flowering dates for varieties on 15 of the subset farms, and those of their immediate neighbours. For Chiro, six of nine farms, and five of the six Miesso farms, had at least one variety at mid-flowering within 5 days of another in the same or an adjacent field, offering possibilities for gene flow. Not surprisingly, flowering time closely overlapped between different varieties in the same field. Flowering times were significantly closer (both within and between fields) for the Miesso sample than for Chiro (Table 5.8). Whether this results from greater conformity in planting time or in germplasm is uncertain. Though farmers in both locations use material with maturity times ranging from 3 to over 6 months, very little is known about whether farmers manage different maturity times as a livelihood strategy; more characterization and *in situ* study of maturity times is needed.

**Table 5.8.**  Average differences (with standard errors) in mid-flowering times of sorghum variety populations, observed in August–September 1998.

| Comparison | Chiro | | Miesso | |
|---|---|---|---|---|
| | N of observed comparisons | Days between 50% flowering | N of observed comparisons | Days between 50% flowering |
| Between different varieties in same field | 3 | 8.0 (5.1) | 3 | 1.0 (1.2) |
| Between varieties in adjacent fields | 13 | 12.1 (1.7) | 25 | 3.2 (1.5) |

When asked in the formal survey if seed from some farmers can be 'better' than others, a minority felt it could, citing that some farmers select seed 'at the proper time', or that some farmers' seeds were free of diseases. Seed health and viability is as important a factor in performance as any genetic improvements. Some farmers also related good storage with better seed. If they did note differences between different farmers' seed lots of the same variety, most related it to different management, and not to any intrinsic (i.e. genetic) differences between seed lots. A few gave statements like 'because our storage system here is the same, our seed is the same'. Interestingly, a few lowland farmers mentioned frequent seed exchange: 'we all give to each other, so nobody's seed is different'. In theory, even small levels of gene flow between populations, from seed exchange or pollen flow, can blur distinctions between populations (Dobzhansky *et al.*, 1977).

Farmers note distinct variants of varieties in their fields, especially those differing in colour. Many explain this with the term 'dikala', an Amharic term – also commonly used by Oromo farmers – for a hybrid or cross of plant or animal breeds; for people, at least, 'dikala' can be translated as 'bastard', in both the literal and the pejorative sense (Leslau, 1976). Dikala is a descriptive term, instead of a variety name *per se* and some use it in place of a name; for example, 'this is not "Wogere", but a dikala'. In discussions, most farmers were aware of dikala sorghum plants in their field, though their reactions to them vary, possibly reflecting how strictly they define their varieties, as well as their attitudes to mixtures. Similar to the 'lumping/splitting' discussion above, a few farmers claim that they carefully avoid any dikala, and carefully select for purity; their seed thus represents the 'old type', or unchanged populations. Closer study and characterization of actual seed selection would be needed to better understand if these farmers' management of purity is distinct from their neighbours'.

Farmers recognize some specific types of hybrids with particular names, and generally shun them. Highland farmers mention 'Jengaa', characterized by brown seeds, vigorous growth and impressive yield. However, they find the grain quality very poor, and never sow it deliberately, carefully removing from their fields any individuals found at harvest. Regardless, a few 'Jengaa' plants still appear in most fields each year, prompting a common explanation for its origins: 'ye beré shint' (ox urine!). Some younger farmers thought this was not literally true, and suspected a type of dikala between different varieties in their field to be the cause.

Highland farmers in Chiro also try to remove and destroy all examples of 'Fechatee', sorghum that shatters before harvest (the name means 'shatter' in Oromo), though it is not detectable before then. While this trait and its seed and glume morphology suggests 'Fechatee' is a hybrid with wild sorghum, wild sorghum types are not seen in the highlands. Interestingly, older farmers in Chiro say 'Fechatee' first appeared during a feud long ago, when one farmer brought it up from the lowlands, to secretly sow in his enemy's field to sabotage the harvest; it has persisted in the area ever since.

Lowland farmers commonly find crop–weed hybrids in their fields, which they name 'Qillee'. Like their highland counterparts, they can only identify these after the flowering stage, but will rogue and burn any 'Qillee' plants then, for fear of further contaminating their field. 'Qillee' is not consciously sown, and some local explanations of its origins include the droppings from birds nesting in adjacent marshy areas. Finally, on one Miesso farm I found several plants that had both red and white seed on the same panicle, usually with entire primary or secondary branches all one colour. Though the sorghum breeder had never seen such a condition, I found that 3 of the 11 subset farmers in Miesso recognized it, calling it 'Tafakuur', a name which I subsequently found to be known among a number of farmers in Miesso and Chiro. Most of the 11 felt they would avoid selecting such a plant if it were to appear on their field, in order to maintain pure colours. Table 5.9 summarizes these terms.

Given the presumed high $V_G$ of these populations, and opportunities for crossing between varieties and with wild types, the range of possible hybrids is potentially enormous. Thus, it is striking that farmers tell of only two or three terms in each location, to describe

**Table 5.9.** Common terms for hybrids in Western Harerghe, in Chiro (highland) and Miesso (lowland), with farmers' usual explanations for their origins.

| Location | Type of hybrid | Term | Comments and views on origins |
|---|---|---|---|
| Both | Any | Dikala[a] | General term for mixture of types: also means 'bastard' in reference to people |
| Chiro | Between varieties | Jengaa | Brown seed and vigorous growth; seeded via 'ox urine' |
|  | With weedy sp. | Fechatee | Shatters before harvest; brought up from lowlands during feud; means 'shatter' |
| Miesso | With weedy sp. | Qillee | Shatters before harvest; from droppings of birds that breed in area |
|  | Different seed colour in clusters on the same panicle | Tafakuur | No farmer comments on origins |

[a]Amharic (though used by Oromo farmers as well), all other terms in Oromo.

them. In general, farmers are aware of hybrids, or describe how sorghum types may 'change characteristics over time'. However, dikala is the term generally used to describe this, if any term is used at all. Occasionally, farmers mention more specific hybrid types, when this is obvious (e.g. 'Sarude–Qillee', 'Muyra–Jengaa'), but in these cases the key traits still come from 'Qillee' (blue-black, shattering seeds) or 'Jengaa' (brown, bitter seeds). The pollution associations with 'Jengaa', 'Fechatee', or 'Qillee' are also striking, and the other meaning of dikala may also be significant. The limited set of names in Table 5.9 may thus be reserved for distinctly negative types, about which there is strongly proscribed behaviour (i.e. immediate destruction, lest you contaminate your neighbours' fields).

## Selection

Most farmers practice mass selection, choosing panicles after harvest from piles in the field or at home, with varieties roughly proportionate to the following year's planting intentions (Fig. 5.2 and Fig. 5.3). Some begin earlier, evaluating the full standing plant (e.g. for signs of stalk sweetness or drought tolerance) before or during harvest, and a few report selecting and multiplying distinct variants or new types from an individual head. Panicles selected for use as seed are handled distinctly from the rest of the harvest, often transported separately to the house for storage.

**Fig. 5.2.**  Harvested sorghum panicles in an individual farmer's field in Chiro district, showing the diversity of varieties in a typical plot. Farmers generally select their seed from piles such as this. The darker panicles in front are 'Jengaa'; this off-type has been separated out so that it is not used for food or seed (photograph by S. McGuire).

**Fig. 5.3.**  Farmer in Chiro district transporting selected seed from the field to her house, where it will be stored until the next season by hanging from the roof (photograph by S. McGuire).

Farmers generally sought disease-free panicles with good grain production. However, the actual traits they seek in selection may differ from those they explicitly describe beforehand. A selection simulation served as a preliminary test of actual selection criteria. Nineteen of the subset farmers (11 in Chiro, eight in Miesso) were shown a group of panicles of one variety (19 'Wogere Diima' for Chiro, 15 'Masugi Diima' for Miesso). These panicles had been randomly collected from the FAs as part of another trial, and were of known measurements. Only one variety was used in each simulation to avoid confounding factors of mixing varieties, hence the small number of panicles used. Farmers were asked to select the best panicles (six in Chiro, five in Miesso), a selection intensity over 30%.

Though the selection intensity was necessarily low (breeders often select less than 10% of a population), some trends were suggested. Farmers' selections in both locations tended to have higher grain weight per panicle (Table 5.10). However, Chiro farmers tended to choose panicles with heavier seeds (1000 grain weight) but with no difference in seed numbers per head, while Miesso selections suggested the opposite trend. Some Miesso selections also had greater height and biomass in the original plants; these traits showed little difference in the Chiro selections.

**Table 5.10.** From selection simulation, average selection differentials (S) for several characters when selecting from a sample of heads at 30% intensity, and the number of selections whose means differ from the overall mean at 0.05 and 0.10 probability levels.

| | Chiro (*n* = 11) | | | Miesso (*n* = 8) | | |
|---|---|---|---|---|---|---|
| | Mean selection differential | No. significant *t*-tests | | Mean selection differential | No. significant *t*-tests | |
| Character | (S) | $P < 0.05$ | $P < 0.10$ | (S) | $P < 0.05$ | $P < 0.10$ |
| Fresh plant biomass | 0.06 | 0 | 1 | 0.72 | 3 | 4 |
| Plant height | 0.07 | 0 | 0 | 0.69 | 3 | 5 |
| Unthreshed head weight | 0.67 | 1 | 6 | 0.71 | 2 | 3 |
| Head length | 0.47 | 1 | 1 | 0.53 | 2 | 3 |
| Head width | 0.61 | 1 | 1 | 0.43 | 0 | 2 |
| Threshed grain weight | 0.72 | 2 | 6 | 0.79 | 2 | 4 |
| Grain no. per head | 0.50 | 0 | 1 | 0.71 | 2 | 3 |
| 1000 grain weight | 0.43 | 2 | 4 | 0.24 | 0 | 0 |

The selection differential (S) expresses the difference of the selected heads from the rest of the population, in terms of standard deviation (Falconer, 1981: 175). S values were slightly lower than those reported by Soleri *et al.* (2000), and averaged between 0.4 and 0.7 for the grain yield traits mentioned above. The degree and direction of selection differentials for given traits varied among farmers, and interviews suggest they probably vary to some degree among varieties. Thus a thorough understanding of farmers' actual selection goals would require far more extensive research than that reported here.

## Environments

### Perceptions of heritability

Even if farmers' selection is directional based on phenotype, it may not produce changes from year to year, given the high environmental variation ($V_E$) under farmers' conditions. The heritability of the traits that farmers select – the proportion of their variance that is genetic – also affects response to selection. In the absence of *in situ* progeny tests to estimate heritability, farmers' perceptions of the relative influence of the environment are of interest, since this also sheds light on the degree of conscious theory behind their selection practices.

Using the approach of Soleri and Cleveland (2001), farmers from the subset sample were asked for their expectations about head size and height. In the first scenario, they were asked if progeny from their selections would have the same head size and height if planted on their own fields: 16 of the 19 farmers expected both traits would vary, due to variation in soil conditions, rain or management. The second scenario asked them to imagine planting on a uniform field with uniform, optimal management: most (15 of 19) felt that there would be little variation. While a more complete exploration of farmers' expectations would also include more heritable traits, these results suggest that farmers perceive variation in environment or management as a large part of total phenotypic variation.

## Environmental zones

As stated above, formal sorghum breeders in Ethiopia define three broad altitude zones for the country. Though these cluster broad sets of rainfall, temperature and stresses, each macro-zone still has considerable internal diversity, especially for rainfall amount and timing. Breeders have in the past in Ethiopia, as in other places, sought wide adaptation, giving attention to mean grain yields across multiple testing sites. Given the limited number of testing stations and personnel, this strategy was understandable. Moreover, it also reflected the widely held view within conventional breeding that selection in centralized, well-managed locations offers advantages due to higher heritability of traits on-station (Simmonds, 1991; Ceccarelli, 1994).

At medium scales (a few hundred metres in mountainous highlands, several kilometres in the lowlands), farmers do associate particular environments with particular varieties considered to perform best in those conditions, which are identified by soil, temperature, topography or moisture. Though farmers note spatial environmental variation within their fields, I saw little evidence that they address variation at this scale by planting different varieties in particular parts of their fields; most mix varieties together. High $V_E$ within a field, irregular patterns of variation in $V_E$, and low heritabilities of desired traits could all limit the opportunities to exploit genotype $\times$ environment ($G \times E$) interactions at this scale (Soleri et al., 2000). If these interactions are low within a field, it may not be worth the effort to designate different variety zones at that scale.

Farmers, therefore, do define environmental zones, though differently from breeders. The study area spans all three altitude classes defined by breeders, though farmers' access and management of varieties may cross breeders' zones, for example, when highland farmers seek faster varieties from mid or low elevations. Farmers are interested in yield stability, in the face of variation over space and time. Since changing (or fallowing) fields is rarely an option, and improved management practices are equally inaccessible, farmers can mainly adjust to this variation with variation within and between varieties ($V_G$). The level and nature of yield stability in FVs warrants much more attention.

These differences in environmental classification raise issues for breeding strategy. Qualitative G × E crossovers have long been apparent among selecting sites within the same breeding zone (e.g. Gebrekidian and Kebede, 1978; Debelo and Gutema, 1997). Recently, formal breeding has started to consider regional trials and releases from individual stations, rather than national trials that contain these G × E crossovers (Z. Gutema, Nazret, Ethiopia, 1998, personal communication). Furthermore, the Ethiopia Agricultural Research Organization (EARO) has been aiming to structure its research strategy to environments more precisely defined by rainfall patterns, temperature and stresses. Both a more nuanced definition of target environment, and decentralizing testing and release could help to bring farmers' and breeders' perspectives closer together. However, the 'optimum' number of zones may differ by crop (Packwood et al., 1998), and these new levels of aggregation may still not reflect variation of environment, genetic resources or management at scales important to farmers (Almekinders et al., 1995). Moreover, capacity to respond to this new level of aggregation needs to be developed for many stations and other institutions, such as seed multiplication and extension, which still reflect a centralized, national approach.

## Implications for Breeding for or with Farmers

Farmers retain an abiding interest in trying new varieties, as most widely used material in the region was first introduced within living memory. This suggests that exposure to a wider range of sorghum diversity could help farmers, such as for varieties adapted to shorter growing seasons. Though current MVs start to fill this need in the lowlands, even these usually meet with low levels of acceptance and use.

Formal breeding does not yet address some important explicit farmer criteria (e.g. stover combustion ability). Yet the methodology proposed to learn more about farmers' criteria (participatory variety evaluation of existing MVs, e.g. Mulatu, 1996) may not reveal farmers' actual or implicit selection criteria. Decentralized selection may be one way to address diverse or high-stress environments, and greater farmer involvement could help to bring farmers' criteria into actual selection goals. Though environments and selection goals may vary by farmer, as this research suggests, this does not necessarily mean that breeding need target all possible social/physical environment combinations. Careful definition of target environments and representative user groups may still be needed, though the end result should be more diverse offerings than conventional breeding currently provides (Sperling *et al.*, 2001). Realistically, decentralization of selection, or direct farmer involvement, will become an institutionalized practice in Ethiopia only when there is a clearer sense of the costs and benefits. Cost-effective CPB methods, definitions of target areas and rigorous assessment of impact in comparison with conventional approaches may all be needed before CPB approaches are seriously applied at the national level.

Genetic gains from selection in farmers' situations may not come easily, given the high levels of environmental variation. Though particular selection methods (such as 'stratified' or 'grid' selection: Roupakias *et al.*, 1997; Sarquís *et al.*, 1998; Bletsos and Goulas, 1999) or analytical approaches (Atlin, 1997) may help to improve impact, we still need to know what traits are priorities for farmers, if there is variation from which to select, and if these traits are heritable under farmers' conditions. Simple, cost effective assessments of heritability of selection under farmers' conditions are still needed (Soleri and Smith, 2002) before much selection should occur under farmers' situations.

This study suggests high levels of gene flow from seed exchange or cross-pollination, making farmers' populations very dynamic. This may actually be important in the adaptation of farmers' materials (and adopted or 'creolized' MVs, cf. Bellón and Brush, 1994) to varying conditions, but broader theoretical frameworks, such as metapopulation biology (Husband and Barrett, 1996) may be needed to assess this. Indeed, such frameworks have been suggested to understand and promote dynamic conservation in breeding (le Boulc'h *et al.*, 1994) and in farmers' seed systems (Louette *et al.*, 1997).

One approach receiving attention, especially from NGOs, is to support farmers' own management and selection of their FVs. There may be scope for extending 'best local practice' in selection, as promoted in northern Ethiopia, through the Relief Association of Tigray (REST) (Berg, 1996). In Western Harerghe, farmers' own statements, and direct observations of selection practice indicate that the degree of attention to seed selection can vary considerably. However, farmers need to perceive practical benefits from increased attention to selection, for this to be worth their additional effort. With variable environments and dynamic gene flow, seed selection on its own may not have as much impact as, say, improvements to seed physical health.

Finally, seed systems for supply and storage may warrant support, both for regular supply of planting material in seed insecure situations, and for supplying new materials to farmers. Key aspects of seed systems to address may include social barriers to seed access, such as for poorer farmers, who regularly need off-farm seed but who often lack the means to purchase it (cf. David and Sperling, 1999 for examples of this elsewhere). Barriers in seed systems for accessing new varieties may also relate to knowledge and exposure to new material, because of the changing and sometimes local nature of naming systems, and the *ad hoc* and opportunistic channels through which farmers acquire new material. Much may come from strengthening the flows of information and new genetic material (FV or MV) to farmers. One approach may be to strengthen social capital among farmers, since seed systems are social systems. A popular approach to enhancing local flows of material and information is community seed fairs, though other approaches may be needed to bring in distant material, and to respond to changes in farming systems.

## Conclusions

Farmers' perceptions of genetic and environmental diversity, and the actions they take to manage these, appear to blend empirical observations, theory, environmental specificities and cultural values (compare Soleri *et al.*, Chapter 2, this volume). While there is potential for closer farmer–scientist collaboration in sorghum development, and certainly a need, CPB work needs a better understanding of both farmers' and formal breeders' perspectives and activities in this light, especially with regard to goals, yield stability, gene flow and seed supply.

## Acknowledgements

This research is dedicated to the farmers in Western Harerghe; without their hospitality, patience and enthusiasm this study would not have been possible. The welcoming support and advice from the Ethiopia Agricultural Research Organization, particularly the Melkassa and Miesso Research Centres and the Sorghum Improvement Programme, was equally valued and appreciated. Further gratitude is extended toward the Agricultural Development Bureaux of Miesso and Chiro Woredas, and of Western Harerghe Zone for their support. This study was supported in part by a Natural Sciences and Engineering Research Council (Canada) Postgraduate Scholarship, and by an O'Brien Foundation Fellowship.

# References

Almekinders, C.J.M., Fresco, L.O. and Struik, P. (1995) The need to study and manage variation in agro-ecosystems. *Netherlands Journal of Agricultural Sciences* 43, 127–142.

Ariola, P.E. and Ellstrand, N.C. (1996) Crop-to-weed gene flow in the genus *Sorghum* (Poaceae): spontaneous interspecific hybridization between johnsongrass, *Sorghum halpense*, and crop sorghum, *S. bicolor*. *American Journal of Botany* 83, 1153–1159.

Ariola, P.E. and Ellstrand, N.C. (1997) Fitness of interspecific hybrids in the genus *Sorghum*: persistence of crop genes in wild populations. *Ecological Applications* 7, 512–518.

Atlin, G.N. (1997) In what situations will participatory plant breeding work best? Presented at the CPRO-DLO/IPGRI workshop. *Toward a Synthesis between Crop Conservation and Crop Development. Baarlo, The Netherlands.*

Bellón, M.B. and Brush, S.B. (1994) Keepers of maize in Chiapas, Mexico. *Economic Botany* 48, 196–209.

Berg, T. (1996) Community Seed Bank Project in Tigray, Ethiopia. Report from a review mission October–November 1995, NORAGRIC, Centre for International Environment and Development Studies, Ås, Norway.

Bletsos, E.A. and Goulas, C.K. (1999) Mass selection for improvement of grain yield and protein in a maize population. *Crop Science* 39, 1302–1305.

le Boulc'h, V., David, J.L., Brabant, P. and de Vallavielle-Pope, C. (1994) Dynamic conservation of variability: responses of wheat populations to different selective forces including powdery mildew. *Genetic Selection Evolution* 26, 221s–240s.

Ceccarelli, S. (1994) Specific adaptation and breeding for marginal conditions. *Euphytica* 77, 205–219.

Cohen, J.M. and Isaakson, N.I. (1988) Food production strategy debates in revolutionary Ethiopia. *World Development* 16, 323–348.

Cromwell, E. (1996) *Governments, Farmers, and Seeds in a Changing Africa.* CAB International, Wallingford, UK.

CSA (1995) *Agricultural Statistics, 1994–1995* (1987 Ethiopian Calendar). Statistical Bulletin 132, Vol. 1. Federal Democratic Republic of Ethiopia: Central Statistical Authority, Addis Ababa.

CSA (1997) Estimate of improved seed, irrigation, pesticide and fertiliser applied area, and their percentage distribution by crop for the main (Meher) season, private peasant holdings 1995–1996 (1988 Ethiopian Calendar). Federal Democratic Republic of Ethiopia: Central Statistical Authority, Addis Ababa.

David, S. and Sperling, L. (1999) Improving technology delivery mechanisms: lessons from bean seed systems research in Eastern and Central Africa. *Agriculture and Human Values* 16, 381–388.

Debelo, A. and Gutema, Z. (1997) *Ethiopian Sorghum Improvement Program*, Progress Report 1996, No. 24. Institute of Agricultural Research, Addis Ababa.

Degu, E. (1996) Sorghum breeding and achievements in Ethiopia. In: Deressa, A. and Seboka, B. (eds) *Research Achievements and Technology Transfer Attempts: Vignettes from Shoa. Proceedings from the First Technology Generation, Transfer, and Gap Analysis Workshop.* Institute of Agricultural Research, Nazret, Ethiopia.

Dobzhansky, Th., Ayala, F.J., Stebbins, G.L. and Valentine, J.W. (1977) *Evolution.* W.H. Freeman and Co., San Francisco.

Falconer, D.S. (1981) *Introduction to Quantitative Genetics*, 2nd edn. Longman Scientific and Technical, London.

Gebrekidian, B. (1982a) The Ethiopia Sorghum Improvement Project. Terminal Report, IDRC, Ottawa, Mimeo.

Gebrekidian, B. (1982b) Overview of Ethiopian Sorghum Improvement Program. In: *Sorghum Improvement in Eastern Africa. Proceedings of a Regional Workshop.* Nazret & Debre Zeit, Ethiopia.

Gebrekidian, Brhane and Kebede, Yilma (1978) *Ethiopian Sorghum Improvement Program*, Progress Report 1977, No 5. College of Agriculture of Addis Ababa University and IAR, Addis Ababa.

Guo, J.H., Skinner, D.Z. and Liang, G.H. (1996) Phylogenetic relationships of sorghum taxa inferred from mitochondrial DNA restriction fragment analysis. *Genome* 39, 1027–1034.

Harlan, J.R. (1989) The tropical African cereals. In: Harris, D.R. and Hillman, G.C. (eds) *Foraging and Farming: the Evolution of Plant Exploitation.* Unwin and Hyman, London, pp. 335–343.

Hassen, Mohammed (1994) *The Oromo of Ethiopia: a History, 1570–1860.* Red Sea Press, Treton, New Jersey.

Husband, B.C. and Barrett, S.C.H. (1996) A metapopulation perspective in plant population biology. *Journal of Ecology* 84, 461–469.

IAR (1995) *Sorghum Production Handbook*, IAR Crop Production Handbook Series. Institute for Agricultural Research, Addis Ababa.

IAR (n.d.) *Mashilla. Melkassa Ersha Mermer Ma'ekel* [Sorghum. Melkassa Agricultural Research Centre]. Institute for Agricultural Research, Addis Ababa.

ICRA (1996) *Supporting Agricultural Innovation in Two Districts in Western Harerghe: the Role of Research, Extension, and Farmers.* Working Document Series No. 52, International Centre for Development Oriented Research in Agriculture, Wageningen and Addis Ababa.

ISNAR (1987) Review of Research Programme Management and Manpower Training at the Institute of Agriculture Research in Ethiopia, Report to Board of Directors of IAR. ISNAR R26, ISNAR, The Hague.

Jalata, A. (1993) *Oromia and Ethiopia: State Formation and Ethnonational Conflict, 1868–1992.* Lynne Rienner, Boulder, Colorado.

Kefyalew, A., Degu, E., Mulatu, T., Regassa, S. and Regassa, T. (1996) *Sorghum Production for the Miesso-Assebot Area*, Report No. 27. Institute for Agricultural Research, Addis Ababa.

Leslau, W. (1976) *Concise Amharic Dictionary: Amharic–English; English–Amharic.* Otto Harrassowitz, Wiesbaden, Germany.

Lewis, H.S. (1975) Neighbours, friends, kinsmen: principles of social organisation among Cushitic-speaking peoples of Ethiopia. In: Marcus, H.G. (ed.)

*Proceedings of First United States Conference on Ethiopian Studies.* Africa Studies Centre, Michigan State University, East Lansing, Michigan, pp. 193–207.

Louette, D. (1994) Gestion traditionelle de variétés de maïs dans la Reserve de la Biosphère Sierra de Manatlán (RBSM, états de Jalisco et Colima, Méxique), et conservation *in situ* des ressources génétiques des plants cultivés. PhD thesis, École Nationale Supérieure Agronomique de Montpellier, Montpellier, France.

Louette, D., Charrier, A. and Berthaud, J. (1997) *In situ* conservation of maize in Mexico: genetic diversity and maize seed management in a traditional community. *Economic Botany* 51, 20–38.

Marcus, H.G. (1994) *A History of Ethiopia.* University of California Press, Los Angeles.

Maunder, A.B. and Sharp, G.I. (1963) Localization of outcrosses within the panicle of fertile sorghum. *Crop Science* 3, 449.

McCann, J.C. (1995) *People of the Plow: an Agricultural History of Ethiopia. 1800–1990.* University of Wisconsin Press, Madison, Wisconsin.

McGuire, S. (2000) Farmer management of sorghum diversity in Eastern Ethiopia. In: Almekinders, C. and de Boef, W.S. (eds) *Encouraging Diversity: the Conservation and Development of Plant Genetic Resources.* Intermediate Technology Publications, London, pp. 43–48.

McGuire, S., Manicad, G. and Sperling, L. (1999) *Technical and Institutional Issues in Participatory Plant Breeding – Done from a Perspective of Farmer Plant Breeding: a Global Analysis of Issues and of Current Experience.* CGIAR Systemwide Program on Participatory Research and Gender Analysis, Working Document No. 2. CIAT, Cali, Colombia.

Ministry of Agriculture (1995) *Western Harerghe Population and Production Data, 1990–96.* Ministry of Agriculture, Oromiya Office, Addis Ababa, Ethiopia.

Ministry of Agriculture (1998) Agriculture extension intervention in Ethiopia. In: Zicke, G., Kebede, Z. and Damte, G. (eds) *Special Issue Prepared on Occasion of the 20th FAO Regional Conference for Africa.*

Mulatu, E. (1996) Participatory on-farm variety evaluation of Recommended Highland sorghum varieties under farmers' management. Project Proposal to Ethiopian Sorghum Improvement Program, Melkassa Research Center, Nazret, Ethiopia.

Nyerges, A.E. (1997) Introduction – the ecology of practice. In: Nyerges, A.E. (ed.) *The Ecology of Practice: Studies in Food Crop Production in Sub-Saharan West Africa.* Gordon and Breach, London, pp. 1–38.

Packwood, A.J., Virk, D.S. and Witcombe, J.R. (1998) Trial testing sites in the All India Coordinated Projects – how well do they represent agroecological zones and farmers' fields? In: Witcombe, J.R., Virk, D.S. and Farrington, J. (eds) *Seeds of Choice: Making the Most of New Varieties for Small Farmers.* Intermediate Technology Publications, London, pp. 7–26.

Pausewang, S. (1983) *Peasants, Land and Society: a Social History of Land Reform in Ethiopia.* Weltform Verlag, München.

PHC (1989) *Population and Housing Census, 1984.* Ministry of Population and Housing, Addis Ababa, Ethiopia.

Rahmato, Dessalegn (1985) *Agrarian Reform in Ethiopia*. Red Sea Press, Trenton, New Jersey.

Roupakias, D., Sachinoglou, A., Lazarou, E., Vafias, B. and Tsaftaris, A. (1997) Effectiveness of two grid systems of mass selection in faba bean. *FABIS Newsletter* 40, 2–7.

Sarquís, J.I., Gonzalez, H., Sánchez de Jiménez, E. and Dunlap, J.R. (1998) Physiological traits associated with mass selection for improved yield in a maize population. *Field Crops Research* 56, 239–246.

Simmonds, N.W. (1991) Selection for local adaptation in a plant breeding programme. *Theoretical and Applied Genetics* 82, 363–367.

Soleri, D. and Cleveland, D.A. (2001) Farmers' genetic perceptions regarding their crop populations: an example with maize in the Central Valleys of Oaxaca, Mexico. *Economic Botany* 55, 106–128.

Soleri, D. and Smith, S.E. (2002) Rapid estimation of broad sense heritability of farmer managed maize populations in the Central Valleys of Oaxaca, Mexico, and implications for improvement. *Euphytica* (in press).

Soleri, D., Smith, S.E. and Cleveland, D.A. (2000) Evaluating the potential for farmer and plant breeder collaboration: a case study of farmer maize selection in Oaxaca, Mexico. *Euphytica* 116, 41–67.

Sperling, L., Ashby, J., Weltzien, E., Smith, M. and McGuire, S. (2001) Base-broadening for client-oriented impact: insights drawn from participatory plant breeding field experience. In: Cooper, H.D., Spillane, C. and Hodgkin, T. (eds) *Broadening the Genetic Base of Crop Production*. CAB International, Wallingford, UK, pp. 419–435.

Stemler, A.B.L., Harlan, J.R. and de Wet, J.M.J. (1977) The sorghums of Ethiopia. *Economic Botany* 31, 446–460.

Stroud, A. and Mekuria, Mulugetta (1992) Ethiopia's agricultural sector: an overview. In: Franzel, S. and van Houten, H. (eds) *Research With Farmers: Lessons from Ethiopia*. CAB International, Wallingford, UK, pp. 9–27.

Tesfaye, Million (1961) Mutual-aid associations among the Kottu-Galla of Harer. *Ethnological Society Bulletin, University of Addis Ababa* 2, 71–80.

Teshome, A., Baum, B.R., Fahrig, L., Torrance, J.K., Arnason, J.T. and Lambert, J.D. (1997) Sorghum [*Sorghum bicolor* (L.) Moench] landrace variation and classification in North Shewa and South Welo, Ethiopia. *Euphytica* 97, 255–263.

Teshome, A., Fahrig, L., Torrance, J.K., Lambert, J.D., Arnason, J.T. and Baum, B.R. (1999) Maintenance of sorghum (*Sorghum bicolor*, Poaceae) landrace diversity by farmers' selection in Ethiopia. *Economic Botany* 53, 79–88.

Yemane, G. and Lee-Smith, D. (1984) Evaluation of IDRC-funded research projects in Ethiopia (1972–1983), Vols 1–2. Ethiopian Science and Technology Commission, Addis Ababa and IDRC, Ottawa.

Zewde, B. (1991) *A History of Modern Ethiopia: 1855–1974*. James Currey, London.

# How Farmer–Scientist Cooperation is Devalued and Revalued: a Philippine Example

# 6

## David Frossard

*418 Sunset Drive, Golden, CO 80401–2536, USA*

## Abstract

Scientific collaboration between crop scientists and developing-world farmers is a growing, but still far too rare, occurrence. Reductionism and the privileging of certain forms of 'scientific' discourse typically produce ontological, epistemological and sociological 'exclusions' by which farmer participation is systematically devalued and dismissed. MASIPAG is an unusual organization of peasant farmers and plant scientists founded in response to the negative economic and environmental impacts of the so-called 'green revolution' in the Philippines. MASIPAG members have worked for more than 15 years successfully developing farmer–scientist collaboration in rice development using scientific breeding methods. Based on extensive fieldwork with MASIPAG farmers and scientists, this chapter details the exclusionary roadblocks these farmers face, and their often successful responses.[1]

## The Setting

This chapter concerns a group of Philippine rice farmers and professional crop scientists who call themselves MASIPAG.[1] The study is the result of my long-term interest in alternatives to 'green revolution' rice farming, an interest originally spawned by 2 years as a Peace Corps volunteer in the mountainous cordillera of northern Luzon, where 'modern' rice varieties are still quite rare. I later spent more than a year in Nueva Ecija and Laguna provinces – 'ground zero' for the introduction of 'modern' rice strains in Asia – conducting doctoral research on

©CAB *International 2002. Farmers, Scientists and Plant Breeding*
(eds D.A. Cleveland and D. Soleri)

the history of institutional crop development in the region. While working on a larger project involving the practical ramifications and political ecology of 'green revolution' in the area, I interviewed approximately three dozen scientists, activists and farmers of the MASIPAG movement (as well as many professional crop scientists from the International Rice Research Institute and similar organizations); observed MASIPAG farmers and peasant plant breeders in action; and analysed available MASIPAG literature. I gradually came to see MASIPAG as both a unique kind of farmer's organization and, in fact, a sophisticated critique of the claims made by 'green revolution' stalwarts. As will become apparent, MASIPAG farmers (and their associated scientists) do not readily fit the neo-modernizationist agricultural development models commonly proposed by those in the development industry.

In more than 15 years of continual rice hybridization and meticulous seed selection, MASIPAG farmers, working with sympathetic academics from the University of the Philippines and local non-governmental organizations (NGOs), have produced dozens of new rice strains. And as will be shown below, MASIPAG farmers consider them better adapted to local soils, climates and cultural preferences than competing 'green revolution' varieties. Though the latter, 'modern' varieties are extremely common in the Philippines, they have been abandoned en masse by MASIPAG farmers.

Whether the MASIPAG rice varieties are in fact superior in the measures most of interest to professional crop scientists – primarily *yields* and the associated quality of *pest resistance* – is a question still being answered, though anecdotal evidence and limited surveys of farmer yields and practices (e.g. MASIPAG, 1991) suggest that MASIPAG seeds are at least equal to 'green revolution' strains in this regard. Clearly, more work needs to be done on the nuts and bolts of the MASIPAG experience, and on the documentation of these results.[2] In 2000 the organization began a major internal project to document its technical (and other) successes over the years. But from a subjective MASIPAG farmer's perspective, benefits of various kinds – including, but by no means limited to the narrow categories of yield and pest resistance – seem quite clear.

By taking charge of its own seed production, by producing new rice varieties using many of the same techniques used by professional crop scientists, MASIPAG challenges development industry stereotypes of peasant farmers irrationally resistant to change, innovation and science. In this case, their attention to the scientific method has allowed MASIPAG participants to pursue important agricultural, ecological, economic and political aims not always available to other smallholder farmers.

Since 1992, using translators when necessary, I have interviewed many people associated with MASIPAG in various capacities, generating hundreds of pages of interview transcripts. Perhaps the most striking thing about these interviews, whether conducted with the organization's peasant leaders, 'ordinary' MASIPAG peasant farmers, or associated crop scientists and social scientists, is the uniformity of opinion within MASIPAG. It is clear from these interviews, and from the organization's own self-published literature (e.g. MASIPAG, n.d., 1991) that there is remarkable consensus on the following organizational goals:

**1.** Produce new high-yielding, disease-resistant rice varieties requiring few (or no) purchased inputs such as chemical fertilizers and pesticides.
**2.** Improve farmer health and reduce environmental impact by eliminating toxic chemicals from the food chain.
**3.** Maintain and increase genetic diversity of rice strains (a factor the MASIPAG farmers and scientists believe is crucial to achieving objectives 1 and 2, above).
**4.** Give other poor farmers the tools to break out of what are seen as dead-end, chemical-intensive, 'green revolution' farming practices.
**5.** Have their work disseminated and taken seriously by the larger scientific and political establishment as a valid new agricultural paradigm.

Developing promising new high-yield, low-input, disease-resistant rice strains in farmer-established MASIPAG rice breeding centres, as detailed below, is a labour-intensive and time-consuming business, but one that has in fact produced scores of new and useful seed stocks, according to the unanimous testimony of MASIPAG members, all of whom actually use these varieties on a day-to-day basis instead of 'green-revolution' strains. Nevertheless, attempts to validate the MASIPAG model to a larger academic and political world have been more difficult still because farmer-derived knowledge remains, to some, a second-class kind of knowledge, inferior to that produced by official, conventional science. By using the methods of science themselves (albeit to somewhat different ends) MASIPAG farmers challenge that view.

## A Brief History of MASIPAG

While I have written about the MASIPAG organization at greater length elsewhere (e.g. Frossard, 1994, 1998a,b), some details are in order here to set the scene.

In mid-1985, a politically active Philippine NGO known as the Agency for Community Educational Services Foundation, or ACES, arranged a nation-wide meeting of peasant rice farmers to take place on the campus of the University of the Philippines at Los Baños (UPLB), south of Manila. ACES had long been involved with agricultural development in Central Luzon, and was even sponsored in its early days by the International Rice Research Institute (IRRI), UPLB's next-door neighbour in Los Baños and the inventor of 'green revolution' rice in the 1960s. Although initially sympathetic to the new agricultural technologies set forth by IRRI and disseminated by Philippine government extensionists, ACES field researchers eventually began to doubt the benefits of IRRI's 'green revolution', especially for smallholders. By 1985, ACES fieldworkers – in close, daily contact with the farmers 'served' by the new technologies – were in full revolt against the agricultural policies of the Philippine government and its top-down imposition of IRRI technologies.

Thus, from the ACES perspective, UPLB was a doubly appropriate venue for a national conference on the problems of the 'green revolution'. The campus not only housed the pre-eminent agricultural research university in South-east Asia, but it was, like other University of the Philippines branches in those days, a hotbed of resistance to the dictatorial regime of Ferdinand Marcos, itself closely associated with 'green revolution' agriculture.

During this 'BIGAS National Consultation',[3] as it was known, 44 farmer participants from 30 farmers' organizations nation-wide compared experiences, strategized and finally issued a document titled 'A Declaration on the Root Cause of Our Problems'. It read, in part:

> We, the farmers from Luzon, Visayas, and Mindanao are currently experiencing a downward spiral in our standard of living. Each day our lives are becoming more miserable and difficult to bear . . .
>
> The misery that we are experiencing has, undoubtedly, been hastened and made worse by a one-sided system of production that involves the use of high-yielding varieties of rice grains created by IRRI (International Rice Research Institute) – an international institution engaged in the research on rice strains that thrive on a package of new technologies involving use of fertilizer, pesticides, machineries, irrigation, etc.
>
> We believe the root cause of our suffering can be traced to the closed conspiracy between giant foreign and local capitalists, and our government. We see evidence of this conspiracy in our lack of control of our lands, in questionable farming technologies thrust to us, in the credit program for farming, in the pricing and marketing of prime farming inputs – e.g. fertilizers, pesticides, etc. – in the pricing and marketing of our farm products, etc. They have largely accomplished and perpetuated these through a well-oiled machinery of deception and repression that

both persuade and force us to accept deceptive programs they created and implement to suit their interests.

(BIGAS, 1985: Annex II)

In their declaration, farmers also demanded 'genuine land reform' to include redistribution of large rice farms, as well as coconut, banana and sugarcane plantations; the launching of a 'nationalist agro-industrial program' not dependent on foreign companies or IRRI technology; subsidized rice prices and irrigation fees; and cancellation of outstanding farm debts.

On the tumultuous third day of the conference, farmers presented their declaration to the cream of the Philippine agricultural establishment, including then-IRRI Director General M.S. Swaminathan and representatives from the Philippine Ministry of Agriculture and Food, among others. The meeting was frank and, at times, tense. When invited guests like Swaminathan were finally confronted by farmers, emotions ran high. Angelina Briones, a UPLB soils scientist and an early participant in MASIPAG, recalls farmers shouting at the now sullen and unresponsive dignitaries, a brave, or perhaps foolhardy, course of action in the repressive atmosphere of the time. 'Oh my God,' she remembers thinking, 'how many of these farmers will not reach home again?' (A. Briones, Los Baños, 1992, personal communication).[4]

While, as far as is known, all conference participants eventually did return home safely, a peasant contingent from nearby Nueva Ecija province first met with the UPLB agriculturalists sympathetic to their criticisms.

'How can we help?' the scientists asked the farmers (a radical act in the context of Philippine agricultural development at that time). 'Teach us to breed rice like you do', the farmers replied. As both farmers and scientists related years later, this was a transformative moment (Frossard, 1994: Chapter 3).

Within a year, the MASIPAG organization took shape as a partnership between a few dozen Nueva Ecija rice farmers, ACES organizers and progressive academics from the UPLB. The farmers were politically active individuals with strong and outspoken views on the conditions under which they laboured, but otherwise not noticeably different demographically from their neighbours. Yet with the help of like-minded individuals from other political sectors a new idea was born in Philippine agricultural development: a genuine marriage of equals between farmers and academics to produce new rice varieties and rice-growing technologies using established scientific methods. Equality was an explicit goal of both farmers and scientists (a goal first suggested, it should be noted, by the social scientists of ACES, with their long experience organizing farmers in the barrio).

In some ways, it was a difficult marriage, at least at first. According to Corazon 'Dinky' Soliman, an ACES representative to MASIPAG at that time, the initial process of institution building was something of a minefield. A particular source of tension was the translation of the idea of a 'farmer–scientist partnership' into concrete terms. The elite and highly educated scientists and development workers,

> found themselves struggling within themselves in situations where the farmers assert their rights to know and decide even sensitive matters like finances. It is power sharing because access to resources can be a tool for control. However, we all went through a process of breaking down traditional relationships and attitudes. While the structure was participatory . . . we were all struggling within ourselves.

(Soliman, 1989: 77)

According to Soliman, a graduate of the University of the Philippines School of Social Work, the farmers had to overcome an 'inferiority complex' about their rice-farming knowledge, always 'reminding themselves they were among equals'. The scientists, in turn, had to 'struggle with their intellectual dilemma': the idea that scientific knowledge was available and valid not only in laboratories, but in the rice fields as well (Soliman, 1989: 77ff.).

Thus, the basis for the MASIPAG partnership is, as MASIPAG's former chief extensionist, Perfecto 'Pec' Vicente, put it, a 'radical belief in farmers' (P. Vicente, Los Baños, 1992, personal communication).

In the end, equality (and, indeed, farmer dominance) was institutionalized as a method of de-emphasizing differences in class and status of participants. The MASIPAG Board of Trustees was made up of five farmer leaders elected from participating *barangays*, with two representatives each from the 'scientist' (UPLB) and 'social scientist' (ACES) camps (Frossard, 1994: 114).

While the founders of MASIPAG came from widely different intellectual backgrounds, with hugely diverse life experiences, they shared an interest both in improving rice farming and in organized resistance to the Marcos dictatorship, which they saw as corrupt, anti-nationalist, anti-democratic and anti-poor. Since the Marcos regime was an early and strong supporter of 'green revolution' technologies, and since these technologies, by the mid-1980s, were becoming increasingly problematic for many small farmers, MASIPAG vowed to contest the Marcos regime and 'green revolution' simultaneously, through a programme of self-directed agricultural development.[5]

While MASIPAG farmers joined the organization in part for political reasons, the practical problems confronting chemical-intensive farmers were also a significant motivation. By that time, these farmers faced historically high input costs and low farm-gate prices (IRRI, 1991a), along with the perceived inferior disease resistance of

IRRI-developed seeds, which seem to require never-ending doses of toxic pesticides.

Equally alarming to virtually every MASIPAG farmer I interviewed were the increasingly obvious adverse environmental and health effects from pesticides and herbicides. 'My carabao drank from the irrigation canal', one farmer told me with a mixture of anger and resignation. 'It died from the toxic chemicals.' Other farmers talked about the rise of what they described as 'high blood pressure', 'cancer' and 'heart problems' among those who practised chemical-intensive farming.

In fact, the farmers' health concerns were merited. According to a 1987 study by Michael Loevinsohn published in the prestigious British medical journal *Lancet*, insecticide use was the chief cause of increased mortality in young males of Central Luzon (young males are usually the ones to apply agrochemicals). Many deaths attributed to 'high blood pressure' or 'heart attack' were in fact due to chemical poisoning, Loevinsohn found.

Interestingly, IRRI researchers eventually reached similar conclusions. In a 1995 study of pesticides and farmer health, IRRI's resident pesticide expert Prabhu Pingali found that when using disease-resistant 'modern' varieties, 'natural control or the 'do nothing' option is the most profitable pest-control strategy under normal circumstances' (Pingali, 1995: 9). Pingali *et al.* (1995) note that pesticide exposure can produce a host of significant heart, lung, nerve, blood and skin disorders and should be avoided. In fact, they conclude, for the farmers they studied in Nueva Ecija and Laguna provinces, 'the positive productions benefits of insecticides were overwhelmed by the increased health costs' to those exposed; health costs 61% higher than for unexposed farmers (pp. 357–358).

In the mid-1980s, however, IRRI opposition to the most toxic pesticides was not yet so forceful, and decades of pronouncements about the absolute necessity of pesticides had been internalized by many farmers. Indeed, what else could farmers think when farm loans were tied to their willingness to buy and use specific pesticides mandated by professional agricultural 'experts'?[6]

Nevertheless, it was clear to MASIPAG participants that a new rice-growing regime was required. They would need to develop their own locally specific disease-resistant rice, to help free themselves from the prevailing agricultural system, to eliminate endless seed and chemical purchases, and to minimize health risks to themselves and their animals. Rather than participate in a crop-growing system that was quickly becoming less economical and more unhealthy, MASIPAG farmers and scientists decided to, as they put it, 'take back' rice production from the Philippine government, from 'foreign' organizations like the International Rice Research Institute, from multinational chemical and seed corporations. Renouncing their role as passive recipients of 'green revolution' technology, MASIPAG farmers would try to put varietal improvement once again directly into the hands of smallholder farmers. Clearly, to the extent that members now rely very little on outside agents for seeds or technology, MASIPAG has succeeded.

The MASIPAG experimental farm would not, farmers and scientist quickly decided, return to pre-'green revolution' 'traditional' methods of growing rice. While others elsewhere had retreated to past production modes, MASIPAG farmers saw that as a dead-end, low-yield solution (and, not incidentally, an action guaranteed to draw unwelcome scrutiny from the Marcos government, which had long staked its political prestige on the success of the 'green revolution'). Politically and philosophically, such a reversal was untenable. Economically, too, farmers would find it hard to continue to live modern lives, as part of a larger market economy, without at least a small rice surplus to sell.

MASIPAG farmers also appreciated the rich social capital inherent in the scientific method. They would pursue science wholeheartedly, by establishing a clearly scientific, IRRI-style plant breeding programme of their own. This was both an unprecedented peasant show of support for science (particularly science in the direct service of the farmer) and an explicit critique of the International Rice Research Institute, which these farmers felt had produced neither rice varieties nor farming systems that they could live with happily.

This 'peasant science', as I have labelled it, is a dramatic refutation of the many professional crop scientists and development theorists who saw peasants as inevitably 'backwards', 'traditional' and a general impediment to 'modern' crop-growing practices (more on which later). And, given the participation of UPLB scientists, the MASIPAG partnership is also a dramatic confirmation that not all crop scientists can be lumped together by peasants or academic critics as tools of 'green revolution' interests.

## MASIPAG in Practice

MASIPAG's institutional structure now in place, farmers and scientists tackled difficult questions: How exactly would an experimental breeding centre be set up? Who would supply the land? Who would staff it? Who would teach farmers the scientific methods they craved? Who would make decisions about which hybridizations (of the near-infinite number possible) would be pursued? Who would decide which crosses merited further propagation and selection? Who would benefit from the organization's efforts, having access to the new seeds? (Unless otherwise noted, the claims made in this section come from research documented in Frossard, 1994, particularly Chapter 3.)

Three peasant farmers volunteered land for the project. A site in Jaen, Nueva Ecija, was eventually chosen for its proximity to a major road (to expedite farmer training) and for a deep well suitable for irrigation (necessary at times because MASIPAG varieties have varied maturation times and cannot be counted on to follow government irrigation schedules). A local farming family lived at the station as full-time caretakers, hosting participating scientists and farmers.

The MASIPAG Central Station, as it came to be known, including scores of experimental plots and a nearby cinder-block building with office, teaching, sleeping and cooking facilities, was financed initially through *piso-piso*, pass-the-hat donations and volunteer labour, and later through a small grant from a German church group.

ACES social scientists provided bureaucratic and fund-raising services for the new organization. More importantly, ACES staff ran many 'conscientization' sessions with farmers, teasing out the causes of their growing disaffection with rice growing and soliciting extensive farmer input (Modina and Ridao, 1987).

In an ongoing series of technical sessions, UPLB crop scientists taught the techniques of rice hybridization, record-keeping and selection that farmers had asked for. Whereas farmers had traditionally *selected* promising seeds over generations, and thus were slowly able to mould the rice phenotype, hybridization allowed farmers to accelerate the process greatly. The scientists were in fact so successful that farmers were ultimately able to conduct training sessions themselves, teaching other farmers the new techniques.

Soon, a farmer seed-selection board was formed to evaluate the potential of new hybridizations. After an initial training period, farmers made these decisions themselves. After all, farmers wryly noted, it was the decision of plant breeders at places like IRRI that produced such varieties as the famous IR-8, well known for its amazing yields . . . as well as quick spoilage after cooking, poor taste and weak insect resistance. This time, farmers would choose the characteristics *they* found most important in a rice plant.

It was also decided that any farmer could partake of new MASIPAG varieties as they were developed, as long as they volunteered a few hours of labour to the experimental station, planting, harvesting and evaluating new strains. Scores, then hundreds, of farmers did so and the best MASIPAG varieties were disseminated widely through the province, then elsewhere.[7]

In the end, was it ironic or inevitable that Nueva Ecija should be the site of the first MASIPAG experimental centre? Was it coincidence that this farmer–scientist agricultural development effort grew so popular with local farmers so quickly? In fact, Nueva Ecija was by this time the centre of 'green revolution' rice farming in the Philippines. In 1967–68, high-yield high-input rice covered only 12% of rice lands in the province, increasing to 41% in 1970/71, 80% in 1975/76, and about 99% in 1978/79. At this time, use of IRRI varieties in Nueva Ecija was as much as 25 percentage points higher than the Philippine average (Kerkvliet, 1990: Table B2; IRRI, 1991a).[8] As penetration of high-input varieties increases nation-wide, other farmers in other places may be expected to find farmer–scientist partnerships more attractive.

Though the ongoing MASIPAG experiment has yet to be fully documented, a few tantalizing observations can be made, suggesting fertile areas for future study.

First, the original MASIPAG experimental farm successfully produced 30 generations (two generations per year) in a programme of careful hybridization and selection to produce new rice varieties, and did so in a fashion overtly similar to large international research centres. Beginning with less than ideal raw germplasm, farmers and scientists used what they considered the most promising strains available to them at that time. Although farmer-developed landraces were rare in Nueva Ecija by 1985, virtually eradicated by the subsidized spread of 'modern' varieties, a few were discovered in remote areas. Other landraces were obtained by MASIPAG scientists themselves, often taking speaking fees in 'native' rice as they travelled the country talking to farmers' groups about MASIPAG. In addition, high-yield strains developed by the UPLB, and even a few IRRI varieties, helped to form the organization's original genetic pool. Today, after many generations of crosses and backcrosses, and the release of scores of homozygous new varieties, MASIPAG seeds are presumably quite different from the 'green revolution' varieties grown nearby by many of their neighbours. Exactly how different (and exactly what practical purposes those differences might fulfil) remains to be documented by plant physiologists and anthropologists.

In any case, MASIPAG farmers are proud of defying the stereotypes of crop scientists outside the organization who may grudgingly concede that while farmers can select for particular traits in their traditional, old-fashioned way, hybridization and other 'modern' techniques are beyond them. 'Mang' Marciano, land donor for the MASIPAG experimental farm and a MASIPAG-trained plant hybridizer in his own right, has this response for the doubting scientists:

> You are wrong. You are underestimating the ability of small farmers. Scientists are limiting the capabilities of farmers. Because they are learned and studied, they belittle small farmers. Now the little farmers are trying to prove they can do the work. They are encouraged by the example of MASIPAG in producing varieties. If farmers are really interested, they can

easily learn ('Mang' Marciano, Nueva Ecija, 1992, personal communication) (compare Ceccarelli and Grando, Chapter 12, this volume).

Second, despite the predictions of some crop scientists, the sky has not fallen on MASIPAG farmers now that they have abandoned the 'security' of officially vetted rice varieties. Favourite MASIPAG varieties, as determined by the farmers themselves, produce yields roughly equal to 'green revolution' varieties, often with fewer purchased inputs. My re-analysis of unpublished (MASIPAG, 1991) survey data gathered by UPLB graduate statistics students suggests that the difference in yields between 195 Nueva Ecija farmers growing either MASIPAG or IRRI strains exclusively was statistically insignificant (averaging 3.82 and 3.87 t $ha^{-1}$ per growing season, respectively)[9] (compare Ceccarelli and Grando, Chapter 12, this volume). With an emphasis on integrated pest management (IPM) and other 'organic' techniques (in addition to directed seed production), some MASIPAG farmers were able to eliminate pesticides entirely and to reduce the use of chemical fertilizers substantially (typically substituting various local biofertilizers such as *Azolla*, other green manures, compost or farm wastes).

Anecdotally, one of my most vivid memories from Nueva Ecija involves a MASIPAG farmer who had used no pesticides or chemical fertilizers whatsoever for several years. His particular field was lush and pest-free, full of beneficial predators, particularly spiders (often among the first organisms to disappear when spraying begins). His was one of those fields of ripened grain that would almost, as the farmers say, support a door laid flat on it. However, the field of his non-MASIPAG next-door neighbour, literally inches away and separated only by a low earth berm, was repeatedly sprayed with large quantities of agrochemicals yet contained virtually no harvestable rice. The crop had been destroyed by leaf-hoppers and stem-borers, which still swarmed over the bare stalks. This was a common sight near MASIPAG fields I visited. A decade later, with many more farmers nation-wide employing MASIPAG methods, it is time for new surveys to assess the ongoing efficacy of the organization's methods.

Third, a final, telling endorsement of MASIPAG strategies is simply the rapid growth of the organization itself. Beginning with a few dozen participants in 1986, MASIPAG now counts thousands of farmers in several hundred Philippine localities, actively developing their own seeds on the Nueva Ecija MASIPAG model. Exactly how many such MASIPAG organizations may exist, the original organizers are unsure; new experimental farms are started frequently, often with the help of other farmers who have received MASIPAG training. MASIPAG crop scientists now take a less central role. In addition, since the end of the Marcos dictatorship, farmers' unions nation-wide, and even elements of the Philippine government itself, have endorsed the organization's methods and goals. MASIPAG's rapid expansion from one experimental farm to so many in little more than a decade indicates that farmers find something very useful in the MASIPAG process and products.

In summary, MASIPAG farmers have for more than a decade successfully hybridized rice and selected new varieties to meet their own particular, local criteria for cost, yields, environmental and health effects, pest resistance and taste. They did so because they perceived that 'modern' crop science would not do so for them. Their achievements seem impressive, worthy of emulation by other poor farmers and worthy of study by scientists. The latter, however, especially those associated with the scientific agricultural development industry, have historically dismissed, even belittled the MASIPAG efforts. Partly this is the fault of MASIPAG itself, which has emphasized practical results over documentation of those results (except in the matter of record-keeping for rice hybridizations). But even with voluminous records, MASIPAG would face sceptics in the larger social marketplace, sceptics who simply dismiss the products of farmers. How, in practice, does this 'devaluation' take place? What are the practical effects of devaluation? And what, if anything, can the farmers do about it?

## Devaluation of Farmer Knowledge

It is only relatively recently that farmer participation in crop development has come to be seen as valuable, though more often by social scientists studying 'indigenous knowledge' than by crop scientists *per se*. As I noted above, at least one highly persistent strain of discourse regarding peasant farmers sees them not as an aid to scientific crop development, but as an impediment. Supposedly stubbornly traditional and resistant to modernity, these farmers are seen, in this view, as 'backward' creatures to be tamed and brought into modernity, kicking and screaming if necessary (compare Schneider, Chapter 7, this volume).

For a (literal) textbook example of this view, we need look no further than the standard reference work on rice science, *Rice*, by Sir Donald H. Grist (1986, its 6th edition). In it, Grist maintains that 'under Asian conditions, the limiting factor to yield improvement through better seed is probably the low educational standard and apathy of cultivators'. Peasants, he allows, while 'not antagonistic to improvements' are nevertheless 'slow to absorb new ideas and to appreciate the possibilities of improvement' (p. 506). While Grist may be politically incorrect by today's standards, he is not alone in his views. Decades of development research among peasants have long painted rural peoples as antagonistic to change (George Foster's 'Image of Limited Good'), apt to retreat behind village walls when confronted with new circumstances (Eric Wolf's 'Closed Corporate Peasant Community') or simply not interested in innovation (the 'laggards' of adoption-of-innovation theory) (Soleri *et al.*, Chapter 2, this volume). Extensive interviews with IRRI crop scientists in 1992 and 1995 reinforced the notion that farmers are often seen there as peripheral to the actual business of seed development. For instance, one senior IRRI crop scientist confided to me that he typically paid farmers to grow 'test' seed varieties, then asked them to comment on the result. This, he said, constituted 'participatory' research. Admittedly, IRRI social scientists have a far more subtle view of farmer–scientist collaboration, and have had some success at making farmers' voices heard by crop scientists (see, e.g. Goodell, 1984; Fujisaka, 1989, 1990; Fujisaka and Garrity, unpublished). Arguably, however, social scientists wield relatively little power in the IRRI hierarchy.

Fortunately for the farmers themselves, these images of benighted peasants antagonistic to science are incorrect in many or most instances, as attested by a growing number of empirical accounts from places like Gaviotas, Colombia, or Kerala, India. The image of peasant farmers as dull, sullen children irrationally resistant to change but in need of a particular kind of scientific 'education' – a picture closely tied to modernization theory – is in fact politically dangerous to farmers (and advantageous to others) because it legitimizes many forms of coercive intervention into peasant life. Certainly the Marcos regime took this view in the 1980s when government extension agents are reported to have collected and burned stocks of 'traditional' seeds, forcing farmers to use 'green revolution' strains (McAfee, 1985: 294).[10] More subtle and more widely practised forms of coercion today involve selective access to credit or irrigation water reserved for government-approved 'green revolution' farmers.

This is why the MASIPAG effort is so potentially counter-hegemonic, and thus so threatening to some within organized crop science, or indeed within Philippine politics. MASIPAG is clearly *not* anti-science, anti-technology or anti-progress. (Its members 'merely' want more say in reaching their own definition of 'progress'.) In these circumstances, it is much harder for agricultural professionals to explain to non-MASIPAG farmers why they should continue to follow the lead of the International Rice Research Institute or the Philippine government in rice farming matters.

Nevertheless, such explanations, some more persuasive than others, have indeed been offered and MASIPAG's efforts have been systematically discounted by actors at various nodes in the development web. IRRI, in particular, spent much of the 1980s in an acrimonious public relations fight with Dr Burton Oñate, an eminent UPLB

statistician and founding member of MASIPAG, who publicly called
IRRI a 'capitalist instrument' and tool of multinational agribusinesses
(a battle I recount at greater length in Frossard, 1994: Chapter 4).
Thomas Hargrove, at that time head of public relations for IRRI,
detailed his relatively successful counter-attack (from IRRI's point of
view) on Oñate, in a paper presented to a convention of agricultural
communications specialists (Hargrove and Pollard, 1990). In this
instance, while Oñate was often successful in taking his case to the
public through the popular press, IRRI's status in the Philippines
changed little. Through an extensive (and expensive) public relations
campaign, Hargrove ultimately convinced enough Philippine policy
makers of IRRI's value, and of the supposed 'irrationality' of IRRI's
opponents. The MASIPAG challenge, even coming from an eminent
scientist like Oñate, rather than a farmer, was ultimately dismissed by
those who counted politically.

The great Indian social theorist Vandana Shiva suggested how
this devaluation might be accomplished, in a seminal paper entitled
'Reductionist science as epistemological violence' (Shiva, 1988).[11]
Shiva bases her argument on the premise that 'modern science is
quintessentially reductionist'. Because it selectively reduces its vision
to a few very specific properties of an object (e.g. a rice seed, or a
rice farmer) it is able to squash alternative forms of knowing very
effectively. While science is said to be a 'value-neutral' and 'universal'
process of truth-finding, the evidence from the rice field suggests that,
in fact, science regularly excludes certain forms of knowledge while
promoting others. Shiva suggests that several types of 'exclusion' (as
she calls them) are regularly practised in agricultural science, 'exclusions' that are familiar to those involved in the MASIPAG project.

The first type, according to Shiva, may be referred to as *ontological
exclusion*, in which particular *intrinsic* properties of the experimental
subject are ignored. In rice farming, ontological exclusions might
include: the comparatively low cash cost of both 'traditional' and
MASIPAG crop-growing systems (and what that implies for incorporation of farmers into the debt-based market economy); the disease resistance inherent in genetically diverse, rather than monogenetic, crops;
even the *taste* of the rice. These factors, if mentioned at all in scientific
literature, are clearly considered more or less irrelevant to the job
at hand. The most 'relevant' factor, in the reductionist scientific
worldview, is simply *higher yields*, nothing less, nothing more. To
judge from a decade of IRRI publications, that focus has not changed,
only the methods of achieving it have evolved (principally, farmer
education in 'optimal' management practices on the farm side, and
biotechnology focused on the production of 'super' rice on the development side). As former Director General Klaus Lampe put it, 'So long as

scientists are unable to alter the fundamental structure of the rice plant itself or of the environment in which it is grown, breaking today's yield ceiling to reach higher levels of productivity will follow the law of diminishing returns' (IRRI, 1991b: 9). The *IRRI Reporter* announced in a headline that same year: 'IRRI Goal: a new rice plant type' (IRRI, 1991c).

Second, says Shiva, is *epistemological exclusion*, in which alternative ways of knowing are ruled out. In this case, one way of knowing – laboratory science – is employed at the expense of the more experiential epistemologies employed by farmers (see Duvick, Chapter 8, this volume; Soleri *et al.*, Chapter 2, this volume). Rather than experimentation in tightly controlled test plots, farmers tend to do their 'testing' of different schemes under rather chaotic, real-world conditions, involving many simultaneous variables. Any knowledge gained from such a process, scientists often claim, would be idiosyncratic at best; not the 'universal' knowledge produced by science. To many professional crop scientists, 'true' science must be done only by those specially trained in its methods. All other 'science' is, therefore, science in name only (and not really worthy of the name, at that). This position is powerfully reinforced by the elaborate systems of scientific accreditation enforced by universities and governments.

The fact that MASIPAG explicitly duplicates the scientific methods of the International Rice Research Institute makes it more difficult, though obviously not impossible, to dismiss the knowledge and products produced by this farmers–scientist partnership. For instance, MASIPAG opponents may try to change the rules of the game. While it may grudgingly be allowed that the organization's 'peasant science' is an adequate crop development strategy today, in the future advanced bioengineering techniques will make it a poor cousin at best, and no longer adequate or competitive with the new biotechnologies (which, finally, must surely be beyond the capacity of peasants to reproduce). IRRI's recent and ongoing emphasis on a promised genetically engineered 'super rice' is a none too subtle reminder that science is powerful, and that IRRI science is the most powerful of all.

Finally, there is a third type of exclusion, *sociological exclusion*, in which the non-scientist (in this case, the peasant farmer) is denied both the right of access to knowledge and the right to judge knowledge claims. Farmers are, by and large, excluded from the 'green revolution' development process, except as passive recipients of finished technology.

Professional crop scientists that I met at IRRI in 1992 and 1995 flatly rejected the possibility that peasant farmers could perform their own rice hybridizations. In interviews with IRRI plant breeders I was constantly asked to define my terms. Were farmers really hybridizing rice or just picking this or that seed to try out next season? Peasant farmers, it was allowed, are able to *select* promising plants from their fields and thus slowly improve varieties over generations, but the much more powerful technique of directed hybridization, the scientists claimed, is unknown and unavailable to farmers. When I described the hybridization process (involving the selection of parental varieties and physical pollination by peasant field technicians), the crop scientists seemed genuinely surprised, even stunned. Social critic Claude Alvares (1992: 227) describes this mindset:

> The state claims its right to 'develop' people and nature on the basis of a vision of progress set out in blueprints supplied by modern science . . . The people have no role other than spectators or cogs in this 'great adventure'. In exchange, they, or some of them at least, are privileged to consume the technological wonders that result from the heady union of development and science. In the eyes of a patronizing state, this is adequate compensation for a surrender of their natural rights. As for those who cannot or will not participate, they must lose their rights.

Indeed, for millions of smallholder farmers in Asia and elsewhere using 'green revolution' technologies, crop growing is not what it used to be, nor in a very real sense are they. Smallholder farmers in these situations have become denatured, devalued, empty vessels into which 'green revolution' technology is poured.

Still, as Galileo is said to have muttered when forced to disavow a heliocentric solar system in the face of torture: 'And yet, it [the Earth] moves.' And yet, despite their supposed inability to do so, MASIPAG farmers demonstrably perform hybridizations and produce new rice strains in an explicitly scientific fashion, varieties that these farmers clearly prefer to institutionally developed 'green revolution' ones.

Critics of MASIPAG thus may thrive only when particular aspects of the MASIPAG programme, such as its explicitly scientific nature and successful products, can be effectively ignored, and other aspects, like farmers' relative lack of (formal) education, emphasized.


## Revaluation of Farmer Knowledge

Revaluation of farmer knowledge is not easily done. Not only is science a powerful trump card when juxtaposed with supposedly 'traditional' and 'uneducated' peasants, but science effectively masks its powerful political effects through equally powerful rhetoric. Science (or so it is often claimed) is 'apolitical'. Thus science becomes a kind of 'anti-politics machine', in the words of Jim Ferguson (1990). Science as a discipline simultaneously defends its particular, reductionist view of the world while claiming to be above political considerations.

'Scientific' pronouncements about the deficiency of farmer knowledge inevitably carry some weight, at least to many Western ears.

Farmer knowledge, and farmer–scientist partnerships like MASIPAG, may nevertheless be defended and revalued through some of the same mechanisms by which they are devalued, and through other methods. Though what follows is no doubt incomplete, let me make a few hopeful observations about farmer and scientist partnerships like MASIPAG.

First, as science takes away, so can it give. Sympathetic, accredited mainstream scientists, social and otherwise, may be able to validate 'peasant science' in scientific terms (Ceccarelli and Grando, Chapter 12; Soleri *et al.*, Chapter 2, this volume). My own comparison of MASIPAG and IRRI yields, above (yields which are apparently rather similar), can be interpreted in this light. In this volume are several chapters attesting to the empirical soundness of farmer knowledge in particular cases (see McGuire, Chapter 5; Soleri *et al.*, Chapter 2; Zimmerer, Chapter 4, this volume). Such chapters may use scientific terminology, statistics and even the odd mathematical formula, to illustrate the empirical soundness of particular farmer practices. MASIPAG itself, recognizing the usefulness of this kind of knowledge both to crop production and to public relations, is hard at work documenting what its members perceive as its stunning success. Expect numerical data on the MASIPAG experience, generated by the organization itself, to follow in the next few years.

In another widely reported case, anthropologist Stephen Lansing and systems ecologist James Kremer were able to model indigenous Balinese irrigation systems and show that they were, in fact, to a statistical certainty, superior to the farming methods of 'green revolution' in producing sustained yields and keeping pests at a minimum (Lansing, 1991). Given enough evidence, enough scientific testimony, enough demonstrably positive results, science can be forced to acknowledge the validity of 'peasant science' in terms of Western science itself, an acknowledgement which can only serve to decrease the typical marginalization of the peasant farmer.

Second, the underlying premises of science, especially as they intersect with modernization theory ('neo' or otherwise), may be contested within the scientific academy. This chapter is one obvious example. Social scientists, particularly, may at times successfully problematize the premises and promises of institutional science. In fact, many social scientists have done so, bringing the pejorative term 'scientism' – fetishized, superficial science – into the epistemological debate on truth and knowledge production.[12] Sceptical social scientists and progressive critics like Ivan Illich (1977), Vandana Shiva (1988, 1991, 1993), Ashis Nandy (1988), Claude Alvares (1988, 1992), Serge

Latouche (1993) and others are uniquely positioned to challenge the twin gospels of 'science' and 'progress', to make supposed 'externalities' more visible, to illuminate the ontological properties of both rice and rice farmers, and show how these properties are intrinsic, not peripheral, both to crop production and to farmers' lives.

Third, let us remember that conventional top-down science is not the only game in town. After the fall of the Marcos regime, the Aquino government was much more positive about ventures like MASIPAG. Delegations from the Ministry of Agriculture (as it was then known) trooped to see the MASIPAG experimental farm. More importantly perhaps, one of the largest farmers' unions in the country endorsed the MASIPAG experiment and encouraged its other members – several million of them – to do likewise. While the reverse is perhaps more common, sometimes politics can trump conventional science in useful fashion. Scientists should not necessarily see this as a threat, but perhaps more as an invitation to try something different. After all, the involvement of (once) 'conventional' plant breeders was a crucial element in the success of MASIPAG and the development of a useful alternative to the conventional top-down model.

Finally, and most tentatively, I suggest that we social scientists may also seek to validate other forms of knowledge such as 'peasant science' by transforming *ourselves*. One of the observations made repeatedly by MASIPAG farmers and scientists alike was the difficult time each had 'reimagining' the other. The MASIPAG farmers often talked about how difficult it was for them to see University of the Philippines scientists as equals rather than superiors to whom they should defer at all times. For their part, the UPLB scientists who originally trained MASIPAG farmers in 'green revolution' techniques were forced to confront their own ideas of what a 'poor' and 'illiterate' farmer might ultimately be capable. Also, the academics were used to dispensing knowledge, not negotiating it with peasant farmers. Not insignificantly, the scientists were also used to, and admittedly fond of, being addressed with great deference. Yet, in the end, when each side tried sincerely to approach the other on an equal footing – a process accelerated by the formation of a MASIPAG board of directors with a farmer majority – both scientists and farmers pronounced themselves transformed and enriched by the experience, better able to understand and appreciate the other.

Today, MASIPAG is thriving. No longer hounded by the national government, the growth of MASIPAG is constrained only by the inability of members to fulfil requests for training in MASIPAG techniques; requests that have come from throughout Asia and the Philippines. From one research centre MASIPAG is now an umbrella organization representing 387 people's organizations, 97 church-based groups, and 46 NGOs, according to Dennis Maliwanag, the group's publications officer (D. Maliwanag, Los Baños, July 2000, personal communication via e-mail). Where MASIPAG once released the odd pamphlet or photocopied position paper, it now produces slick publications attacking 'biopiracy' and the international patenting of rice genes, genetically modified transgenic crops, and the 'green revolution' in general.

The organization has clearly struck a chord among a particular category of farmers. These are the smallholder peasant farmers who, for too long, felt themselves at the mercy of economic and political forces beyond their control. In a sense, MASIPAG methods have led to a most radical product of science: self-empowerment. By increasing their control over rice germplasm, MASIPAG farmers increase their control over their own lives in general. They also validate the promise of science as a discipline and way of knowing, at least when it is used democratically in the service of farmers.

The MASIPAG story apparently has come full-circle. From a small anti-Marcos, anti-'green revolution' protest group in the 1980s, MASIPAG and the ideas it has championed for almost two decades may have officially entered the political mainstream. In late January 2001, after a second bloodless 'People Power' revolution against a corrupt leader, Vice-President Gloria Macapagal-Arroyo assumed the Philippine presidency, replacing disgraced President Joseph Estrada. One of her first two cabinet appointments was an early MASIPAG stalwart, Corazon 'Dinky' Soliman, who becomes Philippine Secretary for Social Welfare and Development.

## Acknowledgements

A version of this chapter was originally presented at the Annual Meeting of the Society for Applied Anthropology, Tucson, Arizona, 20–25 April 1999. I especially wish to thank David Cleveland for his insightful comments on earlier drafts of this chapter. And of course, I would once again like to thank the scientists and 'peasant scientists' of MASIPAG for their insights and their example.

## Notes

[1]    MASIPAG (pronounced 'mah-SEE-pahg') means 'industrious' in Tagalog, but in this case also forms a Tagalog acronym for Mga Magsasaka at Siyentipiko Para sa Ikauunlad ng Agham Pang-Agrikultura, roughly 'Farmer–Scientist Partnership for Agricultural Development'.

[2]    This is one area in which MASIPAG perhaps lags behind institutional science: the documentation of results. In some sense, though, this is explained

by the problematic, even precarious, political position of MASIPAG in the
Marcos era, and afterwards.

[3]    *Bigas* means 'husked rice' in Tagalog, but is in this case also a rough acronym for Bahanggunian Ng Mga Isyu Sa Bigas or 'Advisory Group on Rice
Issues'.

[4]    At least one MASIPAG member, Dr Ruben Aspiras, was briefly imprisoned
in a military detainment camp because of his activism. Aspiras survived his
experience and was later named chancellor of UPLB.

[5]    Perhaps the most devastating critique of excesses in agricultural development before and during the Marcos era can be found in Ernest Feder's neo-
Marxist polemic *Perverse Development* (1983), a book widely available and
widely read by Philippine intellectuals, especially those interested in agriculture, during the period in which MASIPAG was formed.

[6]    This is a complicated issue. Partially in response to farmer complaints,
IRRI has worked diligently for many years to build wide-spectrum disease
resistance into its rice varieties, through both the usual hybridization and
new recombinant genetic techniques. In addition, IRRI officially promotes
integrated pest management techniques to minimize pesticide use. Pingali and
others at IRRI are on record advocating high taxes on the most toxic category
I and II chemicals (typically organophosphates and organochlorines) to
encourage farmers to use less toxic category III and IV pesticides (Pingali *et al.*,
1995).

Unfortunately, in the view of MASIPAG participants, IRRI's attempts at
ecologically sound pest control are largely confounded by the organization's
own success in promoting new, high yield rice strains. By developing new rice
varieties that are quickly and widely adopted throughout Asia, IRRI – and now,
a number of national, university and commercial rice breeding centres mod-
elled on IRRI – helps to eliminate nature's oldest defence against insects and
other pests: biodiversity. Where perhaps 140,000 farmer-bred rice varieties
once grew in Asia, a relative handful of landraces remain, supplanted by the
'green revolution' influx. In such a system, an infestation of one rice variety can
destroy hundreds of thousands of hectares of crops. MASIPAG arguably offers a
superior model for biodiversity maintenance, maximizing genetic diversity and
minimizing risk of widespread crop failure. By producing rice in hundreds of
local research centres, biodiversity – and appropriateness for particular local
climate, soil and pest conditions – is maximized. This biodiversity maintenance is an intrinsic feature of the MASIPAG model, with its hundreds, and
potentially thousands, of dispersed breeding centres; an institutional arrangement that large, centralized research organizations cannot easily duplicate.

Of course, mainstream institutions like IRRI necessarily approach biodiversity and disease control through the application of their own strengths:
educational programmes in integrated pest management (which, incidentally,
find favour among MASIPAG members as well) encourage a wide diversity of
rice pest *predators* to guard the 'green revolution' monocrops; gene splicing
inserts pest-resistant genes from other plants – or even organisms such as
*Bacillus thuringiensis* – into new rice varieties (a development strongly
opposed by MASIPAG); diversity of raw germplasm is maintained *in vitro*
through collection and cold storage of thousands of varieties (MASIPAG notes

that IRRI's stored rice is for all practical purposes unavailable to them or to other small farmers).

[7] An anthropologist friend of mine working in a distant rural area of Borneo came upon farmers who called one of their favourite rice varieties 'Masipag'. Coincidence? More concretely, MASIPAG seeds have been introduced and grown in Thailand, as well as throughout the Philippines. Though particular varieties are not always productive in these disparate environments, their dissemination is still a point of honour with MASIPAG farmers, who often ruefully recount their unsuccessful efforts to acquire samples of nearly extinct landraces from IRRI, the world's primary repository of rare rice seeds.

[8] Nueva Ecija, it should be noted, was also the site chosen by Pingali and others in the 1980s and 1990s to document the effects of pesticide over-exposure on farmers, precisely because it had experienced the most 'green revolution' for the longest time.

[9] Interestingly, the survey (MASIPAG, 1991) found no significant differences in landholding size between MASIPAG adopters and non-adopters (both averaged landholdings of slightly less than 2 ha). Adoption apparently hinged on other factors, including the limited number of seeds available for distribution and the fact that MASIPAG seeds were not certified by the government and were thus ineligible for crop insurance. Nevertheless some highly 'idealistic' farmers, according to the survey, defined security not in terms of crop insurance, but in terms of 'control over the factors of production – primarily seed' and questioned the need for MASIPAG certification by the Philippine national seed board (MASIPAG, 1991: 4–5). The survey also provided subjective responses by farmer-adopters of MASIPAG seeds. Farmers most often cited low production costs, high yields and high resistance to pests as reasons to use the seeds. 'Curiosity' was also cited as a major reason! Finally, the farmers warned surveyors about potential 'sabotage' by those with vested interests in the 'green revolution'. If MASIPAG methods were fully adopted, the report calculated, it would cost multinational chemical companies on the order of 1.5 billion Philippine pesos year$^{-1}$, or about US$75 million at that time (MASIPAG, 1991: 5).

[10] The Japanese took similar actions in Taiwan a century ago (Juma, 1989). During the 1970s, the Khmer Rouge leadership in Cambodia also collected and burned more than 1000 rice varieties deemed ideologically impure, the product of 'foreign' influences. These varieties were restored to Cambodia in the 1980s from the germplasm collection of the International Rice Research Institute.

[11] Interestingly, Shiva is by training a physicist and philosopher of quantum mechanics. She is also an astute observer of the agricultural transformation of India in the 'green revolution' era and is thus uniquely situated to comment on scientific practice.

[12] It is well to note that the very term 'scientist' is of comparatively recent origin. 'We need very much a name to describe a cultivator of science in general. I should incline to call him a Scientist,' wrote William Whewell in his *Philosophy of the Inductive Sciences* (1840). The power and prestige accruing to the person of 'the scientist' *per se* is thus a relatively recent historical phenomenon. While the privileging of particular categories of thought-and-thinker

predates Whewell by millennia, the extraordinary, almost religious, devotion to science in our time remains an extreme case.

# References

Alvares, C. (1988) Science, colonialism and violence: a Luddite view. In: Nandy, A. (ed.) *Science, Hegemony and Violence: a Requiem for Modernity*. Oxford University Press, Delhi, pp. 68–112.

Alvares, C. (1992) Science. In: Sachs, W. (ed.) *The Development Dictionary*. Zed Books, London, pp. 219–232.

BIGAS (1985) *Proceedings of the BIGAS National Consultation, 17–19 July. University of the Philippines, Los Baños.*

Feder, E. (1983) *Perverse Development*. Foundation for Nationalist Studies, Quezon City, Philippines.

Ferguson, J. (1990) *The Anti-Politics Machine: 'Development,' Depoliticization, and Bureaucratic Power in Lesotho*. Cambridge University Press, New York.

Frossard, D. (1994) Peasant Science: Farmer Research and Philippine Rice Development. PhD dissertation, University of California, Irvine, California.

Frossard, D. (1998a) Asia's Green Revolution and peasant distinctions between science and authority. In: Fischer, M. (ed.) *Representing Natural Resource Development in Asia: 'Modern' Versus 'Postmodern' Scholarly Authority*, Centre for Social Anthropology and Computing Human Ecology Series 1. CSAC Monographs, Canterbury, UK. Also available online, at http://lucy.ukc.ac.uk/Postmodern/David_Frossard_TOC.html

Frossard, D. (1998b) 'Peasant science': a new paradigm for sustainable development? *Research in Philosophy and Technology* 17, 111–126.

Fujisaka, S. (1989) *Participation by Farmers, Researchers and Extension Workers in Soil Conservation*, IIED Gatekeeper Series No. SA16. International Institute for Environment and Development, London.

Fujisaka, S. (1990) Rainfed lowland rice: building research on farmer practice and technical knowledge. *Agriculture, Ecosystems and Environment* 33, 57–74.

Goodell, G. (1984) Untying the HYV package: a Filipino farmer grapples with the new technology. *Journal of Peasant Studies* 11, 238–266.

Grist, D.H. (1986) *Rice*, 6th edn. Longman, New York.

Hargrove, T.R. and Pollard, L. (1990) An international center's public awareness program in the host country: IRRI and the Philippines. Presented at the *Annual Meeting of Agricultural Communicators in Education, Special Interest Group on International Programs, Minneapolis, Minnesota, 16 July*.

Illich, I. (1977) *Toward a History of Needs*. Heydey Books, Berkeley, California.

IRRI (1991a) *World Rice Statistics 1990*. IRRI, Los Baños, Philippines.

IRRI (1991b) *IRRI 1990–1991: a Continuing Adventure in Rice Research*. IRRI, Los Baños, Philippines.

IRRI (1991c) IRRI Goal: a new rice plant type. *IRRI Reporter* (September). IRRI, Los Baños, Philippines.

Juma, C. (1989) *The Gene Hunters: Biotechnology and the Scramble for Seeds.* Princeton University Press, Princeton, New Jersey.

Kerkvliet, B.J.T. (1990) *Everyday Politics in the Philippines: Class and Status Relations in a Central Luzon Village.* University of California Press, Berkeley, California.

Lansing, J.S. (1991) *Priests and Programmers: Technologies of Power in the Engineered Landscape of Bali.* Zed Books, London.

Latouche, S. (1993) *In the Wake of the Affluent Society: Explorations in Post-Development.* Zed Books, London.

Loevinsohn, M. (1987) Insecticide use and increased mortality in rural Central Luzon, Philippines. *Lancet* 13 June 1987, 1359–1362.

MASIPAG (n.d.) *MASIPAG,* organizational pamphlet. MASIPAG, Los Baños, Philippines.

MASIPAG (1991) MASIPAG Evaluation Survey. MASIPAG Project Benefit Monitoring and Evaluation Survey (PBMES) Unit, Los Baños, Philippines.

McAfee, K. (1985) The Philippines: a harvest of anger. In: Schirmer, D.B. and Shalom, S.R. (eds) *The Philippines Reader.* South End Press, Boston, Massachusetts, pp. 292–301.

Modina, R.B. and Ridao, A.R. (1987) *IRRI Rice: the Miracle That Never Was.* ACES Foundation, Quezon City, Philippines.

Nandy, A. (1988) Introduction: Science as a reason of state. In: Nandy, A. (ed.) *Science, Hegemony and Violence: a Requiem for Modernity.* Oxford University Press, Delhi, pp. 1–23.

Pingali, P.L. (1995) Impact of pesticides on farmer health and the rice environment: an overview of results from a multidisciplinary study in the Philippines. In: Pingali, P.L. and Roger, P.A. (eds) *Impact of Pesticides on Farmer Health and the Rice Environment.* Kluwer Academic Publishers, Boston, Massachusetts, pp. 3–21.

Pingali, P.L., Marquez, C.B., Palis, F.G. and Rola, A.C. (1995) The impact of long-term pesticide exposure on farmer health: a medical and economic analysis in the Philippines. In: Pingali, P.L. and Roger, P.A. (eds) *Impact of Pesticides on Farmer Health and the Rice Environment.* Kluwer Academic Publishers, Boston, Massachusetts, pp. 343–360.

Shiva, V. (1988) Reductionist science as epistemological violence. In: Nandy, A. (ed.) *Science, Hegemony and Violence: a Requiem for Modernity.* Oxford University Press, Delhi, pp. 232–256.

Shiva, V. (1991) *The Violence of the Green Revolution: Third World Agriculture, Ecology and Politics.* Zed Books, London.

Shiva, V. (1993) *Monocultures of the Mind: Perspectives on Biodiversity and Biotechnology.* Zed Books, London.

Soliman, D. (1989) *The Challenge of Participatory Development: the MASIPAG Experience.* Proceedings of the Regional Workshop on Sustainable Agriculture, 8–22 September 1988, International Institute for Rural Reconstruction, Silang, Cavite, Philippines.

# Selecting with Farmers: the Formative Years of Cereal Breeding and Public Seed in Switzerland (1889–1936)

## 7

JÜRG SCHNEIDER

*Wabernstrasse 43, CH-3007 Bern, Switzerland*

## Abstract

Genetic diversity in the form of domestic landraces was the major source for wheat and spelt breeding in Switzerland from 1910 to 1930. The first Swiss breeding programmes were participatory, involving farmers in on-farm selection and propagation of breeding lines from their own landraces. These selections became the basis of all cultivar releases in the years between the two World Wars, when a formal seed sector promoting the displacement of landraces in lower altitude areas was established. In the 1930s, this loss of landrace diversity was first perceived as 'genetic erosion' and addressed with timid conservation measures. At the same time, farmer participation in breeding was phased out. This chapter describes and analyses context and function of collaborative plant breeding (CPB) during the formative years of Swiss national wheat breeding in terms of biological, institutional and political factors.

## Introduction

In the 1990s, the perception that formal plant breeding and seed supply systems are unable to meet the needs of many farmers became widespread. It stimulated analysis of approaches that involve users more closely in crop development; these were variously termed collaborative or participatory plant breeding. A substantial body of literature on participatory or collaborative crop development analysing a wide range of anthropological and agronomic data has emerged (Eyzaguirre and

Iwanaga, 1996; McGuire *et al.*, 1999). In most of the contexts that are analysed, institutional breeding is an established system. By institutional breeding, I refer to complex institutional arrangements with the specific objective of crop improvement. These arrangements transform crop development from a localized farmers' endeavour – embedded in cultivation and harvesting practices – into a locus of recurrent, institutionalized action which is responsive and accountable to scientific and political discourse. Evidently, we need to look at different socio-political and historical contexts to improve our understanding of how farmer breeding and formal plant breeding relate to each other. Such historical studies are largely lacking so far, and this is also true for the formative period of cereal breeding in Switzerland.

I decided to select this case because cereal crops are the staple crops in the Swiss lowlands; the data collection has been restricted to Switzerland to allow for easy access to documentary sources and resource persons. Initially, I had the assumption that contemporary documentary sources would allow the reconstruction of farmer varietal management practices for those crops that were of interest to formal plant breeders, and give a more precise answer to questions such as 'Have formal plant breeders worked with and benefited from farmers' knowledge and farmers' selection methods?' and if so 'how?'

In terms of information on farmers' practices and their perceptions of landraces, the results from documentary studies have been rather disappointing. This lack of information on local variety selection seems surprising given the fact that – as the chapter will show – a group of farmers were actually very closely involved in the first stages of institutionalized breeding. Yet, most aspects of local practices of crop selection remained uncharted. It seems as though the two knowledge systems, 'art de la localité' of farmers (Van der Ploeg, 1992), and national seed selection of the scientists, have not communicated, at least not in a way that is reflected in written sources (compare Ceccarelli and Grando, Chapter 12, this volume; Duvick, Chapter 8, this volume; Joshi *et al.*, Chapter 10, this volume).

The farmer–plant breeder relationship was shaped and reshaped in the context of a broader movement towards regulated seed. How seed productivity, production and distribution should and could be regulated has been an important issue for agricultural interests, scientists and politicians. To reflect this broad meaning, I will use the notion of 'regulated seed' rather than 'regulated seed production'.[1]

In this regard, the years 1889 and 1936 that bracket the period analysed in this chapter each represent a very different type of institutional arrangement. In 1889, when a first series of state-promoted seed fairs[2] for the broader diffusion of 'good and clean cereal seed' were held, the organizational concepts and expertise for institutional breeding were still lacking. Yet when collaboration between plant breeders and farmers first developed around 1900, the driving force was a similar one: the idea that the crisis in cereal production needed to be addressed with 'good seed'. In the perspective of breeders, this soon came to mean 'selected varieties'. In 1936, a nation-wide exhibition of 'quality cereal seed' was organized that showed the products of the national breeding venture and highlighted also the important role of seed growers' associations in the diffusion of a selected number of wheat cultivars.

Between 1889 and 1936, the few plant breeders that there were in Switzerland had assumed an important role in the selection of varieties and the implementation of a seed sector regulated by the state. They had developed collaboration with farmers in on-farm selection of varieties and with farmer seed growers in the production of commercial cereal seed. Breeders described this collaboration mostly in technical language, justifying it on the basis of local crop diversity held by farmers and the absence of specialized producers of seed grain in the country.

When it came to the political project of a protected national seed market and the role of the state as regulator of varieties, the language of breeders also comprised more ideological elements. The special status of local varieties and the quality of domestic production were invoked by breeders and politicians representing farmers' interests as part of their struggle for national self-sufficiency in grain production. For them, the scarcity of grain supplies during the First World War proved to be a blessing in disguise as it had demonstrated the vulnerability of Switzerland's dependence on foreign seed imports and the viability of domestic seed production.

Thus, the analysis of the formative period of institutional breeding – a period during which farmers have been partly associated with the activities of formal plant breeders – requires us to consider not only technical progress in plant breeding, but also the link with seed regulation and the broader social and political context of the nation-state and organized farming interests. How the link between domestic wheat cultivars, plant breeding and seed regulation by the state was shaped is the central focus of this chapter. I will first look at the context of agrarian politics in the late 19th and early 20th century. Second, I will discuss scientific progress in plant breeding and its significance for crop development in Switzerland. Then I describe forms of collaboration in early cereal breeding, look into organizational developments in the domains of breeding and seed regulation, and address the ideological discourse of the protagonists of regulated seed. Finally, I ask what all this has meant in terms of genetic resources.

## Sources and Method

There is an abundance of material published by agricultural organizations in Switzerland even before the turn of the 20th century. A
comprehensive collection of these materials is held by the Swiss
National Library and can be accessed through card catalogues and
partly through electronic databases. Plant breeding was, however, not
important enough to lead to the formation of special journals. It forms
part of this large body of printed sources, which I have searched for
references on plant breeding and communications or articles by plant
breeders during a period of roughly 50 years (1890–1940). The foremost
breeder of the period, Albert Volkart, left scientific research material
to the archives of the Swiss Federal Institute of Technology Zurich.[3]
This unpublished material has also been consulted. In addition, a
retired plant breeder has been interviewed to cross-check some of my
interpretations.

The knowledge and practices of farmers are largely lacking in these
sources, which reflect the perspectives of researchers and breeders,
not those of farmers. Farmer selection was treated summarily under
the term 'mass selection'. Though breeders were aware that artificial
selection had been practised by farmers in their role as cultivators for
many centuries, natural selection[4] was considered by most contemporary plant scientists a determining factor shaping 'landraces' over the
centuries. As a consequence of this strong focus on 'natural selection',
the farmers' role of 'artificial selection' tended to be overlooked.

## The Context of Agrarian Politics

> Once the great investments involved in the building of steamships and
> railroads came to fruition, whole continents were opened up and an
> avalanche of grain descended upon unhappy Europe.
>
> (Polanyi, 1957: 182)

In the late 19th century, the cultivation of cereals in the Swiss lowlands
was shrinking as a result of cheap imports made possible by new
railway connections with Eastern Europe and liberal trade policies at
the national level. Among farmers and agricultural interest groups,
there was a widespread perception of crisis. Protectionist measures had
a lot of support, but the agricultural lobby was just starting to build up
its influence in the national political arena; it was too weak to press for
protectionist legislation.[5]

As far as seed was concerned, there were some timid attempts
to solve the problem. In response to the crisis, many farmers opted
for new cereal varieties, mostly foreign introductions, and tried to

cultivate these on a greater scale. Agricultural associations had a negative view of this trend and tried to promote 'seed replacement'[6] from domestic sources, which meant the replacement of farm-saved seed with a new seed lot not produced on the farm. Ideally, these replacement lots were varieties obtained from markets or seed fairs. The concepts underlying 'seed replacement' were far from clear-cut or scientifically validated, but based on a belief that yields of local varieties were declining, as these strains had 'degenerated', and new, more productive seed had to be introduced.[7] Plant scientists disagreed that this could be an effective measure to boost productivity (Jahresberichte, 1889–1921; Schellenberg, 1901). However, the prevailing opinion in the Swiss Association for Agriculture (SLV[8]) was that it was worth trying, and that one of the means to achieve broader diffusion of new strains was seed fairs. These were organized seasonally by agricultural associations in various locations of the country where locally produced cereal seed was evaluated, promoted and sold on a small scale. Initially, these markets were private initiatives; in the late 1890s, the associations succeeded in securing subsidies from the federal government for these events. The purpose of seed fairs was simple: good seed should be given an official reward, and farmers from the region should have the opportunity to buy. Characteristic for this period was the aesthetic mode of evaluation. The evaluation criteria for these 'best-seed-awards' were based on a set of traits that were all non-productive.[9] In particular, yield did not form part of the evaluation criteria because farmers brought farm-saved seed themselves to the fairs without any but their own yield estimations, and the associations did not perform independent yield measurements on these seeds.

This approach based on minimal interference with varietal choice and seed production of farmers was still prevalent in the early 1910s, without having shown significant results. Subsequently, the First World War proved to be a turning point. The first seed growers' associations[10] were founded during the war as an essential part of an emergency scheme to boost domestic production of cereal seed. The war had caused a serious shortage in domestic food supplies, including bread wheat, and eventually led the state to declare a monopoly in cereal imports. In 1916, the federal state mandated one of its agricultural officers – who was also a plant breeder – to set up the necessary institutional framework. The breeder's role was that of an organizer developing the contractual arrangement between state institutions, seed growers' associations and farmers. Only in a few cases was he/she providing seed from varieties of his/her own selections for this new organizational set-up. At this point, breeding programmes were still too recent to produce enough seed for multiplication. However, a number of local varieties had been identified in the years prior to 1916 on the basis of both performing and non-performing traits. These would also be accepted for the emergency scheme, which succeeded in contributing grain seed for an expansion of national production.

Later this programme was portrayed as an important 'lesson from the war' (Oehninger, 1936: 57) by the agricultural interest groups which saw a need to have a strong domestic seed sector supported by state intervention if necessary. Indeed, the post-First World War period brought more political leverage for the protectionist view and more emphasis on national self-sufficiency in important food crops. Farmers' organizations opted for import tariffs and a state cereal monopoly. The monopoly option was against the interests of workers and liberals and was defeated in a 1926 referendum in which the major political cleavages of that period became manifest. A more modest article regulating cereal production was accepted by popular vote in 1929. Legislation based on this article was adopted in 1932 and formed the basis of the very strong position of the parastatal seed sector for decades to come.

In sum, the strongly regulated grain production in Switzerland was one of the fruits of a new alliance between liberal mainstream politics, agricultural interests and a peasantry (*Bauernschaft*) that was better organized than before and using the experience of the First World War as a call for self-sufficiency. In the 1920s, ruling liberal political forces, although in favour of unregulated seed and grain production, were anxious of rising left-wing influence and thus tried to find a compromise with agricultural interests. After 1930, world markets were again well supplied with grain. Overproduction led to price drops and eventually to a farm crisis; for example, that in the USA. However, in Europe, the argument for self-sufficiency generally took precedence over the argument for free markets, and protectionist policies continued.

The stronger focus on national self-sufficiency and institutions of regulated seed led to a policy encouraging varietal uniformity and discouraging diversity on the farm level. In the years prior to 1932, a restrictive variety policy was implemented under the heading 'variety purification'.[11] The reasons for the 'purification' policy were as much ideological as political and technical. On a technical and administrative level, a low number of varieties eligible for multiplication certainly made the work of seed production easier. But the concept is also informed by the idea that the variety purification would do away with the current mess or jumble[12] of 'too many and too diverse' varieties. This view could not be justified with the goal to protect national seed production. Often, it reads like a devaluation of varietal diversity a priori. As a policy, it informed the actions of the executive of the Association of Seed Growers, resulting in an increasingly lower number of varieties being accepted for multiplication by official seed growers' associations, and a decreasing number of varieties being actually planted. Moreover, the policy of variety purification ran counter to what the seed fairs had tried to achieve: the promotion of diversity based on an attitude of relative non-interference with farmers' choices.

In the 1920s and 1930s, centralized and authoritarian governments in Germany and Italy favoured even stronger processes of 'varietal purification' by which the state developed a firm grip on seed production and variety selection (Fontanari, 1934; Flitner, 1995: 81–85). Such moves have to be seen in the context of an ideological formation favouring centralized planning which – at least in the case of Nazi Germany – enabled the administration of increased production of important food crops through shortened variety lists while protecting the interests of plant breeders cum seed growers more efficiently. In the absence of an authoritarian power centre in the Swiss case, the 'purification' policy did not make it to the top of the agenda of farmers or the agricultural lobby. However, the underlying concepts of 'unification' or 'purification' were influential.

The concept of 'varietal purification' entered the discourse of agriculturalists in the 1920s and was eventually adopted by the top official of the seed grower association in his drive to 'rationalise' seed production of just the 'very best varieties'. A good example is Schnyder (1936: 34–36) who has the period of 'variety unification' start in 1925 with the objective of reducing the number of cultivated wheat varieties to between two and four for the whole wheat growing part of the country. This target was missed, yet for 1935, he observed that just three varieties accounted for 72% of all seed provided by the seed growers' associations. It is pertinent to note here that the notions of 'unity', 'regularity' and 'harmony' are a fundamental triad for many projects of the young, federated nation-state of Switzerland.[13] The concepts of 'unified' and 'purified' seed thus drew part of their strength also from the 'national landscape' that was being built in many areas of social and economic activity.

Through their policies of variety 'purification', agricultural modernizers were influencing varietal choice; modern plant breeding seemed to open the way to determine varietal forms, to 'manufacture' varieties. This potential for 'rational', engineered varietal forms was a source of fascination, as is evident from a description of the breeding facilities at Svalöf in Sweden – then one of the most advanced plant breeding institutes in the world – which was published in 1925 as a critique of less developed Swiss institutions.

> We are shown the storerooms, shops, and offices where trained personnel are sorting, measuring, weighing. The compound reminds of a manufacturing plant in which precision instruments are built. There is something awesome in the business we witness: living beings are engineered here,

beings which did not exist before, new creations following the will of humankind.[14]

(Seiler-Neuenschwander, 1925: 396)

Seiler-Neuenschwander critiqued the Swiss plant breeding system for lagging behind the Swedish example which for him was 'an organization and a bundling of forces for the whole country which is so far singular in its perfection'.[15] What then had Swiss wheat breeders achieved so far?

## Progress in Genetics and its Significance for Plant Breeding

Breeding is unable to create new forms; but among the forms created by nature, the breeder's hand is able to make a selection for further propagation.[16]

(Schellenberg, 1902: 33)

The years just before and after 1900 were revolutionary for the science of genetics. In 1900, the laws of Mendel were rediscovered almost simultaneously by three scientists (Mayr, 1984: 582–585). At least theoretically it had now become clear that a controlled recombination of certain traits through hybridization is possible. But this possibility still seemed remote or practically not applicable, and this was the state of practical breeding when the plant scientist Schellenberg made the statement quoted above. The idea that 'breeding is unable to create new forms' coheres with the technical approaches[17] adopted originally by wheat breeders in Switzerland: these were first dominated by the method of 'single pure line selection'.[18] It differed from the practices of farmers who would usually plant 'multi-line' mixtures and not go for pure lines. It differed from later breeding techniques in that it made no use of hybridization (it did not make crosses).

The first Swiss breeder who applied this breeding method in a wheat selection programme with national scope was Gustave Martinet in Lausanne. Martinet based his work mostly on landraces[19] of his native canton Vaud, and came up with a few selections bearing place names from related landraces or their respective regions (e.g. Bretonnieres, Vuiteboeuf and Gavillet). He selected for yield and lodging tolerance because he wanted to address the productivity issue of domestic wheat production (Martinet, 1921).

A more advanced approach was adopted by the plant scientist Albert Volkart,[20] the second and more influential Swiss breeder. Volkart started his programme in 1907 with a similar claim to improve landraces. He still considered single pure line selection as the best option for self-pollinating crops like wheat, and based his work on farmer-assisted selection of pure lines.[21] However, contrary to

Martinet, he tried early on to experiment with crosses for the long-term objective of using them in applied breeding. Scientists at the Federal Polytechnical School[22] in Zürich had built contacts with Swedish plant breeders from the renowned institute at Svalöf. There, Nillsson-Ehle had developed a more sophisticated wheat breeding technique early in the first decade of the 20th century. Plant scientists in Zürich were well aware of these developments abroad and considered them the model to follow.

Volkart made his first experimental crosses with poppies before 1910,[23] yet he started using bulk methods from segregating wheat populations[24] only in 1918 (Volkart, 1931: 16–18). Figure 7.1 (1931: 18) illustrates his procedure using a method termed 'hybridization with

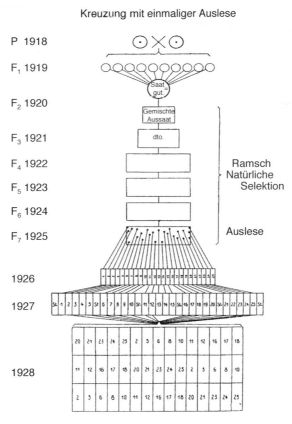

**Fig. 7.1.** A figure from Volkart's book *Kurzgefasste Anleitung zur Getreidezucht, herausgegeben vom Eidg. Volkswirtschaftsdepartement* (1931) illustrating 'hybridization with single selection'; making a wheat cross and cultivating the F$_1$ generation as bulk (the German term used in the figure is 'Ramsch') for seven generations until it had reached the desired degree of homozygous stability.

single selection', during a 10-year period from 1918 to 1928. He started making a wheat cross and cultivating the $F_1$ generation as bulk for seven generations until it had reached the desired degree of homozygous stability.[25] In wheat, this is the case in the $F_7$ generation. Then, in 1925, he made a single selection[26] and proceeded with propagation of this selection. In the period up to 1930, Volkart's use of hybridization techniques in wheat and spelt was still limited and experimental.

In sum we can observe a slow transformation of plant breeding into a scientific field of knowledge, informed by the new paradigm of Mendelian genetics and insights from population genetics. Specialists worked with farmers' assistance on landraces, transforming them within a new context of rational, scientific procedure. In self-pollinating crops, the resulting breeders' varieties[27] were composed of single, pure lines. Rather than any specific breeding technique it was this concept of selection for pure lines by plant breeders that had differentiated the products of plant improvement into *Hochzuchten* (literally high-breeds, referring primarily to pure lines)[28] on the one hand and landraces (consisting of many lines) on the other.

## Selecting with Farmers in Nascent Institutional Cereal Breeding

There had been no institutional cereal breeding in Switzerland previous to 1898, neither in the public nor in the private sector. There were two main reasons for this: wheat production was of lesser importance compared with the dairy sector, and the agrarian structure was based on smallholder operations.[29] In other parts of Europe, for example in northern Germany, large landowners growing cereal crops had traditionally been important and innovative cereal breeders. They engaged in this activity with a view to producing and marketing seed.[30] Private capital, thus, was instrumental in early modern European breeding, before the state assumed a more important role.

It is no coincidence that the beginning of specialized wheat breeding in Switzerland is linked to the establishment of a state agency: Martinet had started his selection programme for wheat at the Federal Station for Seed Experiments in 1898, working mainly with landraces from the western part of Switzerland, traditionally a centre of wheat growing. His methodology essentially followed the ideas of the French breeder Vilmorin. He describes it as follows: on farmers' plots, landraces – usually consisting of between 10 and 20 different morphological types – were evaluated. For each type, relative productivity (seeds, straw) was measured and samples of 20–40 seeds were taken for further evaluation on station. From the $F_2$ generation, one-third of the types was retained on the basis of superior yield performance. This procedure was repeated until very few types remained. Martinet's dominant breeding objectives were yield and tolerance against lodging.

Volkart started his programme in 1907, 9 years after Martinet. His first preferred method was single selection for pure lines:[31] sampling of types from landraces, separate cultivation of selected lines and successive elimination of those lines that perform less well (mostly in terms of yield) until one line remains. Multiplications were run mostly simultaneously to the evaluation of lines. He qualified the method as 'single pure-line selection', because selection took place only once, at the beginning. No crosses were made.[32] Volkart notes, however, that evaluation needs to be done carefully and on multiple plots (as the elimination of lines is proceeding). Yet it was through this method that most of the official varieties of the first two decades of formal plant breeding were produced. Volkart used the term 'formative breeding'[33] to denote the transition from landraces to pure line selections.[34]

Volkart had to organize his programme without an adequate amount of land on an experimental station; thus he also did seed multiplication from selected types within the network of participating farms using plots of participating farmers. Around 1910, Volkart travelled in Switzerland, explaining his wheat and spelt breeding programme to farmers and inviting them to participate. The farmers who participated in Volkart's programme were responsible for the selection procedure, which took place on land they provided (compare Duvick, Chapter 8, this volume). This included the selection of parent material in the field, further selection based on morphological criteria, documentation of relevant morphological characteristics, preparation of propagation plots, observation and data collection during growth, harvesting and second selection. Volkart recommended the elimination each year of half of all planted lines. If farmers made an initial selection of 50 plants for pure line evaluation and propagation, only three lines were left for a final selection after 5 years (Volkart, 1931: 32).

Farmers were assisted by the plant breeder in the selection of lines, and they were given recommendations on the criteria to observe, such as number of growing plants, tillering capacity, ear weight, ear density, kernel weight, straw weight, and 1000 kernel weight (Volkart, 1931: 29). The objective was to obtain higher yielding varieties with tolerance to lodging, resistance to pests and hardiness. Participating farmers were scattered all over the lowlands of Switzerland, each testing and selecting from one or two landraces. Location-specific selection and the diversity of locations represented in the programme ensured specific adaptation to various environments of the lowland wheat and spelt growing regions (Festschrift, 1913: 54–55) (compare Bänziger and de Mayer, Chapter 11, this volume; Ceccarelli and Grando, Chapter 12, this volume). Thus, during the initial phase of his programme, Volkart's goals were not restricted to productivity gains; he also wanted to obtain pure line varieties adapted to various local soil and climatic conditions, mainly in the Swiss lowlands.

The number of farmers involved in the programme grew from 9 (1909) to 64 (1919).[35] As an incentive, small lump sums were paid to participating farmers, but these payments were not commensurate with the work they had to invest. Farmer-breeders had to provide additional labour for careful harvesting of the selected lines.[36] In addition to financial rewards, two other factors seem to have motivated farmers to participate in the breeding exercise: an interest in the knowledge of applied breeding, and the possibility of seeing a wheat selection of their own being spread on a national scale. Starting with many varieties in many locations and reducing them in a sort of pyramidal selection, the programme of the first period resembled a contest instigating farmers to come up with a pure line selection from landraces in their region that would be accepted by the seed board. The decision making for those cultivars that were eventually selected for cultivation on a national scale remained a prerogative of the experts. Yet, Volkart acknowledges that farmers would not easily accept rejection of their final selections by the seed board (1931: 62–64).

Financial constraints and shortage of land for experiments – a national research station for central and eastern Switzerland with sufficient land was built only in the 1940s – was an important reason behind Volkart's drive to include farmers and to use more advanced breeding methods in a collaborative effort. Farmers were trained in the practical tasks of selecting types and preparing plots for selected line evaluation. Volkart developed breeding courses that he conducted for over 20 years. Based on these courses, a breeding manual for a broad audience from farmers to agricultural extensionists was eventually published (1931). In the first part, Volkart explains general principles of plant breeding, including a description of how to make crosses in wheat (1931: 33–36). In the second part, he describes the appropriate methods, in his view, to be applied by non-breeders; thus the practicalities of pure line selection from landraces of various cereal crops.

In his review of 'formative breeding' Volkart (1928: 29) made clear that he associates the collaborative element in breeding with a technical stage based on pure line selection and using landraces. He suggested that in the future farmers would not participate to the same extent they had in the first two decades of Swiss cereal breeding. He recommended establishing breeding as a specialized function in the experimental station, a function led by professional staff who would be able to conduct it with much higher precision. Although he had made

wheat crosses himself since 1918 and included the technique in his
'manual', he did not make the step to include these methods explicitly
in his on-farm selection programme. Implying that this may involve too
much work for farmers, he asked whether 'farmers who have been of
such great service during formative breeding do still have the time
necessary for these tasks of hybridization' (Volkart, 1928: 25). Volkart
referred here to self-pollinating crops, and the task of cultivating the
bulk for a number of generations before a selection could be made. He
would thus postulate a labour constraint for farmers in more complex
breeding methods involving hybridization, not unlike rice breeders in
the Philippines[37] many decades later.

## Promoting 'Good Seed': the Institutions of Regulated Seed

The two early Swiss cereal breeding programmes conducted by
Martinet and Volkart were neither primarily scientific projects in the
search of knowledge nor commercial ventures dominated by an economic rationale. Rather they have to be seen in the context of a concern
for 'improved seed' that developed in the agricultural public, peasant
parties and agricultural research towards the end of the 19th century
(Baumann, 1992; Moser, 1994). The organization of farmers and farming interests took shape towards the end of the 19th century and slowly
brought farmers back into the political arena after 1900.

This was an experimental and formative phase, in which existing
models for and new forms of organization were applied to the problems
perceived in agriculture. Three stages or organizational models can be
discerned with regard to the concern for 'improved seed': seed fairs,
seed control,[38] and seed growers' associations.

I have already referred to seed fairs, the first of these institutional
models. Seed fairs were organized after 1889 by agricultural associations with national support as a measure to promote 'cleaned and
well-sorted seed' of cereal crops, and to stimulate variety turnover
based on the concept of seed replacement. Despite some limitations,
seed fairs became an important instrument of seed politics for more
than two decades (Jahresberichte, 1889–1921). What were the characteristics of these fairs?

First, they were competitions with prize money for the 'best seed'.
The system looks similar to the cattle contests conducted in many
cattle growing areas of Switzerland and could have been inspired by
these. Any farmer could send or bring a seed sample to the market.
There was a closing date after which the samples were ranked by
a price committee. Then the market was officially opened and the
samples put on display.

The ranking was based on a sophisticated set of evaluation criteria relating to quantity, presentation and quality. The 'purity of seed' criterion gave higher values to pure line varieties – in the case of self-pollinated species – compared with seed from mixed line landraces. This standard interfered with farmer practices of seed saving but did not imply a distinction between 'breeder varieties'[39] and landraces, which were treated indiscriminately before 1910.

Seed fairs did not attempt to assess or change farmer practices of seed production. The organizers did not take any seed samples from the fields of producers, who thus became the major source of information about the varieties' performance. The jury rankings were based more on beauty than on performance, because only field visits would have allowed the evaluation of important parameters such as yield potential and seed viability (the percentage of seeds that would germinate). This was considered a shortcoming of the seed fairs, but a system with field visits[40] – envisaged as early as 1895 – was not introduced because it would have been too costly for the organizing associations.

The seed fair model was reviewed by a committee of the Swiss Agricultural Association in 1911, and considered a potentially useful tool for the promotion of certain breeders' varieties. It was then that Volkart proposed to create two classes of samples, one for breeders' varieties and one for landraces. A brochure published by the SLV in 1913 was optimistic that seed fairs would be used increasingly as plant breeding developed and gave no indication that the decline of this institution was imminent (Festschrift, 1913: 49).

The issue that led to the sudden demise of the 'seed fair' model was definitely seed control, which is the second organizational element of regulated seed. Historically, the demand for stronger seed control in European agriculture is linked with more commercial seed being produced and sold in the second half of the 19th century; fraud with seed that would not germinate seems to have been quite common, and the first control mechanisms were set up in Western Europe during this period. These controls were in many cases administered by state institutions, but in some cases also by the seed industry through some kind of self-control mechanism.

From a breeder's perspective, his or her varieties can reach a large 'audience' only if seed is multiplied and if its quality and identity can be guaranteed. Seed control thus is at the heart of applied breeding. In addition to quality concerns, seed control mechanisms are among the best instruments for the regulation of variety choice (compare Ceccarelli and Grando, Chapter 12, this volume; Joshi et al., Chapter 10, this volume). Simultaneously with his breeding activities, Volkart – in his function as chairman of the Committee for Plant Production of the SLV – played a crucial role in the shift from seed display to seed control. Around 1910, the terms of field control of cereal seed were developed, and in 1913 seed certification through field visits was made mandatory in a regulation by the federal department for agriculture (Volkart, 1928: 15–16).

State-controlled seed is the chief outcome of the third and most important instrument of the state's seed institutions: seed growers' associations tied into a contractual regime which could be used to compel farmers to grow those varieties selected by the breeder or a national seed board. It was Volkart's network of selecting farmers that first grew marketable wheat seed (of preferred varieties). In 1916, he also organized some of the first seed growers' associations in order to increase the output of seed of selected varieties.[41] In the 1920s, the seed growers' associations were more and more used to diffuse varieties selected by the agricultural research stations. 'Variety lists' were put together which restricted the number of varieties allowed under this regime (Schnyder, 1936). By contractual arrangement, the seed growers had to grow a variety required by the federal association of the seed growers' associations.

After the First World War had ended, support for a continuation of regulatory measures was far from unanimous. The struggle for legislation supporting a regulated domestic cereal seed complex took much of the 1920s. In 1929, the process led to the so-called 'cereal amendment' (Getreideartikel) in the constitution and brought major social issues to the forefront. In 1926, 3 years before the vote of 1929, one of the most acrimonious debates of 20th century Swiss history led to the rejection of a proposed state wheat monopoly. If the struggle was eventually won by a new alliance of an organized Bauernstand, agricultural research institutions and some liberal allies, this also reflects the victory of a new political formation. Reading the text of article 23[bis] of the 1929 constitutional amendment, we notice the crucial role given to the state for the regulation of all aspects of grain production.[42] The article provides for the protection and regulation of breeding, seed production, domestic prices, milling and stocks to be held by state agencies. The major justification given in the article itself is national self-sufficiency (Landesversorgung). The institutional setting established for the seed and breeding sector until 1935 was to last right into the late 1980s, when new political paradigms would challenge it.[43]

The legislation of 1932 – adopted for the implementation of the 1929 amendment – established a monopoly for domestic seed at subsidized prices. Farmers growing cereal seed were in a favourable position from then on. Now that seed supply was effectively controlled by the federation of seed growers' associations, seed fairs – a valuable instrument of variety policy according to the 1913 scenario – had become obsolete. In 1935, subsidies were discontinued. A 1938 anniversary brochure of the SLV (Festschrift, 1938) omits any reference to the seed fairs.

In sum, seed regulation evolved from a public display at seed fairs to a process regulated by public or parastatal institutions. During that shift seed production and the products of national cereal breeding – initially two different processes with no common framework – were increasingly merged in a complex institutional web.

Issues of seed production and breeding were not only considered an issue of public concern and state intervention, they were also interwoven with the ideological frameworks of actors who tried to tie them into grand political projects such as the self-sufficient *Bauernstand* or the productive and competitive nation-state. Two elements of discourse were of particular importance: one that was emphasizing the idea of territory, soil or nature in relation to wheat varieties, and another that was making the link of domestic production with quality.

Reference to wheat cultivation and bread production within Switzerland could have nationalist undertones. Martinet's notion of *blé du pays* (wheat from our country) evoked the tradition of cultivating local varieties of wheat, producing local types of bread, etc. With a grain of patriotism, Martinet wanted to demonstrate the possibility of a national breeding programme, and the validity of using local varieties. In 1921, he stated 'that the activity of breeders directed by the two federal institutions for varietal tests at Oerlikon and Lausanne has strongly contributed to maintain the cultivation of wheat in honour' (1921: 30).[44] In a more accentuated phrase, Näf (1943: 49) called local varieties an age-old 'product of the tilled soil'.[45] The term 'Scholle' is not only explicit in its reference to cultivated soil, it is also an ideological construct of conservative nationalism obscuring the human agency in the very development of these varieties. Diversity is essentialized as a product of nature within national borders.

However, most phrasings in the programmatic publications of the pioneering state breeders are subtler. Martinet (1921) and Volkart (1928) omit any strong-worded rhetoric of nationalism.[46] If the material for breeding could still have been confined within national borders, the knowledge could not. The scope of breeding was already international or European at least. Volkart in particular had many contacts with plant scientists in Germany, Holland, Denmark and Sweden. Not surprisingly his discourse was a rationalist one, infused with a conviction that the domestic product was, and had to be, of high quality. The term 'domestic quality wheat'[47] was coined, which probably shows best the emphasis of those national programmes devoted to cereal breeding and seed production. This was, in addition, a statement that parallels the emphasis on quality of national production in other areas and resounded with the discourse on national self-sufficiency.

A significant demonstration of this image of quality and national achievement was an 'exhibition of quality cereals'[48] in 1936, shown in Zürich and Burgdorf (Getreideschau, 1936). There, the concept of a regulated cereal seed sector and downstream economic activities was presented physically to the rural and urban public: seed samples of breeder's varieties, the institutional arrangement of seed production, milling and baking.

Among organized seed growers, lobbying farmers and politicians, one is able to detect stronger expressions of a conservative nationalism. If we consider the nature of the transformation that has taken place in the seed sector, the conservative position is ambiguous. In the limited sphere of the seed sector it was promoting a structural transformation or modernization. However, conservatives took a negative attitude to modernization for its potential to affect the traditional ways of life of the rural population, preferring instead to espouse a romantic image of the eternal peasant. On the one hand, self-sufficiency was idealized in the image of farmers producing bread in their family oven from their own wheat. On the other hand, the appearance of regulated seed and the disappearance of landraces was not interpreted as a sign that part of that self-sufficiency had perhaps already been lost. Baumann (1992, 1993) sees this as typical for the rural population in Switzerland in this period: slowly building more political strength in a liberal state, the Swiss Farmers Association[49] under Ernst Laur promoted both selective modernization where it seemed opportune and a conservative ideology projecting the image of a *Bauernstand* (peasantry) with its own identity and eternal values.

## Regulated Cereal Seed and Genetic Resources

I have pointed out above that the first Swiss wheat breeders adopted a breeding strategy based on the selection from mixtures to obtain 'pure lines' with the characteristics they desired. Breeding at this point was thus a function of genetic diversity of landraces cultivated in their respective regions. Landraces offered good levels of environmental adaptation, a condition for rapid progress at a time when the introduction of exotic breeding materials still posed too many technical problems.

The shortcomings of this strategy became evident when its products replaced landraces on a large scale. These pure lines were genetically more uniform than landraces. If no effort is made to conserve landraces, genetic erosion is inevitable. This scenario was not a surprise to the more conscientious breeders. As early as 1914, the German geneticist Erwin Baur (1914: 109) warned of the consequences of the spread of pure lines and proposed to collect landraces and to make seed of these collections available to breeders through a centralized institution.

In the Swiss case, the very success of Volkart's approach had exactly the consequences anticipated by Baur in his 1914 paper. Within a short period of time, the transfer of selected genotypes through a strong seed sector into farmers' fields led to a considerable loss of intraspecific diversity, at least in major cereal crops in the lowlands of Switzerland (Popow, 1992). In the 1920s and early 1930s, no efforts for conservation such as collection of landraces were made. The only samples remaining today of the diversity that once existed are Volkart's selections from landraces that had been screened in his effort for collaborative plant breeding, which today form part of national collections. These selections already represented a greatly reduced diversity because they consisted mostly of pure lines that had gained wide acceptance and were conserved later in continuing breeding programmes.[50]

A more elaborate notion of the function of genetic resources developed around 1930. If breeders in Switzerland followed international developments in genetics – and we have good reason to believe that Volkart and his colleagues did – they must have taken notice of the theory developed in the 1920s by the Russian plant geneticist Vavilov (1951). Vavilov's theory on the global distribution of genetic diversity postulated a high degree of diversity in centres of domestication. This implied a high value for landraces in their centres of cultivation. Not surprisingly, in 1935 we see a working group of the SLV looking at ways to conserve landraces. In its deliberations, one finds for the first time the very modern notion of 'genetic reserve' which, by definition, understands genetic diversity not merely as a resource for immediate use, but takes a long-term view.

Interestingly the group came up not with a proposal for collection of landraces, but with a modest scheme for *in situ* conservation. It had compiled lists of endangered varieties and farmers who cultivated them, and thus decided to compensate these farmers in exchange for their effort and a seed sample that they had to send each year to an Agricultural Research Station. The programme existed until 1962, but remained very small: between 1936 and 1939, for example, only 12 landraces of various cereals were included (Popow, 1992). This was justified by a lack of funds; however, it led to the formulation of very restrictive criteria for a variety to enter the programme and limited its function as an efficient tool of *in situ* conservation.[51]

Simultaneously with this programme, efforts to collect and conserve *ex situ* began. Landraces of *Zea mays* were collected in southern Switzerland (1930s, Kauter) and spelt (*Triticum spelta*) in the lowlands and pre-alpine hill zone (after 1933, Wagner). Spelt is a good example of the extent to which landraces in the lowlands had been lost: Wagner could collect only four landraces in the lowlands. In the hill zone, where spelt was still an important crop, he obtained an additional 30 accessions.

Opportunities for collection were no longer missed: for example in 1939 during the compilation of a land register which covered the total cultivated area of the country, the staff of the project received orders to collect landraces as they visited farmers' fields. If the crop was not mature, they had to report on the exact location for later sampling. A substantial part of today's national collections of landraces of maize, wheat, spelt and barley was collected during the work for the register.[52]

In sum, early breeding strategies were selections from landraces in farmers' fields. Once these selections had received the status of official varieties in the seed grower programme, they endangered the very diversity from which they had been selected. The appearance of the term 'genetic reserve' in 1935 is significant as a theoretical watershed with regard to the conservation of genetic resources. Yet considerable erosion of diversity found in landraces had already taken place on the ground.

## Conclusion

Collaborative plant breeding represents an essential component of the formative phase of Swiss plant breeding. It is triggered in the 1880s by concerns about seed quality and the overall productivity of cereal agriculture in Switzerland. Initially, these concerns are addressed with seed fairs (since 1889 on a federal level), an instrument developed to facilitate variety exchange among farmers, thus making farmer varieties more widely available. Seed fairs represent an approach based on the introduction of novel germplasm mainly through farmer-to-farmer exchange, supplemented with an evaluating mechanism for the 'best seed'.

The second instrument focused on participatory variety selection with selected farmers. The first two wheat breeders in Switzerland developed breeding programmes (after 1898) for wheat relying on on-farm selection from landraces in multiple sites. During the first two to three decades of formal plant breeding efforts, access to multi-line populations grown on farmers' fields was essential for breeders who had little funds and almost no land for experimentation. CPB was obviously the most viable and cost-effective option. In part these programmes also included the development of farmers' skills in plant breeding techniques, such as pure line selection.

Initially, the impact of these instruments was limited as they only marginally interfered with seed supply systems. On the one hand, seed fairs left the final choice of seed entirely to farmers. They were also lacking an instrument – such as field controls – to ensure the quality of seed from those varieties promoted through their juries. On the other hand, selections from wheat breeding programmes could not rely on a domestic operation to produce seed for farmers. After 1910, these constraints were addressed with new institutional mechanisms eventually leading to regulated seed production. Field controls to check seed quality were declared mandatory for any seed certified by state agencies in 1913. And the rapid spread of seed growers' associations during the First World War allowed for effective propagation of selected varieties.

Both seed fairs and collaborative pure line selection in wheat were used during the formative period of formal breeding. Support for seed fairs was phased out in the 1930s as the policy shifted towards exclusive support for specialized breeding programmes, rather than a farmer-driven germplasm exchange mechanism. Similarly, the involvement of farmers in participatory variety selection began its decline in the late 1920s, when centralized breeding and formal seed supply systems for wheat had been established. A restrictive policy on variety releases and professional specialization in wheat breeding were becoming institutional obstacles to the approach of CPB.

Historically, the formative period of Swiss cereal breeding is closely linked to an agrarian movement for protection of domestic grain production and for national food self-sufficiency. The protectionist approach manifest in grain policy during this period represents the dominant policy approach in Western Europe. It was the liberal answer to the agrarian question of how the economic future and political stability of the peasantry in the nation-state was to be maintained (McMichael, 1994). In this perspective, the Swiss CPB effort in the first three decades of the 20th century was part of a larger political shift towards a protected national cereal production and a peasantry loyal to the state. In a more technical sense, CPB was the framework within which farmers contributed their skills and genetic resources to the creation and propagation of wheat varieties for nation-wide distribution. This was a necessity in a context where neither state financing nor scientific institutions were strong enough to do cereal breeding without farmer support.

As a consequence of the new law adopted in 1998,[53] agricultural research institutes had to withdraw from most breeding activities in wheat. The new agricultural policy of Switzerland (Bundesrat, 1996) purports an ecological turn by discontinuing product-based subsidizing and channelling subsidies to plots used less intensively.[54] It does not address the issue of crop development and plant breeding in the national context. However, CPB might be considered as an option for knowledge-intensive crop development. Indeed, organic farming movements have begun small breeding programmes to develop niche products for green markets. Whether this context – and the premium prices that green markets offer – is a viable environment for collaborative plant breeding remains to be seen.[55]

## Acknowledgements

Research for this chapter was supported by the Stiftung zur Förderung der Forschung an der Universität Bern. I would like to thank Peter Moser, Heinzpeter Znoj, David Gugerli, Wolfgang Marschall, David Cleveland and Peter Schmiediche for their advice and for commenting on draft versions of this chapter.

## Notes

[1]    There are numerous examples of such developments towards seed regulation; see also my analysis of the transition towards a new regime of rice seed production in Indonesia starting in the late 1950s (Schneider, 1998).

[2]    Ger. *Saatgutmarkt*. The English 'seed fair' will be used subsequently to refer to this special institution which was an open market cum seed fair.

[3]    Eidgenössische Technische Hochschule (ETH) Zurich.

[4]    The distinction between natural selection and artificial selection goes back to Darwin. Before his *On the Origin of Species*, 'selection' was mostly understood as 'artificial selection' (by humans in their relation to plants and animals) (Hodge, 1992: 213).

[5]    See McMichael (1994) for a discussion of the agrarian crisis of the late 19th century.

[6]    Ger. *Samenwechsel*.

[7]    See for example the annual report of the Swiss Agricultural Association (SLV) for 1891, p. 44, quoted in Festschrift (1913: 49).

[8]    Schweizerischer Landwirtschaftlicher Verein.

[9]    Kloppenburg (1988) describes a similar evaluation system based on aesthetic criteria for the 'corn shows' in the USA.

[10]    Ger. *Saatzüchtervereinigungen*.

[11]    Ger. *Sortenbereinigung*.

[12]    Ger. *Wirrwarr*. 'Sortenwirrwarr' is a term often used in contemporary agricultural policy documents.

[13]    For example, the first national mapping project for Switzerland started in 1832: a case of a 19th century national survey enabling and fostering the cartographic (re-)production of the nation. Gugerli (1998) analyses the symbolic power emanating from this project and its contribution to nation-building.

[14]    'Wir werden . . . durch Lagerräume, Oekonomiegebäude aller Art, Institutsräume geführt, in welchen geschultes Personal sortiert, zählt, misst, wiegt. Das

Ganze gleicht einer grossen Fabrik, die Präzisionsinstrumente baut. Dem Getriebe, dem wir zuschauen, haftet aber fast etwas Unheimliches an: Hier werden Lebewesen 'fabriziert', die früher in dieser Ausbildung nicht bestanden, Neuschöpfungen nach des Menschen Willen!'

[15]   '. . . eine Organization und eine Zusammenfassung aller Kräfte für das ganze Land erfolgt ist, die in ihrer Vollkommenheit vorerst noch einzig dastehen dürfte'.

[16]   'Neue Formen vermag keine Züchtung hervorzubringen; die Hand des Züchters kann nur innerhalb der von der Natur erzeugten Formen die Auswahl für die weitere Vermehrung treffen.'

[17]   For technical plant breeding terms see Becker (1993).

[18]   Ger. *einfache Stammzucht* or *Linienzucht*, characterized by single (one-time) selection of plants (Ger. *Stammpflanzen*) based on external traits, separate cultivation of the grains of each plant, comparison and selection from the progenies until one line remains (Volkart, 1931: 13).

[19]   The characteristic of landraces which is most prominent in contemporary definitions is their adaptation to regional climatic variation, i.e. 'types which grow best under the respective climate' (Schellenberg, 1906: 9). In a lecture, Volkart defined landrace as a 'variety which has evolved in the course of time through natural selection (soil) in a region that often gives it its name', emphasizing again natural over human selection. (Nachlass Volkart, Box 13, Lectures on Plant Breeding. Wissenschaftshistorische Sammlung, ETH Zürich.) In contrast, modern definitions emphasize the multi-line or population character of landraces. Following Keller (1990: 5), 'landraces are populations comprising a varying number of different genotypes. Landraces of self-pollinating crops include a smaller or larger number of homozygous forms (phenotypes) the offspring of which are called pure lines.' (Landsorten sind Formengemische (Populationen), die eine unterschiedliche Anzahl verschiedener erblicher Typen enthalten. Landsorten von Selbstbefruchtern bestehen aus einer mehr oder weniger grossen Zahl homozygoter [reinerbiger] Formen, deren individuelle Nachkommenschaft als reine Linien bezeichnet werden.)

[20]   Albert Volkart (1873–1951), from a rural background in the Zürich region, was first trained as a farmer. He then studied agronomy from 1891 to 1894 and received a PhD in systematic botany from the ETH Zürich in 1899. He pursued his career at a seed control station, became lecturer and eventually full professor (1925) for plant production and plant pathology at the ETH. From 1919 until 1929, he also led the federal plant research station in Zürich-Oerlikon.

[21]   Volkart gave only minor attention to cross-pollinating cereals such as rye which was not an important cereal in Switzerland. In the German plant breeding tradition, the term 'Hochzucht' was used for the method of 'pedigree breeding' and its varietal products (Flitner, 1995: 295). It involved recurrent selection of pure lines and was in many cases also applied for self-pollinating crops. Volkart correctly thought that the method was valid only for outcrossing species.

[22]   Past denomination of today's Swiss Federal Institute of Technology.

[23]   In his fieldnotes from June/July 1908, Volkart notes frequent visits to experimental plots on a farm near Zürich, where he both made selections from

wheat and experimented with making crosses from poppies ('On July 11, the first white poppies were flowering, on July 12, several white and brown poppies which I crossed on July 14'). (Wissenschaftshistorische Sammlung der ETH Zürich, Nachlass Volkart, Box 20, Notes on phenology, experiments, official travel etc. 1902–39.)

[24]  Ger. *Kreuzung mit einmaliger Auslese* (Volkart, 1931: 15).

[25]  Ger. *Beständigkeit.*

[26]  Ger. *einfache Auslese.*

[27]  Ger. *Zuchtsorten.*

[28]  The term *Hochzucht* became popular in German plant breeding around 1900. For a discussion of the term see endnote 21.

[29]  On the changes in Swiss agrarian structure in the 19th century, see Pfister (1995: 250–255).

[30]  (Volkart, 1928: 17.)

[31]  Ger. *einfache Stammzucht.* The term 'Stammzucht' should not be confounded with the 'pedigree method' translated by Volkart (1931: 13–16) as 'fortgesetzte Auslesezucht' or 'Stammbaumzucht'.

[32]  Volkart started to work more systematically with segregating populations in 1918. On the classic pedigree method for handling segregating populations from crosses – used by most plant breeders in the early 20th century – see Allard (1999: 65–66, 175–176).

[33]  Ger. *Begründungszucht.*

[34]  Because of the importance of this method, the full German original is given here. 'Es wird auf dem Felde eine grössere Zahl Stammpflanzen wiederum nach äusseren Merkmalen (Leistungsmerkmalen) ausgewählt, diese Pflanzen . . . einzeln entkörnt und ihre Körner getrennt ausgesät. Man erhält dadurch eine Reihe von Nachkommenschaften (Stämmen), die man in allen Entwicklungsstadien und namentlich auch im Ertrag miteinandern vergleicht. Die ungeeigneten und weniger geeigneten werden nach und nach beseitigt, die mit jedem Stamm bestellte Fläche von Jahr zu Jahr vergrössert, bis zuletzt noch ein einziger Stamm, der beste, übrigbleibt. Er stammt von einer einzigen Pflanze ab. Die Auslese erfolgt also nur einmal. Die Vermehrung beginnt gleich von Anfang an und wird nicht neben der Zucht, sondern in der Hauptsache in der Zucht bei der Prüfung der verschiedenen Stämme durchgeführt. Der Unterschied und der grosse Fortschritt gegenüber der Massenauslese besteht in der getrennten Prüfung der Nachkommenschaften . . . Die einfache Stammzucht ist das gegebene Zuchtverfahren für selbstbefruchtende Getreidearten, bei denen die einmalige Auslese genügt, und zwar bei der Begründungszucht, d.h. wenn aus den vorhandenen Landsorten passende Züchtungssorten geschaffen werden sollen. Durch dieses Verfahren sind bei uns die meisten älteren Züchtungen (Plantahofweizen etc.) entstanden. Es wird auch in Zukunft noch angewendet werden müssen, wenn bei selbstbefruchtenden Getreidearten auf die Landsorten zurückgegriffen werden muss' (Volkart, 1931: 13).

[35]  Jahresbericht des schweizerischen landwirtschaftlichen Vereins für 1918, pp. 38–41 (SNL V Schweiz 1970).

[36]  Jahresbericht des schweizerischen Landwirtschaftlichen Vereins pro 1910, p. 51 (SNL V Schweiz 1970).

[37]   See Frossard (Chapter 6, this volume) on the MASIPAG movement in the Philippines, a farmers' breeding initiative which demonstrated that farmers can build their own competence in technically more complex breeding methods.

[38]   Ger. *Saatgutkontrolle*.

[39]   Ger. *Zuchtsorten*.

[40]   Ger. *Feldbesichtigungen*.

[41]   Jahresbericht pro 1916 des SLV (Jahresberichte, 1889–1921).

[42]   With the qualification of *Brotgetreide*, i.e. grain earmarked for human consumption. For fodder cereal crops, the regulations were less protective of domestic producers.

[43]   On 29 November 1998, the Swiss electorate had to vote on the abolition of the 1929 constitutional amendment, and on the transition to a fully liberalized wheat market after 2003. The message of the federal council for the electorate calls the 1929 amendment obsolete because in 1929 a major objective of domestic agricultural production was to secure food supply in times of crises and war. The abolition was accepted with a majority of 79.4% yes votes. The virtual absence of a public debate before the vote is significant for the fundamental differences between the historical contexts of the events of 1929 and 1998. This vote formalized the succeeding phasing out of regulated seed and state-sponsored plant breeding.

[44]   'On peut . . . dire sans crainte que l'activité des selectionneurs dirigée par les deux etablissements fédéraux d'essais de semences d'Oerlikon et de Lausanne . . . a fortement contribué à maintenir la culture du blé en honneur.'

[45]   Ger. *Produkt der Scholle*.

[46]   That such rhetoric existed is exemplified by a short polemic on the modernity of Swiss breeding programmes compared with those of other nations, Sweden in particular. The critique was mounted by Seiler-Neuenschwander (1925) and countered by Volkart (1928).

[47]   Ger. *inländischer Qualitätsweizen*.

[48]   Ger. *Qualitätsgetreideschau*.

[49]   Ger. *Schweizerischer Bauernverband*, SBV.

[50]   Information on the early thinking on landrace conservation among Swiss plant breeders was mostly obtained from Georg Popow, retired plant breeder at the Federal Research Institute for Agroecology and Agriculture (FAL) in Zürich-Reckenholz.

[51]   See also Zeven (1996).

[52]   Figures for 'landraces' of these crops in national collections are as follows (based on Derron *et al.*, 1993): *Zea mays* 560 accessions; *Triticum aestivum* 2000; *Triticum spelta* 2208 + 560 (in two gene banks, with some amount of duplication); *Hordeum vulgare* 791. It has to be noted that these collections include material both from Switzerland and from other countries. The *Triticum spelta* collection, for example, was assembled in other European countries such as Belgium and Spain (Popow, 1998, personal communication).

[53]   Bundesgesetz vom 29 April 1998 über die Landwirtschaft.

[54]   Land used under the title of 'ecological compensation area'.

[55]   For example the 'Forschungslaboratorium am Goetheanum' in Dornach near Basel which, based on the philosophy of Rudolf Steiner, supports a CPB

programme for wheat. This institution also holds a substantial collection of spelt (1300 landraces) and *T. aestivum* (150 landraces).

# References

Allard, R.W. (1999) *Principles of Plant Breeding.* John Wiley & Sons, New York.

Baumann, W. (1992) Bauernstandsideologie und Rolle der Bauern in der Schweizer Politik nach der Jahrhundertwende. In: Tanner, A. and Head-König, A.-L. (eds) *Die Bauern in der Geschichte der Schweiz.* Heft 10 der Schweizerischen Gesellschaft für Wirtschafts- und Sozialgeschichte. Chronos, Zürich.

Baumann, W. (1993) Bauernstand und Bürgerblock: Ernst Laur und der Schweizerische Bauernverband 1897–1918. Orell Füssli, Zürich.

Baur, E. (1914) Die Bedeutung der primitiven Kulturrassen und der wilden Verwandten unserer Kulturpflanzen in der Pflanzenzüchtung. *Jahrbuch der Deutschen Landwirtschaftlichen Gesellschaft* 29, 104–109.

Becker, H. (1993) *Pflanzenzüchtung.* Eugen Ulmer Verlag, Stuttgart.

Bundesrat (Swiss Federal Council) (1996) *Botschaft zur Reform der Agrarpolitik: Zweite Etappe (Agrarpolitik 2002): vom 26 Juni 1996.* Federal Chancellery, Bern.

Derron, M., Kleijer, G., Corbaz, R. and Schmid, J.E. (1993) Die Erhaltung der genetischen Ressourcen von Kulturpflanzen in der Schweiz. *Landwirtschaft Schweiz* 6, 217–232.

Eyzaguirre, P. and Iwanaga, M. (eds) (1996) *Participatory Plant Breeding: Proceedings of a Workshop on Participatory Plant Breeding 26–29 July 1995, Wageningen, The Netherlands.* International Plant Genetic Resources Institute, Rome.

Festschrift (1913) *Festschrift zur Feier des 50-jährigen Bestehens des schweizerischen landwirtschaftlichen Vereins 1863–1913.* Effingerhof AG, Brugg.

Festschrift (1938) *Festschrift zur Feier des 75-jährigen Bestehens des schweizerischen landwirtschaftlichen Vereins 1863–1938.* Verbands-druckerei AG, Bern.

Flitner, M. (1995) *Sammler, Räuber und Gelehrte: Die politischen Interessen an pflanzengenetischen Ressourcen 1895–1995.* Campus, Frankfurt/Main, New York.

Fontanari, G. (1934) Die Entwicklung des italienischen Weizenbaues unter der faschistischen Regierung. PhD thesis, University of Fribourg, Fribourg, Switzerland.

Getreideschau (1936) Erste schweizerische Qualitätsgetreideschau 1936 durchgeführt vom schweizerischen Landwirtschaftlichen Verein. Zürich, Tonhalle-Pavillon, 24–26 April Burgdorf, Markthalle, 30 April–3 Mai. Swiss National Library V Schweiz 2584.

Gugerli, D. (1998) Kartographie und Bundesstaat. Zur Lesbarkeit der Nation im 19. Jahrhundert. In: Ernst, A. (ed.) *Revolution und Innovation, Die konfliktreiche Entstehung des Schweizerischen Bundesstaates von 1848.* Chronos, Zürich, pp. 199–215.

Hodge, M.J.S. (1992) Natural selection: historical perspectives. In: Fox Keller, E. and Lloyd, E. (eds) *Keywords in Evolutionary Biology*. Harvard University Press, Cambridge, Massachusetts, pp. 212–219.

Jahresberichte (1889–1921) *Jahresberichte des Schweizerischen Landwirtschaftlichen Vereins*. J. Rüegg, Zürich.

Keller, L. (1990) Anbau- und erntewert von Getreidelandsorten im Vergleich zu neuen Zuchtsorten unter Berücksichtigung gewisser ökophysiologischer Parameter. PhD thesis. Swiss Federal Institute of Technology, Zürich.

Kloppenburg, J.R.J. (1988) *First the Seed: the Political Economy of Plant Biotechnology 1492–2000*. Cambridge University Press, Cambridge.

Martinet, G. (1921) *La Question du Blé en Suisse*. Lausanne, Switzerland.

Mayr, E. (1984) *Die Entwicklung der Biologischen Gedankenwelt: Vielfalt, Evolution und Vererbung*. Springer, Berlin.

McGuire, S., Manicad, G. and Sperling, L. (1999) *Technical and Institutional Issues in Participatory Plant Breeding – Done from a Perspective of Farmer Plant Breeding: a Global Analysis of Issues and of Current Experience*. CIAT, Cali, Colombia.

McMichael, P. (1994) Global restructuring: the agrarian question revisited. Paper prepared for the Program in Agrarian Studies, Yale University, 14 October 1994.

Moser, P. (1994) *Der Stand der Bauern: Bäuerliche Politik, Wirtschaft und Kultur gestern und heute*. Huber, Frauenfeld, Switzerland.

Näf, A. (1943) Die Entwicklung des Saatzuchtwesens in der Schweiz und seine Bedeutung für den inländischen Getreidebau. In: Koblet, R. (ed.) *Festgabe zum Siebzigsten Geburtstag von Prof. Dr. A. Volkart*. Berichte der Schweizerischen Botanischen Gesellschaft, Vol. 53a. Büchler, Bern, pp. 44–61.

Oehninger, J. (1936) Grundsätzliches zur bäuerlichen Selbstversorgung mit Brot. (ed.) *Erste Schweizerische Qualitätsgetreideschau 1936 Durchgeführt vom Schweizerischen Landwirtschaftlichen Verein*. Zürich, Tonhalle-Pavillon, 24–26 April, Burgdorf, Markthalle, 30 April–3 Mai. Swiss National Library V Schweiz 2584, pp. 56–58.

Pfister, C. (1995) *Im Strom der Modernisierung: Bevölkerung, Wirtschaft und Umwelt 1700–1914*. Historischer Verein des Kantons Bern, Bern.

Polanyi, K. (1957) *The Great Transformation. The Economic and Political Origins of our Times*. Beacon, Boston, Massachusetts.

Popow, G. (1992) Erhaltung und Nutzung der genetischen Vielfalt in früheren Jahrzehnten. *Paper presented at the Swiss Federal Institute of Technology Zurich, 6 January 1992*.

Schellenberg, H.C. (1901) Der Wert des Samenwechsels. *Schweizerisches Landwirtschaftliches Centralblatt* 20, 182–184.

Schellenberg, H.C. (1902) Ziele und Aufgaben der Pflanzenzüchtung. *Schweizerisches Landwirtschaftliches Centralblatt* 21, 33–40, 74–81.

Schellenberg, H.C. (1906) Die Ergebnisse der experimentellen Vererbungslehre und ihre Anwendung in der Landwirtschaft. *Mitteilungen der Gesellschaft Schweizerischer Landwirte* 2, 3–18.

Schneider, J. (1998) The making of 'new seed': ritual, politics and rice seed production in Indonesia. *Asiatische Studien* 52, 611–634.

Schnyder, A. (1936) Die Entwicklung des Saatzuchtwesens und der Saatgutversorgung im schweizerischen Getreidebau (ed.) *Erste Schweizerische Qualitätsgetreideschau 1936 Durchgeführt vom Schweizerischen Landwirtschaftlichen Verein. Zürich, Tonhalle-Pavillon, 24–26 April, Burgdorf, Markthalle, 30 April–3 Mai.* Swiss National Library V Schweiz 2584, pp. 31–36.

Seiler-Neuenschwander, J. (1925) Die praktische Ausnützung der Ergebnisse der Erblichkeitsforschung in Schweden: ein Musterbeispiel planmässiger, moderner Kulturpflanzenforschung. *Der Kleine Bund* 6, 394–397.

Van der Ploeg, J.D. (1992) Knowledge systems, metaphor and interface: the case of potatoes in the Peruvian highlands. In: Long, N. (ed.) *Encounters at the Interface: a Perspective on Social Discontinuities in Rural Development.* Wageningen Studies in Sociology 27, Wageningen, pp. 145–164.

Vavilov, N.I. (1951) The origin, variation, immunity and breeding of cultivated plants: selected writings of N.I. Vavilov. In: Vavilov, N.I. (ed.) *Chronica Botanica.* Waltham, Massachusetts, pp. 1–54.

Volkart, A. (1928) *Die Getreidezucht in der Deutschen Schweiz. Ein Rückblick und Ausblick.* J. Rüegg Söhne, Zürich.

Volkart, A. (1931) *Kurzgefasste Anleitung zur Getreidezucht, Herausgegeben vom Eidg. Volkswirtschaftsdepartement.* Verbandsdruckerei Bern, Bern.

Zeven, A.C. (1996) Results of activities to maintain landraces and other material in some European countries in situ before 1945 and what we may learn from them. *Genetic Resources and Crop Evolution* 43, 337–341.

# Theory, Empiricism and Intuition in Professional Plant Breeding

**8**

## Donald N. Duvick

*Iowa State University and Pioneer Hi-Bred International, Inc. (retired), PO Box 446, Johnston, IA 50131, USA*

## Abstract

Professional (full-time) plant breeders usually speak of their profession as a science, and their publications typically discuss only the scientific basis of plant breeding. But they know that successful variety development depends on a combination of art and science, with art 'skill in performance, acquired by experience, study, or observation' (*Webster's Collegiate Dictionary*, 1941) often playing a larger part than science. This chapter is based on my own long-term experience as a maize breeder and supporter/administrator of professional plant breeders in the USA and abroad. I discuss the ways in which professional plant breeders combine theory, empiricism and intuition, and show how such combinations enable them to collaborate with farmers to produce useful new varieties.

## Introduction

Plant breeding broadly defined is as old as plant domestication, but the concept of deliberately planned plant breeding is much younger. It arose in the late 18th and early 19th centuries when botanists first described sexuality in plants and learned how to make planned crosses to provide new materials for selection (Smith, 1966; Harlan, 1992; Smith, 1995; Evans, 1998).

The concept of scientific plant breeding is even younger. It developed in the first decades of the 20th century, simultaneously with evolution of the science of genetics and of methods for statistical manipulation of data (Smith, 1966). These sciences provided the stimulus for the development of the plant breeding profession; that is full time professional plant breeders, producing new varieties for use by a variety of customers including crop farmers in production agriculture. The professional plant breeders worked (and still work) in the public sector and in the private sector.

Typically, the public sector engages in research to develop theory and basic breeding materials for plant breeding and the private sector devotes its energies to variety development and distribution, aided by the contributions of knowledge and germplasm from the public sector. The two sectors thus form a plant breeding team. Public sector breeders perform the first task, development of theory and basic breeding materials, and the commercial sector performs the second task, variety development and distribution (Frey, 1996). However, one should note that public sector breeders are often also the chief source of new varieties for crops and in countries (especially developing countries) where commercial seed markets are too small to support commercial plant breeding. In recent years, some of the commercial breeding firms have undertaken increasing amounts of basic research, particularly that related to uses of biotechnology in plant breeding.

The following discussion describes breeding of field crops by professionals in the public and private sectors. It also examines the participatory role of farmers in professional plant breeding in industrial countries. The history of professional crop plant breeding in industrial countries is a history of close and continuing interactions among these three groups: farmers, public sector breeders and private sector breeders.

## Theory

Faith in professional plant breeding rests in part on faith in the contribution of theory to success of the enterprise. Rediscovery and elaboration of Gregor Mendel's principles of genetics at the start of the 20th century sparked interest in genetic manipulation of plants, to satisfy intellectual curiosity about principles of heredity. The subsequent increase in knowledge about genes and their effects also gave promise for practical use in plant breeding.

But in the early decades of the century little use could be made of genetics for plant breeding because most of the important traits in farm crops seemed to be governed by 'quantitative traits', traits whose phenotypes show more or less continuous distributions in segregating populations (Baker, 1984). 'Mendelian' genetics did not seem to apply to quantitative traits. The pioneering plant genetic studies of the early decades of the 20th century dealt primarily with simply inherited traits, that is, with single genes that governed clearly expressed but economically unimportant traits such as anther colour.

However, also during the first part of the 20th century, scientists began to study the inheritance of quantitative traits, and gradually they developed the science and theory of quantitative genetics and its application to manipulation of quantitative traits (Smith, 1966; Sprague, 1966). The early work of Johannsen and Nilsson-Ehle and then of Fisher, Wright and Haldane (see Sinnott and Dunn, 1932) showed that quantitative traits could be explained with assumptions based on Mendelian genetics. 'Characters involving size or quantity, which for the most part do not show sharp and simple mendelian assortment and which in their inheritance were long thought to be exceptions to Mendel's laws, have been definitely brought into line with the mendelian explanation by the Multiple-factor Hypothesis' (Sinnott and Dunn, 1932: 11). The new science of genetics could be applied to all traits, not only those with simple, monogenic inheritance. This understanding enabled plant breeders (among others) to design breeding schemes that dealt with multifactorial inheritance and interactions of genotype with environment. New and more efficient schemes of population improvement gave breeders greater speed and precision in development and enhancement of breeding populations, and in designing appropriate selection procedures (e.g. Frey, 1968; Branson and Frey, 1989).

A specific example: the maize (*Zea mays* L.) Stiff Stalk Synthetic (BSSS) was formed at Iowa State College (now Iowa State University) in the 1930s. BSSS was then improved in several successive cycles of breeding, using methods based on the continually evolving knowledge and theory of quantitative genetics (Lamkey, 1992). BSSS was the source of several important inbred lines (B14, B37, B73, B84) that in their day were widely used by commercial seed companies as parents of maize hybrids and as breeding material for new generations of elite inbred lines (Russell, 1991).

Maize breeders have pointed out that the need to improve the rate of progress in hybrid maize breeding was the stimulus for the formation of BSSS (Sprague, 1966). Hybrid maize breeders needed new sources of superior inbred lines and they conjectured that intercrossing the best inbreds then on hand, with emphasis on important traits such as resistance to stalk lodging, could produce a series of 'synthetic open pollinated varieties' that could be sources of superior new inbreds. With one or two exceptions (such as the first generation of BSSS) the first-generation synthetic varieties were not as useful as had been hoped, so breeders such as G.F. Sprague applied some of the new and evolving principles of population improvement to these 'synthetics'.

BSSS was one of the more successful results. It produced not only valuable inbred lines but also valuable data about possible new ways to effect population improvement.

This example illustrates the productive interactions that can take place between practical plant breeding and genetic theory. Empirical breeding needs stimulated development of investigations in quantitative genetics which in turn provided valuable knowledge and germplasm to practical plant breeders.

The BSSS example also illustrates the productive interactions that often occur between public and private sector plant breeding. The public sector developed and sequentially improved BSSS. The private sector used its products, superior inbreds, as an important part of their empirical hybrid breeding programmes. The two sectors operated as a team in the service of plant breeding as a whole.

Despite the contributions of quantitative genetics to applied plant breeding, the precise genetics of multifactorial inheritance remained (and still remain) a mystery. Breeders could not identify the individual genes that govern traits with multifactorial inheritance and then move them quickly and precisely from one genotype to another. They could not assign genes for quantitative traits to positions in linkage maps, as is done routinely for genes that govern simply inherited traits. And since most of the economically important traits in crop plants are quantitative, it follows that breeders could not easily manipulate the majority of the traits they wanted to modify.

However, as the years passed, breeders began to identify a few simply inherited traits that were economically important (Walker, 1966). Experiments in England, for example, showed that resistance to yellow rust (*Puccinia striiformis* West.) in wheat (*Triticum* spp. L.) was conferred by a single dominant gene (Biffen, 1905). Experiments at Kansas State University, beginning in 1915, established the genetic basis in wheat for resistance to Hessian fly (*Mayetiola destructor* Say). Starting from this base, breeders have found numerous instances of useful single-gene pest resistance; for example, to Hessian fly in wheat, to crown rust (*Puccinia coronata* Corda var. *avaenaie* W.P. Fraser & Ledingham) in oat (*Avena sativa* L.), and to greenbug (*Schizaphis graminum* Rondani) in sorghum (*Sorghum bicolor* (L.) Moench) and wheat (Forsberg *et al.*, 1991; Ratcliffe *et al.*, 1996; Porter *et al.*, 1997).

Unfortunately, however, many of the single genes (also called major genes) for pest resistance had undesirably short useful lifetimes. In a very few years after release of a resistant variety, selection among genetically diverse insect and/or pathogen populations brought forth forms that overcame the resistance genes (National Research Council (US) Committee on Genetic Vulnerability of Major Crops, 1972). The new genotypes of insect or disease organism multiplied rapidly, and soon the 'resistant' varieties were (from the farmers' point of view) no longer resistant. Plant breeders had to scurry to find new and hopefully more durable (long lasting) forms of resistance. By the end of the 20th century their chief quest was for durable resistance (Simmonds, 1991). Durable resistance usually seemed to be governed by several genes of minor effect rather than single genes of major effect. Manipulation of these gene groups was more difficult than manipulation of single genes. Precision and speed had to be sacrificed in favour of better outcomes.

Simple Mendelian genetics has been useful and dependable for a few traits in some crops. Manipulation of two or three genes for height and maturity enabled sorghum breeders in temperate climes to use high quality landraces from the tropics as more or less locally adapted breeding material (Stephens *et al.*, 1967). Sweetcorn breeders greatly improved the quality and taste of sweetcorn by introducing two or three new genes that affect starch and carbohydrate development in the maize endosperm (Tracy, 1997).

During the past decade a new laboratory-based technology – biotechnology – has added a few more useful single-gene traits to the portfolio of plant breeders (e.g. Brunke and Meeusen, 1991; Dyer *et al.*, 1993; James, 2000). Genetic transformation has brought in genes from biologically distant species (e.g. from bacteria) that confer resistance to insects such as European cornborer (*Ostrinia nubilalis* Hübner) or to specific herbicides. (The technology of genetic transformation is based, ultimately, on the theory that DNA carries genetic information (Watson *et al.*, 1983).)

These single genes of major effect have made spectacular changes in variety performance and in the efficiency of weed control. There is no reason, however, to expect that they will have useful lifetimes of greater length than lifetimes of the major resistance genes found within a crop species or its close relatives (Benbrook, 1996; Snow and Moarán-Palma, 1997). Laboratory experiments, for example, have shown that the European cornborer can produce genetic variants that overcome transgenic resistance derived from *Bacillus thuringiensis*. Researchers speculate that the same change can occur in the field. Maize breeders therefore will need to transfer genes repeatedly, trying to keep up with changes in borer genotypes. They will duplicate the experiences of their predecessors who worked with non-transgenic major resistance genes. Like their predecessors, breeders who use transgenics will have to search for durable (relatively long lasting) resistance, perhaps multi-factorial, perhaps governed by single genes, with action that is not easily bypassed by genetic variants of the pest organism.

And at the moment transgenes of all kinds have an additional problem. Use of genetic transformation as a plant-breeding tool has come under severe attack from several quarters, and its use is effectively stopped in some countries (*AgBiotech Reporter*, 2000a,b,c). Objections to the technology are based on normative as well as biological reasons. As stated by one writer (van Dommelen, 1999: 192), 'the larger biotechnology debate . . . is riddled with ideological, ethical, and other normative evaluations'. The status and thus the utility of this new plant breeding technology will remain in doubt for several years to come.

Despite the criticisms about its safety and utility, biotechnology, in theory, could raise genetics-based plant breeding to much higher levels of achievement and efficiency. Some improvements are in place already. Molecular marker technology helps breeders to track and retain important but hard to identify genes as they are backcrossed into elite germplasm (Staub *et al.*, 1996; Prabhu *et al.*, 1999). Soybean (*Glycine max* (L.) Merr.) breeders, for example, use molecular markers to speed up and simplify backcrossing programmes to insert new genes for cyst nematode (*Heterodera glycines* Ichinohe) resistance into elite soybean varieties. Marker technology may be useful in the near future (when costs are brought into line with the value of results) to help identify and then assist in purposive movement of gene blocks (linkage groups) that affect important quantitative traits (Stuber and Polacco, 1999).

The new field of genomics eventually will help breeders to identify genes and gene interactions with much greater depth and precision than has been possible up to the present time (Lee, 1998; Stuber and Polacco, 1999). Classical and transgenic manipulations will be used to identify, understand and then improve important genes and also complex genotypes that control key traits such as drought and heat tolerance (and ultimately yield) in crop varieties.

An example: investigation of some of the genes that may have been critical to domestication of maize from its probable parent, teosinte (*Zea mays* ssp. *parviglumis* or spp. *mexicana*), has shown that 'the evolutionary switch from teosinte to maize involved changes in the regulatory regions of *tb1*' (Wang *et al.*, 1999). The gene *tb1* is one of several that are involved in the evolution of the compact maize ear from the many-branched flowering structure of teosinte. Breeders someday may be able to alter this particular regulatory region in ways that enhance the grain yield or production stability of maize. But such potential achievements are primarily for the future, perhaps decades hence. They represent possible future contributions of theory to plant breeding.

Another way in which theory has impacted plant breeding is through attempts to implement the 'ideotype' concept, a description of an ideal phenotype with certain assumptions about physiology (Donald, 1968). This concept was used in the 1960s and 1970s as an aid in development of new high-yield rice and wheat varieties ('green revolution' varieties). Wheat and rice breeders purposely bred varieties with reduced plant height (semi-dwarf) and upright leaves as well as other characteristics which were intended to allow them to take advantage of high plant (or stem) density and heavy applications of nitrogen fertilizer (Peng *et al.*, 1999; Reynolds *et al.*, 1999). The idealized ideotype concept (considering all its traits) had some utility for breeding rice and wheat but was not applicable in all respects (Evans, 1993: 267; Wallace and Yan, 1998: 253). As in the example of maize BSSS, rice and wheat breeders' empirical experience tended to precede and stimulate development of theory and, subsequently, theory and empirical breeding interacted to the benefit of both.

In recent years, rice breeders have devised another ideotype, the 'new plant type'. In working towards its realization they again have found that certain modifications of theory are required for practical success in the breeding programme (Khush, 1995; Peng *et al.*, 1999). Wheat breeders, also, have experimented recently with a new ideotype (*Science*, 1998) and they too are now working to find the best blend of theory and empiricism. In both crops, breeders have intended to increase yield primarily by increasing the size of the panicle or the head (and by changing other traits to compensate for this enlargement). Once breeding got underway they found that other unanticipated changes were needed to enable the larger flowering structure to fill with well-developed grains. Theory gives good ideas for future actions and can help breeders to understand what they have done in the past, but it is not a substitute for empirical breeding. An eminent plant physiologist, speaking for physiology, has given advice that can apply as well to genetics:

> Selection for greater yield potential has not, could not and never shall wait on our fuller understanding of its functional basis, despite the pleas of [some] physiologists. In that sense crop physiology may be retrospective, but the purpose of its backwards glance is to discern some of the ways forward and to provide at least a partial map for plant breeders.
>
> (Evans, 1993: 266)

In summary, genetic theory has given assistance to professional plant breeders but has not been their primary means of support (see Soleri *et al.*, Chapter 2, this volume). Manipulation of individual genes has been useful, but such genes have not had a major effect on yield and yield reliability of crop varieties. Use of quantitative genetics theory has been helpful but has its limits (see Ceccarelli and Grando, Chapter 12, this volume; Bänziger and de Meyer, Chapter 11, this volume; Joshi *et al.*, Chapter 10, this volume). The breeders do not know the precise genetics of important traits such as multifactorial resistance to insect pests and diseases, tolerance of cold and wet growing conditions, tolerance to heat and drought, or tolerance of low or unbalanced amounts of soil fertility elements such as nitrogen. They do not know the precise genetic basis for heterosis, that mysterious phenomenon that is the rationale for development and use of hybrid crop varieties.[1] And most important of all, they do not know the genetics of 'yield', whatever that might be. They have not identified and located 'yield genes' as such. Perhaps the greatest surprise, to those who have watched the development of quantitative genetics over the past 75 years, is that the pedigree system – essentially, crossing successful varieties to make new breeding populations – is the backbone of nearly all successful professional breeding programmes today (Hallauer *et al.*, 1988; Sneller, 1994; Mercado *et al.*, 1996; Rasmusson and Phillips, 1997).[2] Although pedigree breeding can be (and is) guided by theory from population genetics, it essentially is cut and try (empiricism), and many breeders use little or no quantitative genetics in design or operation of their pedigree breeding programmes.

In sum, the well-developed science of genetics gives professional plant breeders strong faith in the genetic basis of the major traits they wish to change, but it does not give them sufficient knowledge to identify and then precisely manipulate the genes and consequent physiological systems that govern those traits.

## Empiricism

The above statement of course is not news to practising professionals, plant breeders who are charged with turning out an unending succession of improved varieties for a multitude of uses and growing conditions. But lack of major help from genetics and/or physiology has not stopped them from carrying out their mission. Breeders of most of the major field crops, for example, can point to well-documented data that show genetic yield gains of 1–3% per year (non-compounded) during at least the past half century (Fehr, 1984; Russell, 1991; Duvick, 1992a; Ortiz-Monasteriao R. *et al.*, 1997; Voldeng *et al.*, 1997). And other traits such as stability of performance across seasons and environments, adaptation to specific environments, or enhancement of product quality, have been improved continuously as well, all in the absence of precise knowledge of the genetics that determines those traits.

The professionals have made their advances by using the tactics that (presumably) have been used since humans first domesticated favoured wild species. They have examined genetically diverse populations, chosen individual plants that caught their fancy, and used seed from those plants as the foundation of a new cycle of hybridization, segregation and selection. Selection of desired new genotypes is practised in the environments and with use of farmers' cultural practices that are typical for the crop. As one breeder (R.D. Riley, Pioneer Hi-Bred International, Inc., Iowa, 1999, personal communication) has put it, 'You cross best by best and then select the best and do it over and over in the place where you want to grow the crop'.[3]

Or in more erudite language, the professionals practise empiricism; their plant breeding is based primarily on observation and experience.[4] They depend heavily on trial and error, they adopt the techniques that seem to give the best results, and then they continually modify those techniques to produce (hopefully) even better results. The need for repetitious trial and error in practical plant breeding was emphasized recently in results of an informal survey of 35 practising plant breeders and five allied professionals. The survey respondents placed the trait, 'patience and persistence' at the head of a list of personal qualities required for success in plant breeding, both now and in the years to come (Duvick, 1999).

## Powerful new tools

But this explication simplifies rather too much. Today's professionals may use some of the same tactics as their predecessors, but without doubt professional breeders have made faster progress during the past century than was made by the farmer breeder/selectors of the previous ten millennia. (Perhaps one should qualify this by saying that the professionals have made faster progress in the development of varieties suited to the demands of commercial agriculture.) This means they must have made some changes or additions to the empirical practices of their forebears. And they have. The empiricism of the full-time breeders has been aided in ways that were not available or even imaginable to farmer breeders. Professional breeders have had access to a kit of new tools that help them to move faster and with more precision than was possible for even the most skilled of their farmer predecessors.

A few examples of the new tools are as follows:

- greatly increased knowledge of sexuality in plants and therefore the ability to make planned crosses of selected genotypes rather than to depend on chance outcrossing;
- knowledge of the role of accurate measurement, replication and statistical analysis in providing increased precision to ratings for yield and other traits, compared with selection based on results in one or two locations per season;
- the ability to bring in germplasm from anywhere in the world compared with dependence primarily on the genotypes in one's own farm or in neighbouring farms;

- the ability to inflict planned and repeated infections and infestations of important plant diseases and insect pests thereby giving the ability to make stronger selection for tolerance or resistance to these biotic stresses, compared with waiting for sporadic pest outbreaks;
- the ability to grow performance trials in critical environments (such as drought, nutrient imbalance, or cold and wet soils) thereby providing increased opportunity to select for tolerance to those abiotic stresses;
- in recent years,
  - mechanization of planting and harvesting, making it possible for breeders to gather great quantities of performance data (sometimes more than may be useful);
  - use of computers to summarize and analyse great quantities of data;
  - off-season nurseries to enable several breeding generations per year.

And supporting all of these tools is the sure knowledge of the breeders that they are working with biological organisms that obey the rules of nature – especially those of genetics – even if the particular applications of those rules are understood only imperfectly.

One also must acknowledge that in many cases (see examples in the 'Theory' section above), breeders have designed new tools on the basis of scientific theory. In the absence of theory the tools might not have been designed and tried. But as also noted in the 'Theory' section, the harsh taskmaster of empiricism always winnows and alters the design and use of the fanciest of tools. In the end, 'art and experience – not precision genetics – are the key to successful use of these [new] tools' (Duvick, 1996).

## Yield Trials: Breeders and Farmers in Partnership

Professional plant breeders and farmers collaborate in the development of new varieties, sometimes in ways they may not realize. Breeders make breeding crosses, design selection experiments and make initial selection of experimental varieties. Farmers participate in the final stages of selection by conducting trials on their own farms to supplement those conducted by the breeders. They also inform the breeders, often in response to planned surveys, of any changes in farming practice that might require new variety characteristics. (An example: no-till farming can cause cooler soil temperatures in the early part of the growing season. Breeders therefore need to develop and/or identify varieties with strong germination and vigorous seedling growth in such conditions.)

The most important partnership contribution of the farmers is one they may not realize. Breeders have a very strong tendency to choose the most popular varieties (or their parent lines in the case of a hybrid crop) as parents of new breeding populations. Breeders make the reasonable assumption that popular varieties must have a good selection of the genes needed for best performance according to farmer standards. Thus, as farmers collectively popularize certain varieties, they also prescribe the genotypes used by breeders for the next cycle of breeding. The farmers, in a sense, control the direction of professional plant breeding.

Perhaps the plant breeders' greatest use of empiricism – of practical experience – in the industrial countries has been in the development of the yield trial technique (Fig. 8.1). Starting with a few plots at the local worksite, yield trials, often called 'performance trials' because they measure much more than yield, have developed over the years into highly organized, wide-ranging enterprises. Modern performance trials

**Fig. 8.1.** Maize breeders weighing ears from a small-plot performance trial, c. 1937. Samuel Goodsell and James Weatherspoon were new employees at a young hybrid maize breeding company, Pioneer Hi-Bred (started in 1926, in Iowa). Technology for conducting performance trials was relatively simple in the early years. Plots were harvested by hand, ears were weighed on a 'dairy scale' and a hand-shelled sample was set aside for testing with a moisture meter at the central 'laboratory'. Goodsell and Weatherspoon each worked for Pioneer Hi-Bred until they retired many years later. Goodsell set up and led the company's seed germ-ination department (essential for delivering sound seed to the farmer), and Weatherspoon set up and led the company's field trial department (essential for choosing superior, dependable hybrids for sale to the farmer). Both departments were pace-setters for the USA hybrid maize breeding industry (photograph courtesy of Pioneer Hi-Bred International, Inc.).

often encompass thousands of plots per breeder and intricate systems of information and data exchange among breeders. Well-conducted trials require careful testing in a diversity of well-defined soil and weather ecosystems, a diversity of appropriate agronomic cultural practices, and most important of all, they are conducted in partnership with the customers, the farmers who will plant the seed of the new varieties, if they like them. The overwhelming majority of the breeders' test plots are planted on farm fields and tended by the farmers in the same way as they tend the rest of their crop.[5]

A second tier of trials has been layered over the breeders' trials. These trials are usually conducted by government agencies, and are intended to give farmers an unbiased report of variety performance. In some countries varieties cannot be released unless they have equalled or exceeded certain standards in the official government trials, sometimes called 'national release trials'. In other countries the trials serve as sources of information for farmers and other interested parties, but they are not used as a means of controlling variety release. In the USA, such informational trials are conducted by university extension department personnel, or by semi-autonomous organizations with close ties to state and university agencies (e.g. Robinson and Knott, 1963; Kalton, 1992).

In addition to the breeders' trials in which thousands of experimental varieties are compared in replicated small-plot trials, and the government trials in which scores of finished varieties are compared in similar fashion, farmers themselves (primarily in the industrialized countries) make their own field-scale yield trials (compare Schneider, Chapter 7, this volume). Such comparisons are principally fostered by the commercial seed companies, but they are also (at least in the USA) sponsored by public interest groups such as a local high school agriculture class. These farmer-run trials, often called 'strip tests', date back at least 50 years (primarily for maize) but they have become much more precise and widely used in most of the important field crops since the 1980s (e.g. Duvick, 1992b; Duvick and Cassman, 1999). Combine harvest and more recently widespread use of computers has allowed great increase in the use and utility of the farmer-run strip test comparisons.

Several varieties – new, experimental varieties, relatively new, commercial varieties and older, well-known varieties – are planted side by side in farm fields using the farmer's equipment and cultural practices. The farmer can choose which varieties to compare, but eager seed companies are also willing to make suggestions and sometimes even furnish free seed of their newest varieties and hybrids for the farmer to grow and evaluate. At harvest time measured strips of each variety are harvested by the farmer, and grain weights are obtained by using a special 'weigh wagon' that pulls alongside the combine to receive and weigh the grain from the measured strip. Each sample is tested for grain moisture, and yields are calculated to a standard moisture level for all varieties. More recently, farmers have begun to use in-combine yield monitors for direct observation of grain yield at any point in the harvest field. They can make yield comparisons of all varieties grown on their farm and, furthermore, compare those grown under different growing conditions, such as on a droughty hill-top or in a well-watered site with high yield potential.

Farmers can compare hybrids by examining strip test (or yield monitor) data that they have gathered on their own farms and they can also obtain summaries of similar trials, performed by dozens or hundreds of other farmers in their locality. Farmers typically do not use the summaries, however, perhaps because they suspect bias on the part of the commercial firms that make them, but also because they tend to believe that their own farm fields and farming practices are unique. They prefer to use their own on-farm data for help in making decisions about which varieties to plant in the coming year.[6]

But the commercial breeders and sales staff make good use of the data summaries. The summaries often involve thousands of farmer comparisons that can be sorted according to such categories as geographical location, seasonal weather pattern, severity of disease or insect attack, general yield level or type of farming system. The results add greatly to the information from the small-plot trials, helping the breeders to interpret the small-plot trials in greater depth. Perhaps more importantly, study of strip-test results helps the breeders to invent ways to improve the organization and operation of their small-plot trials, to make them more accurately predictive of on-farm performance. For example, comparison of data from strip trials and small-plot trials convinced one group of maize breeders that they should plant certain small-plot trials as four-row plots and use data from the centre two rows only, to avoid the deleterious shading effect of tall hybrids on short ones. Alternatively, small-plot trials could be subdivided, with subgroups containing only hybrids of similar phenotype. In another example, the breeders changed their performance trials from multi-replication per location at a few locations to one replication per location at numerous locations. Study of strip-test data (which were one replication at many locations) convinced them that varieties in such a test design (one replication/many sites) could be differentiated with a high degree of statistical precision. The tests would also give better prediction for stability of performance under the multiplicity of conditions to be expected with widespread planting after variety release.

Breeders have set up no balanced experiments to test the hypothesis that the recent great increase in use of data from farmer-run trials has improved breeding progress. They have used the data simply because they were there (courtesy of the variety comparisons promoted by the sales arms of commercial firms) and the test designs were easily adapted for breeder analysis. The breeders assumed that as long as the data were sound (an easily tested premise), the addition of more test sites (especially on-farm sites) could only improve the accuracy of their evaluations of variety performance.

In addition to their strip-test interactions, breeders and farmers confer directly, either in the course of field days when farmers visit the breeders, or by individual visits of breeders to the farm fields in company with the farmers. This is not a new phenomenon; it has been common practice in industrial countries since the first years of professional breeding.[7]

Farmers directly influence breeding direction (as noted earlier) by using their own data to make decisions on what varieties to plant. Breeders note carefully the traits of favoured varieties compared with those that the farmers reject. But farmers also give more direct help to breeders, as they set their breeding goals. They confer with breeders or (more often) with extension and sales personnel who in turn counsel the breeders. They talk about desired improvements or specific faults of present varieties, and they may discuss changes in agronomic practices that could dictate the need for new traits in future varieties. The farmers often initiate new agronomic practices as they search empirically for new ways to increase yields or decrease costs of production or both. The breeders then are challenged to find out about such changes and rebreed their crops in timely fashion to fit the new agronomic practices.

An example of a farmer-initiated change that in turn forced breeders to alter breeding goals is the constantly increasing plant density of maize in the USA. Planting rates have increased every year since at least 1930, going from about 30,000 plants ha$^{-1}$ in 1930 to about 75,000 plants ha$^{-1}$ at the present time in the heart of the USA corn belt. Farmers increase plant density with the hopeful intention (often successful) of getting more yield per hectare. But high density planting – crowding the maize plants together – increases the deleterious effects of biotic and abiotic stresses. Plants are more likely to lodge because of poorly developed roots, more prone to suffer from certain leaf and stalk rot diseases, to suffer from drought, or to have problems with seed set or kernel fill. Breeders therefore have been forced to breed for increased tolerance to these stresses, and have done so with remarkable success (Duvick, 1992a, 1997; Duvick and Cassman, 1999). But each time the breeders make an advance, the farmers push the new varieties beyond the expected limits (as noted, planting rates have increased steadily for the past several decades), and breeders must again develop hybrids with even greater tolerance to crowding.

The net result is that today's hybrids grown at today's planting rates produce about the same weight of grain per plant as was produced by maize varieties in 1930, but they do so on two to three times as many plants per hectare. And yields per hectare are increased by a similar ratio (Duvick, 1997).

Interactions between farmers and breeders are not always perfect. At a time of large and rapid changes in agronomic practice for maize growing in the 1960s and early 1970s, contacts between farmers and breeders were not as close as they might have been, and farmer-conducted strip-test yield comparisons were not yet widespread or highly developed. Breeders sometimes were slow to change their small-plot yield trial conditions to fit the new cultural practices, particularly those utilizing higher plant density and higher rates of nitrogen fertilizer (Troyer, 1996; compare Ríos Labrada et al., Chapter 9, this volume). Hybrids selected for good performance at relatively low densities were prone to exhibit undesirable amounts of stalk and root lodging, premature death and dropped ears when farmers grew them at high densities. They were not suited to the new cultural practices. But eventually the breeders recognized the need to change their breeding goals, as farmers rejected the breeders' 'low density' hybrids in favour of new-model 'high density' hybrids from other sources. The laggard breeders changed their testing procedures to fit the new cultural conditions and in time brought forth hybrids with the needed new characteristics (e.g. Troyer and Rosenbrook, 1983).[8]

The above examples show how farmers, extension/sales personnel and professional breeders work together as an informal team (even though this may not always be apparent to the team members) in a constantly evolving programme for interactive change in crop culture and breeding. Empirical methods – observation and experience – are their chief means of effecting change. The scientific method adds precision and logic to the interpretation of their experience. And especially for the breeders, theory aids and prods the practice of empiricism. But there is more to plant breeding than theory and empiricism.

## Intuition

*Intuition, insight* and *perceptivity*: these words characterize the breeders who are recognized by their peers as the best. These are the breeders who bring forth uniquely improved varieties, or ways of developing them, or ways of introducing valuable traits from unlikely sources of exotic germplasm. Although these top-ranked professionals can often rationalize their successes after the fact, their initial insights seem to come without conscious or painstaking rationalization (see

Soleri *et al.*, Chapter 2, this volume). Rather, they are based on intimate knowledge of the plants that they breed.

In the words of the Nobel Laureate Barbara McClintock, 'You have to be aware . . . You need to know these plants well enough so that if anything changes . . . you [can] look at the plant and right away you know what this damage you see is from . . . [You must have] a feeling for the organism' (Keller, 1983: 198). McClintock was a cytogeneticist and not a plant breeder *per se* but she spoke the same language as top-flight breeders. She and they, together, understood the essentiality of 'a feeling for the organism'.

Successful professional plant breeders live with their plants, constantly looking, taking systematic notes, or just wandering through their plantings, communing with their 'children'. Those who do not do this pay the price. The eminent maize breeder, George Sprague, noted this fact in explaining the reasons for poor output by a particular maize breeder. Sprague said,

> he was not a field organized man. He seldom went to the research fields to see what was going on. He laid out a set of detailed plans for his field technicians to carry out. But by and large they carried them out without much supervision and often times without too much background information as to what they were doing and what they were trying to do. I don't think you can do corn breeding [if you have that] approach
>
> (G.F. Sprague, Oregon, 1996, personal communication)

Intuition works for the future as well as for the present. A contemporary maize breeder says, 'Intuition also plays an important role in knowing what to breed for. The ability to speculate on what . . . traits will be important 5–10 years from now seems critical to successful plant breeding' (M.E. Smith, Cornell University, New York, 2000, personal communication). (One wonders if the first plant domesticators also had this vision?)

Perhaps the function of intuition in plant breeding is not too different from the function of instantaneous reflexes in sports. Constant immersion in the object of study, constant repetition of a particular motion, both of these actions may provide a reflexive reaction that is more direct and to the point than the most carefully reasoned course of action. As the baseball player Yogi Berra is supposed to have said, 'You can't think and hit at the same time.'

Of course the successful use of intuition and insight, of 'a feeling for the organism'; is not limited to professional plant breeders. Accounts of outstanding farmer breeders from the past show that they had the same deep familiarity and insights into the potential of the crop they were manipulating (Wallace and Brown, 1988).

## Synthesis and Comment

Plant breeding is a singular undertaking. It is more than theory, empiricism and intuition. It has a single name but it is not practised uniformly. Starting with the visionary plant selections of the initial domesticators, continuing with the increasingly guided selection practices of farmers through the millennia, and moving on to purposive hybridization and selection of the present time, plant breeding has been an evolving discipline.

The human tendency to arbitrarily classify and separate leads us to speak of different types of plant breeding, such as farmer breeding, scientific breeding, professional breeding, commercial breeding or public breeding. Such categorizations can be helpful for understanding the diversity of plant breeding styles and goals but they also can hinder when they blind us to the common characteristics of the many kinds of plant breeding and plant breeders.

For example, all breeding requires selection within segregating populations derived from hybrids of genetically dissimilar individuals. The fact that farmer breeding typically works with products of accidental hybridization and contemporary professional breeding usually works with products of planned hybridization does not decrease the fundamental importance of genetic segregation for advances in crop breeding. All plant breeding depends on and is limited by the underlying rules of nature.

Thoughtless categorization can hinder us from learning how different types of breeding are related, how they may grade imperceptibly from one type to another. If we can rise above categorization we can consider the prospect that techniques and ideas can be transferred further than has seemed possible. Can farmer breeders adopt some of the tools used by today's professionals? Can professionals work with the same deep understanding of local environmental requirements that is held by smallholders? How about comparisons of farmer breeders and professional breeders, private sector breeders and public sector breeders, 'corporate' breeders and 'small company' breeders, or public sector breeders from developing countries and public sector breeders from industrialized nations?[9]

Theory, empiricism and intuition are essential parts of professional plant breeding. Proportionate use of the different ingredients can and should vary with the opportunities and goals of the practitioner. Successful plant breeding, appropriate to the time, location and customer, depends not only on the ingredients but also – and especially – on their proper integration and execution.

## Acknowledgements

I give my deep thanks to the many professional plant breeders with whom I have worked over the years. Their common sense, ingenuity and single-minded devotion to improvement of crop varieties as desired by their customers – the farmers – have instructed and inspired me.

## Notes

[1]     'Heterosis has been used successfully even though its genetic basis has not been for the most part determined. It seems it will remain so in the near future' (Hallauer, 1999).

[2]     This statement is based primarily on written and oral comments made to me over the years by senior (and junior) breeders of the major crop plants worldwide, and on my own experiences as a plant breeder and plant breeding director. The importance of pedigree or 'empirical' breeding was affirmed recently by rice (*Oryza sativa* L.) breeders at the International Rice Research Institute (IRRI): 'Empirical breeding consists of hybridization between the best adapted inbred parents and selection of progeny on the basis of yield per se . . . This approach has been the basis of indica cultivar development at IRRI' (Peng *et al.*, 1999). A prominent maize geneticist has made a similar statement for maize breeding: 'Pedigree selection is the most widely used breeding method to develop inbred lines for use as parents of hybrids' (Hallauer *et al.*, 1988).

[3]     He notes, however, that modern breeders (in comparison with their predecessors) have access to a greater diversity of genotypes and a broader range of target environments. They therefore have more scope for action even though they may use antique tactics.

[4]     My own experience and observations reinforce this statement. This experience was gained during the course of plant breeding and administrative duties in the private sector (involving the major field crops in developed and developing countries), service on boards and advisory committees for public sector plant breeding, and five decades of informal interactions with farmers and plant breeders worldwide.

[5]     A breeder typically will plant a performance trial at the home station and then replicate it at 6 to 12 on-farm sites. The trials are fully tended (except for planting and harvest) by the farmers.

[6]     During my years in commercial plant breeding this statement ('farmers primarily trust only their own data') was made to me repeatedly by sales people and by farmers. Since my retirement, I have occasionally asked farmers and sales people whether there has been a change. I find no evidence for a change.

[7]     I recall an autumn day in the 1950s when I took routine notes on the appearance of hand-harvested ears of maize yield trial entries in company with the farmer who furnished land and cultural practices for the trial. We discussed good points or bad points of entries that caught our attention, as well as comparing notes on field performance of certain hybrids in the test, as we had

seen them from our separate vantage points. Combine harvest has eliminated the possibility for interactions like this 1950s experience, but in recent years I have ridden in combines with farmers as they harvested strip-test trials or their ordinary production fields. We watched the yield monitor and discussed strong points and weak points of maize hybrids or soybean varieties, as we examined them first-hand from the vantage of the combine seat. We educated each other in about the same way as was done 50 years earlier. Farmer/breeder or (more likely) farmer/sales representative and farmer/extension agent interactions on the farm and in the crop are still the norm in the USA.

[8]     In the 1970s, as a direct result of such experiences in 'cultural lag', I helped to initiate a programme whereby farmer-run strip trials were used for pre-release testing of experimental hybrids (Duvick, 1992b). The programme was initiated in collaboration with the sales director of the company for which I then served as director of plant breeding. His sales personnel were helping farmers to conduct combine-harvested strip-test comparisons among commercial hybrids of the farmers' choice. The sales director and I agreed that free seed of experimental hybrids in the last year of testing before release would be provided for farmers to plant and compare with current commercial favourites. Hybrid comparison data from such trials could then be compared with data for the same hybrids grown by breeders in small-plot trials. If the two data sets agreed, one could be reasonably confident about predictions of performance of the experimental hybrids in the farmers' hands. If they did not agree, breeders examined their trial and selection procedures with the intention of modifying them to better predict on-farm performance. In either case, the combination of farmer trials and breeder trials was used to help determine whether or not an experimental hybrid would be released for widespread distribution and planting.

[9]     An example of commonality: new varieties of common bean (*Phaseolus vulgaris* L.), bred for use in an East African country, will not be accepted if they require a long cooking time even though genetic yield potential may be high. Women have to carry firewood long distances, thus short cooking time is important. The East African bean breeders (of any category) must be sure that their products meet the requirements of their customers (the cooks), just as professional maize breeders in the USA must be sure that their products meet the requirements of their customers (modernizing, high yield farmers). In either case, participatory plant breeding (in these examples, involvement of the customer in goal setting) is essential to avoid misdirected plant breeding.

# References

*AgBiotech Reporter* (2000a) Germany wants GMO moratorium. *AgBiotech Reporter* 17(7), 11.

*AgBiotech Reporter* (2000b) Wales adopts anti-GM resolution. *AgBiotech Reporter* 17(6), 8.

*AgBiotech Reporter* (2000c) Japan manufacturers halt biotech development. *AgBiotech Reporter* 17(6), 20.

Baker, R.J. (1984) Quantitative genetic principles in plant breeding. In: Gustafson, J.P. (ed.) *Gene Manipulation in Plant Improvement*. Plenum Press, New York, pp. 147–176.

Benbrook, C.M. (1996) *Pest Management at the Crossroads*. Consumers Union, Yonkers, New York.

Biffen, R.H. (1905) Mendel's law of inheritance and wheat breeding. *Journal of Agricultural Science* 1, 4–48.

Branson, C.V. and Frey, K.J. (1989) Recurrent selection for groat oil content in oat. *Crop Science* 29, 1382–1387.

Brunke, K.J. and Meeusen, R.L. (1991) Insect control with genetically engineered crops. *Trends in Biotechnology (TIBTECH)* 9, 197–200.

Donald, C.M. (1968) The breeding of crop ideotypes. *Euphytica* 17, 385–403.

Duvick, D.N. (1992a) Genetic contributions to advances in yield of U.S. maize. *Maydica* 37, 69–79.

Duvick, D.N. (1992b) Participatory on-farm research: an agricultural industry perspective. In: Clement, L.L. (ed.) *Participatory On-farm Research and Education for Agricultural Sustainability*. Illinois Agricultural Experiment Station, University of Illinois at Urbana-Champaign, pp. 3–10.

Duvick, D.N. (1996) Plant breeding, an evolutionary concept. *Crop Science* 36, 539–548.

Duvick, D.N. (1997) What is yield? In: Edmeades, G.O., Bänziger, M., Mickelson, H.R. and Peña-Valdivia, C.B. (eds) *Developing Drought- and Low N-tolerant Maize. Proceedings of a Symposium, 25–29 March 1996, CIMMYT, El Batán, Mexico*. CIMMYT, México, DF, pp. 332–335.

Duvick, D.N. (1999) The profile of a plant breeder for the third millennium. In: Borém, A., Del-Giúdice, M.P. and Sakiyama, N.S. (eds) *Plant Breeding in the Turn of the Millennium*. Federal University of Viçosa, Viçosa, MG, Brazil, pp. 133–182.

Duvick, D.N. and Cassman, K.G. (1999) Post-green revolution trends in yield potential of temperate maize in the North-Central United States. *Crop Science* 39, 1622–1630.

Dyer, W.E., Hess, F.D., Holt, J.S. and Duke, S.O. (1993) Potential benefits and risks of herbicide-resistant crops produced by biotechnology. In: Janick, J. (ed.) *Horticultural Reviews*, Vol. 15. John Wiley & Sons, New York, pp. 367–408.

Evans, L.T. (1993) *Crop Evolution, Adaptation, and Yield*. Cambridge University Press, Cambridge.

Evans, L.T. (1998) *Feeding the Ten Billion: Plants and Population Growth*. Cambridge University Press, Cambridge.

Fehr, W.H. (ed.) (1984) *Genetic Contributions to Yield Gains of Five Major Crop Plants*. Crop Science Society of America, Madison, Wisconsin.

Forsberg, R.A., Brinkman, M.A., Karow, R.S. and Duerst, R.D. (1991) Registration of 'Centennial' oat. *Crop Science* 31, 1086–1087.

Frey, K.J. (1968) Expected genetic advances from three simulated selection schemes. *Crop Science* 8, 235–238.

Frey, K.J. (1996) *National Plant Breeding Study – I: Human and Financial Resources Devoted to Plant Breeding Research and Development in the United States in 1994*. Iowa Agriculture and Home Economics Experiment

Station, and Cooperative State Research, Education & Extension Service/ USDA cooperating, Ames, Iowa.

Hallauer, A.R. (1999) Heterosis: What have we learned? What have we done? Where are we headed? In: Coors, J.G. and Pandey, S. (eds) *Genetics and Exploitation of Heterosis in Crops*. American Society of Agronomy, Crop Science Society of America, Soil Science Society of America, Madison, Wisconsin, pp. 483–492.

Hallauer, A.R., Russell, W.A. and Lamkey, K.R. (1988) Corn breeding. In: Sprague, G.F. and Dudley, J.W. (eds) *Corn and Corn Improvement*, 3rd edn. American Society of Agronomy, Crop Science Society of America, Soil Science Society of America, Madison, Wisconsin, pp. 469–565.

Harlan, J.R. (1992) *Crops and Man*, 2nd edn. American Society of Agronomy, Crop Science Society of America, Madison, Wisconsin.

James, C. (2000) *Global Review of Commercialized Transgenic Crops: 2000*. ISAAA, New York.

Kalton, R.R. (1992) Public variety testing as viewed by private breeders. *Twenty-Second Soybean Seed Research Conference*, Vol. 22. American Seed Trade Association, Chicago, Illinois, pp. 65–71.

Keller, E.F. (1983) *A Feeling for the Organism: the Life and Work of Barbara McClintock*. W.H. Freeman and Company, San Francisco.

Khush, G.S. (1995) Breaking the yield frontier of rice. *GeoJournal* 35, 329–332.

Lamkey, K.R. (1992) Fifty years of recurrent selection in the Iowa Stiff Stalk Synthetic maize population. *Maydica* 37, 19–28.

Lee, M. (1998)Genome projects and gene pools: new germplasm for breeding? *Proceedings of the National Academy of Sciences, USA* 95, 20001–20004.

Mercado, L.A., Souza, E. and Kephart, K.D. (1996) Origin and diversity of North American hard spring wheats. *Theoretical and Applied Genetics* 93, 593–599.

National Research Council (US) Committee on Genetic Vulnerability of Major Crops (1972) *Genetic Vulnerability of Major Crops*. National Academy of Sciences, Washington, DC.

Ortiz-Monasteriao R., J.I., Sayre, K.D., Rajaram, S. and McMahon, M. (1997) Genetic progress in wheat yield and nitrogen use efficiency under four nitrogen rates. *Crop Science* 37, 898–904.

Peng, S., Cassman, K.G., Virmani, S.S., Sheehy, J. and Khush, G.S. (1999) Yield potential trends of tropical rice since the release of IR8 and the challenge of increasing rice yield potential. *Crop Science* 39, 1552–1559.

Porter, D.R., Burd, J.D., Shufran, K.A., Webster, J.A. and Teetes, G.L. (1997) Greenbug (Homoptera: Aphididae) biotypes: selected by resistant cultivars or preadapted opportunists? *Journal of Economic Entomology* 90, 1055–1065.

Prabhu, R.R., Njiti, V.N., Bell-Johnson, B., Johnson, J.E., Schmidt, M.E., Klein, J.H. and Lightfoot, D.A. (1999) Selecting soybean cultivars for dual resistance to soybean cyst nematode and sudden death syndrome using two DNA markers. *Crop Science* 39, 982–987.

Rasmusson, D.C. and Phillips, R.L. (1997) Plant breeding progress and genetic diversity from *de novo* variation and elevated epistasis. *Crop Science* 37, 303–310.

Ratcliffe, R.H., Ohm, H.W., Patterson, F.L., Cambron, S.E. and Safranski, G.G. (1996) Response of resistance genes H9-H19 in wheat to Hessian Fly (Diptera: Cecidomyiidae) laboratory biotypes and field populations from the eastern United States. *Journal of Economic Entomology* 80, 1309–1317.

Reynolds, M.P., Rajaram, S. and Sayre, K.D. (1999) Physiological and genetic changes of irrigated wheat in the post-green revolution period and approaches for meeting projected global demand. *Crop Science* 39, 1611–1621.

Robinson, J.L. and Knott, O.A. (1963) *The Story of the Iowa Crop Improvement Association and Its Predecessors.* Iowa Crop Improvement Association, Ames, Iowa.

Russell, W.A. (1991) Genetic improvement of maize yields. *Advances in Agronomy* 46, 245–298.

*Science* (1998) Wonder wheat. *Science* 280, 527.

Simmonds, N.W. (1991) Genetics of horizontal resistance to diseases of crops. *Biological Reviews* 66, 189–241.

Sinnott, E.W. and Dunn, L.C. (1932) *Principles of Genetics.* McGraw-Hill Book Company, New York.

Smith, B.D. (1995) *The Emergence of Agriculture.* Scientific American Library, distributed by W.H. Freeman and Company, New York.

Smith, D.C. (1966) Plant breeding – development and success. In: Frey, K.J. (ed.) *Plant Breeding: a Symposium Held at Iowa State University.* The Iowa State University Press, Ames, Iowa, pp. 3–54.

Sneller, C.H. (1994) Pedigree analysis of elite soybean lines. *Crop Science* 34, 1515–1522.

Snow, A.A. and Moarán-Palma, P. (1997) Commercialization of transgenic plants: potential ecological risks. *BioScience* 47, 86–96.

Sprague, G.F. (1966) Quantitative genetics in plant improvement. In: Frey, K.J. (ed.) *Plant Breeding.* Iowa State University Press, Ames, Iowa, pp. 315–355.

Staub, J.E., Serquen, F.C. and Gupta, M. (1996) Genetic markers, map construction, and their application in plant breeding. *Hort-Science* 31, 729–740.

Stephens, J.C., Miller, F.R. and Rosenow, D.T. (1967) Conversion of alien sorghums to early combine genotypes. *Crop Science* 7, 396.

Stuber, C.W. and Polacco, M. (1999) Synergy of empirical breeding, marker-assisted selection, and genomics to increase crop yield potential. *Crop Science* 39, 1571–1583.

Tracy, W.F. (1997) History, genetics, and breeding of supersweet (*shrunken2*) sweet corn. In: Janick, J. (ed.) *Plant Breeding Reviews,* Vol. 14. John Wiley & Sons, New York, pp. 189–236.

Troyer, A.F. (1996) Breeding widely adapted, popular maize hybrids. *Euphytica* 92, 163–174.

Troyer, A.F. and Rosenbrook, R.W. (1983) Utility of higher plant densities for corn performance testing. *Crop Science* 23, 863–867.

van Dommelen, A. (1999) *Hazard Identification of Agricultural Biotechnology: Finding Relevant Questions.* International Books, A. Numenkade 17, 3572 KP, Utrecht, the Netherlands.

Voldeng, H.D., Cober, E.R., Hume, D.J., Gillard, C. and Morrison, M.J. (1997) Fifty-eight years of genetic improvement of short-season soybean cultivars in Canada. *Crop Science* 37, 428–431.

Walker, J.C. (1966) The role of pest resistance in new varieties. In: Frey, K.J. (ed.) *Plant Breeding*. The Iowa State University Press, Ames, Iowa, pp. 219–242.

Wallace, D.H. and Yan, W. (1998) *Plant Breeding and Whole-system Crop Physiology: Improving Crop Maturity, Adaptation and Yield*. CAB International, Wallingford, UK.

Wallace, H.A. and Brown, W.L. (1988) *Corn and Its Early Fathers*, revised edn. Iowa State University Press, Ames, Iowa.

Wang, R.-L., Stec, A., Hey, J., Lukens, L. and Doebley, J. (1999) The limits of selection during maize domestication. *Nature* 398, 236–239.

Watson, J.D., Tooze, J. and Kurtz, D.T. (1983) *Recombinant DNA: a Short Course*. Scientific American Books, New York.

*Webster's Collegiate Dictionary* (1941) 5th edn. G. & C. Merriam Co., Springfield, Massachusetts.

# Conceptual Changes in Cuban Plant Breeding in Response to a National Socio-economic Crisis: the Example of Pumpkins

9

HUMBERTO RÍOS LABRADA,[1] DANIELA SOLERI[2] AND DAVID A. CLEVELAND[3]

[1]National Institute of Agriculture Sciences (INCA), San Jose de Las Lajas, La Habana, Cuba CP 32700; [2]Centre for People, Food and Environment, Santa Barbara, California, and Environmental Studies Program, University of California, Santa Barbara, CA 93106, USA; [3]Department of Anthropology and Environmental Studies Program, University of California, Santa Barbara, CA 93106-4160, USA

## Abstract

Major changes in the economy of Cuba since 1989 have led to reduced inputs in agricultural production resulting in changes in some of the goals of Cuban plant breeding and a search for new, more appropriate methods of plant breeding for pumpkin, including consideration of participatory plant breeding (PPB). This chapter reports the results of a pumpkin breeding project that explored the benefits of plant breeders' collaboration with farmers in Habana province. The research focused on two issues: (i) the use of landraces as sources of genetic diversity under low-input conditions; and (ii) farmers' and plant breeders' methods of seed production. In this chapter we describe why these issues are important for cross-pollinated crops in the Cuban context, and why PPB might be a useful response in terms of the biological results, energy efficiency and economic returns achieved.

## Cuban Agriculture, 1959–1989

### Objectives and development of Cuban agriculture after 1959

The three most important objectives for the transformation of Cuban agriculture after the Cuban revolution in 1959 were to: (i) meet the growing food requirements of the Cuban population; (ii) create export funds in order to obtain raw materials and empower the food industry; and (iii) eradicate poverty from the countryside (Funes, 1997). An industrialized or conventional agricultural model, typified by the 'green revolution' approach to agricultural development was pursued in order to achieve these objectives.

In Cuba the conventional approach to agricultural modernization in the style of the green revolution was possible because of the strong relations that had been built with the Socialist countries of Eastern Europe and, in particular, with the former Soviet Union (USSR) (Enríquez, 2000). This relationship permitted pursuit of high-input agriculture, focused on labour efficiency and productivity. As such, this reflected an industrialized world trend of progressive substitution of labour with capital as a method for increasing land productivity (Funes, 1997). The management of agricultural processes through high use of external inputs, large-scale production, specialization, mono-culture and mechanization, was adopted in Cuba at a national level, with the intention of producing sufficient food for all. Indeed, this model, supported by high input levels, was successful in achieving the prescribed objectives over its 25-year duration (Table 9.1).

However, throughout the 1980s, 87% of Cuba's export trade was carried out at preferential prices with socialist countries, and only 13% at world 'market' prices with Western countries (Enríquez, 2000). This meant that national goods were sold at high prices, and imports were purchased at low prices. Principal export cash crops, primarily sugar, tobacco and citrus, covered 50% of the agricultural land (Pérez Marín and Muñoz Baños, 1992; Rosset and Benjamin, 1993). The high level of importation of energy (petroleum), fertilizer, herbicides and livestock concentrates was favourable for Cuban food production, but not for Cuban self-sufficiency (Table 9.2). The success of this model was based on a strong dependency on external inputs.

## The plant breeding model

A centralized plant breeding model was a component of the high-input agriculture being used particularly for the country's cash crops (Begemman *et al.*, 2000). Introduction of foreign varieties, hybridiza-tion, landraces and mutation breeding were the principal sources of genetic variation used for varietal development in Cuban plant breeding programmes (Ríos, 1999). At the end of the 10–12 years typically spent in varietal development for a specific crop, the breeding programmes usually released only one or two varieties for the entire country, emphasizing geographically wide adaptation. Wide geograph-ical adaptation was encouraged by policy makers, with most Cuban

**Table 9.1.** Successes obtained with the conventional agricultural model in Cuba. (Sources: Deere, 1992; Funes, 1997; Oficina Nacional de Estadísticas, June 1997.)

| | Year | |
|---|---|---|
| Indicators of success | 1965 | 1988–89 |
| Calorie consumption (kcal per person day$^{-1}$) | 2500 | 2834 |
| Protein consumption (g per person day$^{-1}$) | 66 | 76 |
| Life expectancy (years) | 55 | 75 |

**Table 9.2.** Principal agricultural imports in Cuba in relation to total national requirements at the end of the 1980s. (Sources: Deere, 1992; Pastor, 1992; Funes, 1997.)

| Agricultural import | Percentage of Cuba's national needs met by imports |
|---|---|
| Edible oils and grains | 94 |
| Wheat | 100 |
| Rice | 50 |
| Beans | 99 |
| Meat | 21 |
| Milk and derivatives | 38 |
| Fish | 44 |
| Sugar | 0 |
| Root and tuber crops | 0 |
| Fruits and vegetables | 1–2 |
| Fertilizers | 94 |
| Herbicides | 96 |
| Livestock concentrates | 97 |

governmental organizations providing incentives to scientists involved in releasing a variety for use over a large area (Ríos, 1999).

Ambitious plant breeding programmes were developed in sugarcane, roots and tubers, rice, tobacco, coffee, horticultural crops, pastures, grains, fibres and some fruit trees, undertaken by 15 research institutes and an experimental station network that spread over the island in the 1980s (Begemman *et al.*, 2000).

As a part of the varietal release process, each new variety had to pass through a hierarchical series of steps:

> When a research institute has a significant result they send this to the Scientific Forum (Consejo Cientifico) at the national level. This Forum checks its scientific validity and, if it is approved, they send it to an Expert Group (Grupo expertos) consisting of researchers, teachers and

production directors. If this Group approves the result it is then sent to
the Vice-Minister of Diverse Crops (Vice-Ministro Cultivos Varios). This
Minister will send the results to the provincial delegations, who implement them into their production plans, and this means that producers
have to adopt them as an order. This procedure takes a top-down
approach without consulting the producers. Some researchers do visit
farms, but still the research areas and problems come from the decisions
of the researchers. Further, the Scientific Forum only evaluates the
scientific content and not its applicability, and therefore they may reject
good techniques and accept those that are unsuitable for producers.

(Trinks and Miedema, 1999: 116)

## Landraces and plant breeding

Since professional plant breeding in Cuba was officially established in
April 1904 through the Estación Experimental Agronómica in Santiago
de las Vegas, landraces have been widely used in crop improvement
programmes for different species. Some landraces of food crops such as
garlic, maize, pepper, cowpea, groundnut, tomato, common beans and
pumpkin were released as commercial varieties after roguing the
population of undesirable off types. In others, plant breeders selected
favourable off types within landrace accessions (Esquivel *et al.*, 1994).
Both procedures produced important varieties with good qualities and
acceptable yields throughout the history of professional plant breeding
in Cuba. For tomato, hybridization of Cuban landraces with materials
of temperate origin had a good impact in Cuba (Moya, 1987). Some of
the materials of diverse crop species collected in Cuba with characteristics such as disease resistance, short growing cycles and good food
qualities were not used by the formal plant breeding sector due to their
low yields under high-input conditions (e.g. for beans see Castiñeiras,
1992).

In the 1980s, the peak of industrial agriculture in Cuba, an investigation of plant genetic resources in Cuba was undertaken as a collaboration between the Cuban Fundamental Research Institute in Tropical
Agriculture (the former Estación Experimental Agronómica) and
the Gene Bank of Gatersleben, Germany. This study documented the
diversity of plant genetic resources within the country and found that
the majority of this diversity was being maintained and managed by
farmers and in non-industrial systems (Hammer *et al.*, 1992). Esquivel
and Hammer (1992) describe the 'conucos', traditional home gardens
characteristic of subsistence agriculture in Cuba, as the central ecosystems for maintaining plant genetic resources in the country, and for
maintaining materials to satisfy farmers' needs (medicinal, religious,
economic, food) that were not met by industrial agriculture.

# Cuban Agriculture After the Collapse of the Socialist Block: 'The Special Period'

The collapse of the socialist countries of Eastern Europe and the USSR highlighted the vulnerabilities of the agricultural model being followed by Cuba. Cuba lost its principal export markets and guarantees, which had been provided by these countries. Foreign purchase capacity was dramatically reduced from US$8100 million in 1989 to US$1700 million by 1993, a decrease of almost 80% (Rosset and Benjamin, 1993). Of this new figure, US$750 million was required solely for the purchase of fuel, and US$440 million for basic foods. The ability to purchase agrochemicals was also greatly affected (Table 9.3).

As a result of this situation, agricultural production was immediately and severely reduced. All Cuban farmers suffered in these difficult circumstances; however it was most acutely felt in the large, high-input enterprises; small- and medium-scale farmers were less affected (Funes, 1997). The lack of financial resources also had a severe effect on the seed industry in Cuba. For example, before 1989, seed production capacity for maize and beans had reached over 3180 t year$^{-1}$ but in 1998 only approximately 50% of that amount was produced (Ríos and Wright, 1999).

## Strategic changes and popular responses to the crisis in agricultural production

To address the crisis, the Cuban government implemented changes in all sectors to reduce the negative impact on the national economy. During the early 1990s, severe social and economic changes were made in order to maintain the social guarantees of the government while simultaneously reconstructing the Cuban economy (Rosset and Benjamin, 1993; Enríquez, 2000).

Since the beginning of the crisis, the government attempted to reduce the negative impact of the lack of inputs for agriculture. Accordingly, national strategies were implemented to accelerate research and

**Table 9.3.** Comparison of agrochemical imports pre and post collapse of the socialist block. (Source: Rosset and Benjamin, 1993.)

| Agrochemical | 1989 imports | 1992 imports | Reduction (%) |
|---|---|---|---|
| Petroleum (Mt) | 13,000,000 | 6,100,000 | 53 |
| Fertilizers (Mt) | 1,300,000 | 300,000 | 77 |
| Chemicals (US$ millions) | 80 | 30 | 62 |

application in areas including biological control (Estrada and López, 1997; Fernández-Larrea, 1997), crop rotations and polycultures. Legumes as green manure, biofertilizers and other organic fertilizers were also considered crucial themes for researchers and national institutions (Funes, 1997). A dramatic change in mechanization was quickly implemented; the total number of tractors in the country was reduced to less than 30,000, while the number of ox teams rose to 300,000 (Trinks and Miedema, 1999).

There were also spontaneous responses among both urban and rural populations to the crisis in agricultural production and food availability. For example, urban and peri-urban agriculture reached unanticipated levels of production. In 1994, it was reported that 4200 t of vegetables were produced by urban and peri-urban agriculture, increasing to 480,000 t by 1998, with further increases predicted (Grupo Nacional de Agricultura Urbana, 1999). Similarly, between 1992 and 1993 small-scale production of rice also became significant. The National Rice Research Institute (Instituto de Investigaciones del Arroz, 1999) reported production of more than 150,000 t of rice on smallholdings disseminated over 100,000 ha nation-wide, estimated to contribute more than 53% of the nation's production of that crop.

It became apparent that a substantial amount of the food that people were consuming was not being accounted for in the official agricultural statistics. Plant breeders and policy makers recognized that small-scale producers, using their own knowledge and genetic resources from the informal system, were filling important gaps in agricultural productivity (J.R. Martin Triana, San José de las Lajas, 2001, personal communication). This realization was one of the factors that laid the foundation for some changes in Cuban plant breeding that were to occur over the coming years.

## Refocusing plant breeding efforts during the agricultural input crisis in Cuba

Pumpkin (*Cucurbita moschata.* Duch. ex Lam; hereafter 'pumpkin' always refers to this species) in Cuba is very popular for culinary and medicinal properties, taste, β carotene content as well as use in ceremonies in African religions. A prostrate growing habit with numerous, large lateral branches is a characteristic feature of pumpkin, with branches sometimes reaching lengths of more than 10–12 m (Lira, 1992; Wessel-Beaver, 1995). This growth pattern, as well as being a monoecious plant (separate female and male flowers borne on the same plant) favours cross-pollination (Guenkov, 1969). Pollination is carried out primarily by insects, especially honey bees (*Apis melifera*) since pumpkin pollen grains are too heavy to be transported by the wind (Withaker, 1962).

As a result of the input crisis, no chemical products were applied to Cuban pumpkin fields and artificial irrigation was greatly reduced. At the beginning of the special period the government planned to satisfy the demand for pumpkin by increasing the areas sown to this crop, more than 40,000 ha in 1993. However, yields dropped from 2–3 t ha$^{-1}$ in 1987, to 0.2–0.4 in 1993 (Ministerio de la Agricultura, 1987, 1988, 1989, 1992, 1993) along with decreasing input levels. The abrupt reduction in productivity resulted in pumpkins disappearing from every market and becoming an exotic vegetable in Cuba (Ríos *et al.*, 1994, 1996). The circumstances of the economic crisis made it evident that the conventional breeding model had assumed that varieties should be highly responsive to external inputs (agrochemicals and irrigation). When such inputs were no longer available, the yields of those varieties dropped substantially.

To address this situation a governmental committee composed of researchers and representatives of the Ministry of Agriculture analysed the causes of the decreased yield in pumpkin. Two reasons for the low yields were identified: (i) a lack of chemical inputs including pesticides and fuel for irrigation; and (ii) the degeneration of the commercial variety RG (a crookneck fruit shape), potentially capable of yielding 18–20 t ha$^{-1}$ with intensive application of external inputs (Ministerio de la Agricultura, n.d.). Because of its culinary quality and high yield potential under the conditions before the special period, RG was sown on more than 70% of the area planted with pumpkin across the whole island (Ríos, 1999).

In response to the conclusions of the governmental committee, a plant breeding programme was established in the National Institute of Agricultural Sciences (INCA) in collaboration with the Higher Pedagogical Institute for Professional and Technical Education. The principal goal was to provide seed of new varieties of pumpkin to the national seed company. Initially many efforts were made by the first author's multidisciplinary team to provide fertilizer, pesticides and artificial irrigation to their research plots on the experimental station. However those inputs were simply not available and pumpkins on research stations and in farmers' fields had to grow under abiotic and biotic stresses reflecting the reality of the new Cuban condition.

Plant breeding for abiotic and biotic stresses is common, but these stresses have usually been studied in isolation (Ríos, 1997) with plant breeders typically selecting the best genotypes on the basis of one type of stress with all other factors controlled. However, low-input growing conditions, such as those brought about by the economic crisis in Cuba, result in multiple stress interactions (Ríos, 1997, 1999). Therefore, a central research question was: Is it possible to select pumpkin varieties capable of good yields under the multiple stresses present in the growing environments of the special period in Cuba?

## Pumpkin Landraces: Breeding Material for the New Cuban Agriculture

Two approaches were taken to seeking appropriate pumpkin genetic material for the new Cuban conditions. The first was to explore the possibility of purchasing modern variety (MV) seed from international seed companies. Policy makers acquired different varieties from such companies for testing under Cuba's new, low-input conditions. The second approach was to test, under low-input conditions, a sample of landraces from the collaborative collecting mission of the Gene Bank of Gatersleben, Germany, and the Cuban Fundamental Research Institute in Tropical Agriculture. While evidence suggests that landraces of pumpkin and other species were important as sources of variability with adaptation to high-input Cuban agriculture in the 1980s (Moya, 1987; Hammer *et al.*, 1992), it was rare to find reference to landrace utility for plant breeding for low-input conditions.

Overall, the 20 MVs tested had very low yields, high pest and disease infestations and poor culinary quality (Ríos *et al.*, 1996). In contrast, within the pool of 33 landraces evaluated, favourable variation was found for some characteristics important to the new breeding programme.

### Landrace variability analysis

#### *Materials and methods*

Thirty-three pumpkin landraces from diverse sources collected during the 1980s (Table 9.4) were sown in the Fundamental Research Institute in Tropical Agriculture research plots in San Jose de Las Lajas, La Habana, Cuba, in 1988. The variety RG was included as a control because it was widely sown and well known (Ríos *et al.*, 1994). Growing conditions were in most ways typical of the special period with no agrochemical use and farmers' labour and their traditional management practices. Organic fertilizer in the form of readily available sugarcane processing by-product was applied, at the rate of 3 kg per plant. To avoid any confusion when harvesting the fruit, landrace plots were sown 8 m apart, all having a within-plot sowing distance of

**Table 9.4.** Thirty-three pumpkin landraces[a] and one control variety evaluated in the INCA breeding programme during the special period.

| Landrace | Collection site |
|---|---|
| H-909, H-900, H-1411, H-1345, H-1188, H-1388, H-1291, H-1377 | Holguin |
| IJ-1523 | Isla de la Juventud |
| LH-21, LH-1175, LH-30, Crisostomo | La Habana |
| SS-1130, SS-550 | Sancti Spíritus |
| PR-1189, PR-130, PR-40, PR-39 | Pinar del Río |
| C-1027, C-1029, C-1043 | Cienfuegos |
| G-1442 | Guantánamo |
| CA-711 | Ciego de Avila |
| VC-1118 | Villa Clara |
| LT-826, LT-828 | Las Tunas |
| U-1974, U-357, U-1461, U-2002, U-278, U-1665, RG (control) | Unknown |

[a]Collected as part of the collaborative programme between the Fundamental Research Institute in Tropical Agriculture and Gatersleben Gene Bank (Hammer *et al.*, 1992).

1 m between hills. For this evaluation each landrace was represented by 21 plants all located in one unreplicated plot.

In order to cluster the landraces' phenotypic variability, different characters such as percentage dry matter, placental zone diameter, leaf shape, leaf lobules, number of fruits per plant, and fruit characteristics including weight, calculated yield ha$^{-1}$, length, diameter, flesh diameter, shape, skin colour, flesh colour and skin texture were evaluated according to Esquinas and Gulick (1983). A matrix of landraces based on average values for quantitative and qualitative characteristics was analysed using factorial analysis.

### Findings

Among the landraces, 23 were capable of producing fruits in 80–110 days. The rest fruited after 120 days, and were therefore eliminated from further testing. Earliness is desirable because it allows farmers to produce two crops per year from the same piece of land instead of just one as is the case with longer cycle material. The materials producing ripe fruits in less than 120 days were seen as an important genetic source for earliness because some genotypes can take longer than 160 days to mature in the tropics (Lira, 1992).

Among the quantitative and qualitative characters evaluated in the pumpkin collection, fruit yield and fruit skin colour were predominant in classifying these materials under low-input conditions according to the factorial analysis. The contributions of these two traits allowed formation of six ranked clusters ranging from light to dark fruit skin colour and with fruit yields of less than 1.5 t ha$^{-1}$ to 15 t ha$^{-1}$ (Table 9.5).

Based on the morphological variability observed, the landrace collection seemed to have potential as a starting point for pumpkin breeding for the new Cuban conditions. However, there remained the question of whether landraces in and of themselves could contribute to increasing yields (e.g. Ceccarelli *et al.*, 1998). Our next research question addressed this, asking: Is it possible to increase pumpkin yields under the new Cuban conditions by developing new varieties based solely on landraces?

## Yield as a Selection Criterion

The economic crisis and its impact on research funds and facilities forced one Cuban plant breeder (HRL) to take promising pumpkin landrace families identified in the evaluation trial at the Fundamental Research Institute in Tropical Agriculture experimental station to farmers' fields much earlier than had originally been intended. Once there, he decided to enlist farmers' help in doing family selection within those landraces. Thus, to test the potential for increased fruit production by selecting families originating from landraces, some

**Table 9.5.** Pumpkin landrace clusters produced by a factorial analysis of the 1988 evaluation trials in the Fundamental Research Institute, Santiago de la Vegas, Cuba. (Source: Ríos, 1999.)

| Cluster description | Landraces |
| --- | --- |
| Very high yielding (15.7–7.0 t ha$^{-1}$) genotypes and fruits with green skin colour | H-1388 and LT-828 |
| High yielding genotypes (6.4–6.5 t ha$^{-1}$) and fruits with green skin colour | SS-550 and PR-130 |
| Medium yielding genotypes (3.4–1.6 t ha$^{-1}$) and fruits with brown, yellow or grey skin colour | U-1461, LH-30 and VC-1118 |
| Medium yielding genotypes (3.4–1.6 t ha$^{-1}$) and fruits with green skin colour | H-1411, IJ-1523, C-1043 and C-1029 |
| Low yielding genotypes (less than 1.5 t ha$^{-1}$) and fruits with brown, yellow or grey skin colour | LH-1175, H-909, H-900, H-1377 and LT-826 |
| Low yielding genotypes (less than 1.5 t ha$^{-1}$) and fruits with green skin colour | H-1188, CA-711, U-1974, C-1027, PR-1189, G-1442, PR-39, RG |

experiments, including farmers' participation in selection on farm, were carried out in the middle of the most difficult period of Cuba's input crisis.

## Materials and methods

In the summer of 1989 an experiment that developed into a collaboration between farmers and plant breeders was sown at the 28 de Septiembre Cooperative in Batabanó Municipality, Habana province. Plant breeders brought 13 half-sib families, each one representing a landrace that had been chosen for overall yield under low-input conditions at the experimental station as described above, and planted the experiment at the cooperative. The variety RG was again included as a control. The families were distributed in a randomized complete block design with three replications permitting free cross-pollination aided by the presence of a beehive 100 m from the experimental field. Each half-sib family plot was sown 8 m distant from the other plots, and all had a within-plot sowing distance between hills of 1 m, with 21 plants per half-sib plot as in the original landrace evaluation described above.

During the summer season characterized by heat stress, two farmers from Batabanó (males 58 and 70 years old) chose among the 13 half-sib families according to yield and fruit shape and selected individual fruits within those chosen families. Afterwards the chosen families, represented by the selected individuals, were sown at the same cooperative and at the INCA experimental station under low-input conditions during the cool, dry season (winter) and again in the hot, rainy season (summer).

Components of variation for half-sib progenies were calculated, including additive genetic variance. This permitted estimation of heritability and predictions of genetic response to selection for fruit weight, yield and number per plant, determined according to Galvez (1985), assuming a selected proportion of 40%.

## Findings

Among the 13 half-sib families presented, the farmers chose nine families under the low-input conditions in Batabanó in the summer season. Average estimations of predicted genetic response across families for yield and its components (fruit weight and number) tended to be superior in the winter season (Table 9.6), probably because the relatively low temperatures favourably influence fruit set (Guenkov,

**Table 9.6.**   Average predicted genetic response of nine half-sib pumpkin families to two cycles of selection estimated under low-input conditions. (Source: Ríos, 1999.)

| Selection environment (season and average fruit yield, t ha$^{-1}$) | Predicted genetic response for | | |
| --- | --- | --- | --- |
| | Fruit yield (t ha$^{-1}$) | Number of fruit per plant | Fruit weight (kg per fruit) |
| Hot (summer, 6.1) | 0.6 | 0.3 | 0.7 |
| Dry (winter, 3.2) | 1.5 | 1.5 | 1.1 |

1969). The response of number of fruits per plant tended to be low in the summer season, presumably due to the effects of high temperature on flowering and early fruit abortion (J. Casanova, La Habana, 1998, personal communication).

The genetic response to selection predicted in this experiment was inversely associated with environmental yield, so high genetic response occurred in the low-yielding environment of the winter season and vice versa. In both growing environments, fruit yield had a genetic response superior or similar to its components, indicating that in half-sib families, some genetic advance could be obtained by direct selection for yield under low-input conditions.

Working on farm, with farmers and Cuban pumpkin landraces provided two important insights to pumpkin breeding for the input crisis: (i) wide phenotypic variability of useful traits exists and has been documented among Cuban landraces grown under low-input conditions; and (ii) it is possible to increase production by selecting directly for fruit yield under low-input conditions (see Ceccarelli and Grando, Chapter 12, this volume).

## Participatory Plant Breeding for Multiple Stress Tolerance

To specifically evaluate the pumpkin material selected by farmers in terms of response to the multiple stresses of the agricultural conditions in Cuba during the special period, a further experiment was conducted.

### Materials and methods

The performance of the nine half-sib families chosen by the two farmers and the control variety (RG) were evaluated by the same two farmers and two breeders in the summer and winter seasons as described above. The farmers' selection criteria were obtained through informal interviews at the end of each harvesting period (Ríos, 1999). For each of the nine half-sib families, breeders measured variation in fruit yield in relation to drought and heat stress (Fig. 9.1, Table 9.7). These data permitted estimation of a stress tolerance index (STI) according to Fernández (1992): STI = [(highest yield under stress A) × (highest yield under stress B)]/[yield average under stress most limiting yields (either A or B)]$^2$.

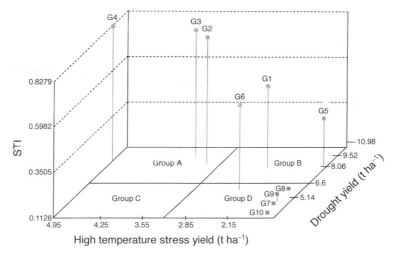

**Fig. 9.1.**   Response to abiotic stresses (high temperature and drought) of lines selected from landraces as measured by their stress tolerance index (STI) (Fernández, 1992). (Source: Ríos *et al.*, 1998a.)

**Table 9.7.**   Codes used in Figs 9.1 and 9.2 for selected lines and their landrace source. (Source: Ríos, 1999.)

| Code | Landrace source of lines selected from half-sib family |
|------|--------------------------------------------------------|
| G1   | IJ-1523                                                |
| G2   | IJ-1523                                                |
| G3   | PR-130                                                 |
| G4   | PR-130                                                 |
| G5   | LT-828                                                 |
| G6   | PR-1029                                                |
| G7   | LT-828                                                 |
| G8   | SS-550                                                 |
| G9   | SS-550                                                 |
| G10  | RG (control, commercial variety)                       |

## Findings

Based on their STIs, the nine half-sib families were classified into the following groups:

- Group A: perform favourably under both high temperature and drought stresses.
- Group B: perform favourably under high temperature stress and poorly under drought stress.
- Group C: perform poorly in high temperature stress and favourably under drought stress.
- Group D: perform poorly under both stresses.

During the drought stress trial in the winter season when diseases are most limiting in pumpkin, the infection indices of silvering leaves (Paris *et al.*, 1987) and downy mildew (*Pseudoperonospora cubensis* Ber.Cur) (CIBA-GEIGY, 1984) under highly infectious natural conditions were plotted against half-sib family average yield, to identify any variation in their responses. Based on this criterion families were classified as follows (Fig. 9.2):

- Group A: half-sib families having high infection index for both downy mildew and silvering leaves.
- Group B: half-sib families having high infection index for downy mildew and low for silvering leaves.

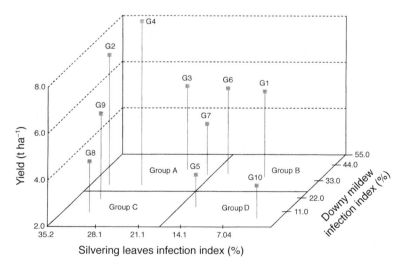

**Fig. 9.2.** Response to biotic stresses (silvering leaves and downy mildew) of lines selected from landraces as measured by yield under infection. (Source: Ríos *et al.*, 1998a.)

- Group C: half-sib families having low infection index for downy mildew and high for silvering leaves.
- Group D: half-sib families having a low index for both diseases.

Farmer-chosen families included the response group A or B (Fig. 9.1), except for one half-sib from PR-130 (see Table 9.4) that had high yield under high abiotic stress conditions but its shape and quality did not satisfy farmers' criteria. The farmers were particularly interested in materials with a good response under summer conditions because production is more economical when summer rainfall eliminates the need for irrigation.

The inclusion of a control variety recommended by breeders was an important component of this experiment. Because of this, farmers were able to compare the new material chosen by them with the material that they usually sow in the area. Sometimes extension workers complain about the difficulties of convincing farmers to adopt new materials; however, providing a clear comparison between what they are currently using and potential new materials provided a valuable discussion point for farmers and extension workers.

A superior performance under biotic stress was shown by another PR-130 half-sib and one from IJ-1523, giving higher fruit yield, despite presenting the highest leaf infection index for powdery mildew and silvering leaf (Fig. 9.2). Indeed, the farmers did not pay attention to leaf infection, because their selection criteria focused on fruit yield and shape, stating that neither disease is important in terms of the quality and yield of fruit.

Formal researchers have two possible explanations for the performance of these two half-sib families. First, it may be that the yield potential of these genotypes is so high that they produce superior phenotypes despite the presence of adverse conditions (Blum, 1988; M. Hernandez, San Jose de las Lajas, 1998, personal communication). Second, these genotypes may be using alternative physiological strategies that allow them to tolerate disease infestation and still produce desirable harvests (Seidel, 1996). As of yet the actual mechanism(s) for the success of these genotypes has not been identified. Still, in response to the new circumstances in Cuban agriculture, Cuban breeders and other scientists accepted both genotypes for release through the formal system, representing an agreement between farmers and scientists on the most desirable pumpkin material.

Evaluation in the target growing environment made it possible to make use of both farmers' and scientists' capacity to identify families tolerant to drought, powdery mildew and silvering leaf interaction. It was also observed that the two participating farmers recognized the possibility of identifying materials with long necks and satisfactory yield under conditions of low external inputs.

The research described above indicated ways in which to improve the breeding process for pumpkin varieties more appropriate for the new conditions of Cuban agriculture. However, the limitations of typical seed production methods developed under the high-input conditions of the 1980s suggested another area that might require rethinking in order to be better adapted to the current realities of Cuban agriculture. An investigation concerning this is the subject of the following section.

## Seed Production and its Relationship to Plant Breeding

One of the fundamental distinctions between formal and informal or traditionally based systems of 'conservation' and plant breeding in terms of their biological consequences is the organization of these processes (de Boef and Almekinders, 2000). As noted by Smale *et al.* (1998) and Soleri *et al.* (Chapter 2, this volume), in formal systems genetic resource conservation, plant improvement and seed production are physically and temporally isolated from one another, while in traditionally based systems these tasks are accomplished simultaneously within the same populations and growing environments. The differences between these two systems could potentially lead to different genetic effects. A similar contrast exists between formal and informal pumpkin breeding and seed production in Cuba. Another part of this research has been testing the hypothesis that one aspect of the organization of seed production (varietal isolation and selfing) has negative biological consequences in comparison with the process of seed production used by the informal system.

A central focus of the formal seed production system in Cuba is maintaining the principal varietal characteristics on release of the variety. The formal sector attempts to maintain the varietal characteristics developed by breeding through continued roguing of off types and complete isolation to maintain trueness to type during seed production (Ministerio de la Agricultura, n.d.). Indeed, early in the special period it was believed that pumpkin yield depression was caused by carelessness in maintaining seed purity during the multiplication phase due to pollen contamination, which resulted in negative deviations from the expected variety response trend. To achieve the desired 'purity' and to ensure high seed production, large amounts of seed are multiplied in optimum conditions outside the target environment made possible through high rates of application of external inputs.

To test the hypothesis that isolation and selfing might in fact have a negative impact on varietal productivity compared with the seed production methods used by the informal system, the two models were examined. The formal model, typically used by the formal seed system in Cuba, separates recombination and selection temporally and spatially from seed production, and the informal model, used by farmers in Cuba, combines recombination, selection and seed production into one simultaneous process (Ríos *et al.*, 1998b).

## Methods

To represent the formal model, seeds from two half-sib families, Marucha and Fifí (the commercial names of the half-sib families G3 and G4, respectively, see Table 9.7), were each multiplied in isolation without any other pumpkin genotypes within 1000 m. Seeds obtained from individuals within each variety characterized by favourable yield and fruit quality were sown separately, avoiding any mixing with extraneous pollen, for two multiplication cycles, one in winter and the other in the summer, with a population size of 2500–2700 plants for each variety in both seasons per cycle. Typical fruits of each variety were selected as seed sources in each cycle.

The informal model was represented by a common approach among Cuban farmers (e.g. Ríos *et al.*, 1998b, for pumpkins; Ríos and Almekinders, 1999, for maize) in which different varieties are grown in close proximity and their seeds saved. In this experiment, Marucha and Fifí half-sibs were grown among eight other pumpkin half-sibs selected from landraces (three half-sibs from the same landrace that Marucha and Fifí originally came from, PR-130, the other five were derived from different landraces). Again, typical fruits of each of the two varieties were selected as seed sources for two multiplication cycles, one in winter and the other in summer, with a population size of 50 plants for each variety in both seasons in each cycle.

The seeds of Marucha and Fifí obtained by simulating the two models of seed production were compared in a randomized complete block design under low-input growing conditions during the summer and winter, with a population size of 100 plants for each model in each season.

## Findings

The pumpkin seed multiplied using the informal model showed higher yields than those produced using the formal model across summer and winter seasons (Table 9.8). As was found in the earlier selection experiment (see Table 9.6), the yield averages are lower during the

**Table 9.8.** Comparison of yields of two pumpkin varieties when seeds are produced using the formal vs. informal seed production models. (Source: Ríos *et al.*, 1998b.)

| | Varietal yield (t ha⁻¹) | |
|---|---|---|
| Seed production model | Marucha | Fifí |
| Summer sowing period (heat stress) | | |
| Formal model | 5.2 | 3.1 |
| Informal model | 7.9 | 4.1 |
| Yield gain using informal model[a] | 35% | 26% |
| Winter sowing period (drought stress) | | |
| Formal model | 3.2 | 1.1 |
| Informal model | 5.9 | 2.6 |
| Yield gain using informal model[a] | 46% | 58% |

[a]Yield gain = [1-(yield from formal model/yield from informal model)] × 100.

winter; however, response to selection for increased yield is higher compared with the summer.

A lineal scheme of varietal development and seed production and dissemination for cross-pollinated crops has been followed in Cuba. This scheme divides the process into two parts: in the first part breeders play a key role in selection and recombination at the experimental station, the second part is accomplished by the Cuban Seed Company – through its seed growers' network – which promotes strict isolation in seed production.

Initially, according to the formal model, recombination is encouraged through cross-pollination between families at the experimental station. However, at the end of the varietal development process on the experimental station, isolation is rigorously enforced with cross-pollination only permitted within families. Once a variety has been released from the research institute the Cuban Seed Company conducts a first multiplication and evaluation, receiving feedback from the seed growers.

In contrast, in Jalisco, Mexico, the informal system for another cross-pollinated crop species (maize) has been able to maintain the population characteristics of greatest interest to farmers by selection while still tolerating continual recombination of populations in the field (Louette and Smale, 2000). Similarly, farmers in a community in central Chiapas, Mexico, were reported to manage more than nine races of maize using the informal system (Bellon and Brush, 1994). Indeed there are some communities in Cuba where maize varietal maintenance and seed production are integrated as described above, and the populations produced have shown promise in terms of yield under low-input conditions (Ríos and Wright, 1999).

It will be important to investigate more carefully the potential for the informal system to improve, as opposed to maintain, their crop varieties. Still, it seems likely that continual recombination in the informal system has played an important role in genetic resource conservation on farm for some cross-pollinating species, and as such may have maintained the genetic potential to respond to ongoing or novel stresses.

Population improvement as a plant breeding approach attempts to improve the mean performance of the population while maintaining genetic variability in the population to facilitate long-term selection (Simmonds, 1979). The informal system may represent population maintenance or perhaps improvement that in itself could offer a more viable alternative for pumpkin improvement in Cuba than the current strategy.


## Conclusions

As emphasized throughout this chapter, the economic crisis in Cuba after the socialist countries collapsed has been the key factor favouring low consumption of conventional agrochemicals. This change has stimulated discussion of the efficiency, advantages and weaknesses of chemical compared with organic inputs (Altieri, 1998). This discussion has also been applied to approaches to plant breeding in the country. For example, Table 9.9 shows a comparison of plant breeding practices under high-input agriculture and those described here, carried out under low-input conditions. The comparison is made in terms of energy consumption, inputs used on farm and farmers' participation.

The collaborative effort towards crop improvement under low-input conditions was more efficient in terms of energy use. And, notably, the yield obtained under the low-input system was comparable to yields under the conventional, high-input technological package.

The economic impact of varieties selected under high- compared with low-input conditions was evaluated by growing the most popular variety disseminated in Cuba (RG) under low-input conditions, as well as two varieties selected collaboratively with farmers under those low-input conditions (Table 9.10).

The situation that occurred in Cuban pumpkin breeding seems to be an example of the possible negative economic effects when varieties are selected in an environment not representative of the target area. The occurrence of a cross-over response (Ceccarelli, 1994; see Ceccarelli

**Table 9.9.** Comparison of input use and results of pumpkin breeding strategies in Habana Province. (Source: Ríos, 1999.)

| Indicators | Before the special period (1980s) | Under conditions of the special period (1990s) |
|---|---|---|
| Mineral fertilization (kg ha⁻¹) | Nitrogen: 42 Phosphorus: 39 Potassium: 62 | 0 |
| Organic soil amendments | Rarely applied | Typically 6–7 t ha⁻¹ |
| Frequency and amount of artificial irrigation (summer season) | 9–11 times per season, 2000 m³ ha⁻¹ | 2–4 times per season, 200 m³ ha⁻¹ |
| Number of varieties released in 10 years | 1 | 2 |
| Varietal maintenance and seed multiplication | Isolation | Cross-pollination |
| Pest and disease control | Agrochemical intensive | Biological |
| Use of honey bees | Sporadic | Frequent |
| Yield | 6–8 t ha⁻¹ | 6–8 t ha⁻¹ |
| Farmer participation | Contracted seed production | On-farm selection of half-sib families |
| Researcher participation | Screening germplasm and varietal evaluation and selection. Cross-pollination control | Screening germplasm, facilitating availability of new germplasm, evaluation of variety with farmers |
| Energy requirements (kcal ha⁻¹)[a] | Fertilization:  679,000 Irrigation:  10,160,000 Pesticides:  6,160,000 Total:      16,999,000 | Fertilization:  42,000 Irrigation: 3,697,200 Pesticides:   88,000 Total:     3,827,200 |

[a]Pimentel, 1984; Masera and Astier, 1993.

and Grando, Chapter 12, this volume) suggests the importance of having a realistic view about who will be using the products of plant breeding. In the case of pumpkins, the effect of sowing in low-input conditions varieties selected under high-input conditions, with the genetic consequences of that selection strategy perhaps reinforced by restrictions on recombination early in the breeding cycle, has meant an inefficient use of energy as well as an economic loss.

During the collaborative pumpkin breeding described here, the farmers and the informal sector they are a part of offered: (i) pumpkin landraces that showed wide phenotypic variability for useful traits under low-input conditions; and (ii) selection skills and a real

**Table 9.10.** Economic impact of pumpkin breeding under low-input conditions. (Source: Ríos, 1999.)

| Indicators (calculated as averages) | Varieties bred under high-input conditions sown in low-input conditions | Varieties bred and sown under low-input conditions |
|---|---|---|
| Cost ha$^{-1}$ under low-input conditions (Cuban pesos) | 702.3 | 708.3 |
| Fruit yield (t ha$^{-1}$) | 1.5 | 6.7 |
| Total income ha$^{-1}$ (0.16 Cuban pesos kg$^{-1}$) | 240 | 1080 |
| Net income ha$^{-1}$(Cuban pesos) | −462[a] | 372 |
| Benefit/cost ratio | 0.34 : 1 | 1.5 : 1 |

[a]Average net loss.

low-input situation including interacting biotic and abiotic stresses and socio-economic constraints and thus the possibility of obtaining advances in characters with complex inheritance such as yield.

In the pumpkin experiences described here, plant breeders offered: (i) a bridge between the plant genetic resources conserved in gene banks and farmers, and the opportunity to screen those resources; and (ii) experimental design, an estimation of plant genetic variability, and estimations of potential genetic advance in a complex character (yield), monitoring variability in time and space.

The advent of low-input agriculture in Cuba, necessitated by the collapse of the Soviet Union, led to some new initiatives in Cuban plant breeding. Clearly, farmers' agricultural knowledge and skills, including those relating to seed management under the pressure of adverse environmental conditions, were an inspiration to develop a new, collaborative approach towards a more efficient use of inputs such as energy, more profitable crop production and maintenance of greater genetic diversity *in situ*.

# Acknowledgements

Ríos Labrada thanks the two farmers who supported the collaborative work described in this chapter, Pedro Romero, the former president of 28 de Septiembre Cooperative, and Tata who was in charge of the experimental area; and Guadalupe Rodriguez Gonzales for her help as cooperative engineer during the entire experimental period.

# References

Altieri, M. (1998) Farmers, NGOs and lighthouses: learning from three years of training, networking and field activities. UNDP-Sponsored-project report INT/INT/93–201:73.

Begemman, F., Oetmann, A. and Esquivel, M. (2000) Linking conservation and utilization of plant genetic resources in Germany and Cuba. In: Almekinders, C. and de Boef, W. (eds) *Encouraging Diversity*. Intermediate Technology, London, pp. 103–112.

Bellon, M. and Brush, S. (1994) Keepers of maize in Chiapas. *Economic Botany* 48, 196–209.

Blum, A. (1988) *Plant Breeding for Stress Environment*. CRC Press, Boca Raton, Florida.

Castiñeiras, L. (1992) Germoplasma de *Phaseolus vulgaris* L. en Cuba: colecta, caracterización y evaluación. PhD thesis, Universidad Agraria de La Habana, Cuba.

Ceccarelli, S. (1994) Specific adaptation and breeding for marginal conditions. *Euphytica* 77, 2005–2019.

Ceccarelli, S., Grando, S. and Impiglia, A. (1998) Choice of selection strategy in barley for stress environments. *Euphytica* 103, 307–318.

CIBA-GEIGY (1984) *Manual para ensayos de campo en protección vegetal*. CIBA-GEIGY, Basel, Switzerland.

De Boef, W. and Almekinders, C. (2000) Introduction. In: Almekinders, C. and de Boef, W. (eds) *Encouraging Diversity*. Intermediate Technology, London.

Deere, C. (1992) *Socialism on One Island? Cuba's National Food Programme and its Prospects for Food Security*. Institute of Social Studies, The Hague, Working Paper Series No. 124.

Enríquez, J.L. (2000) Cuba's New Agricultural Revolution. The Transformation of Food Crop Production in Contemporary Cuba. Development Report 14, Department of Sociology, University of California.

Esquinas, J.T. and Gulick, P.J. (1983) *Genetic Resources of Cucurbitaceae*. A Global Report. IPBGR, Rome.

Esquivel, M. and Hammer, K. (1992) Contemporary traditional agriculture – structure and diversity of the Conuco. In: Hammer, K., Esquivel, M. and Knüpffer, H. (eds) '. . . y tienen faxones y fabas muy diversos de los nuestros . . .' [Origin, Evolution and Diversity of Cuban Plant Genetic Resources] Vol. 1. Institut für Planzengenetik und Kulturplazenforschung, Gatersleben, Germany.

Esquivel, M., Castiñeiras, L., Shagarodsky, T., Moreno, V., Perez, E. and Barrios, O. (1994) The Agronomic Experimental Station of Santiago de las Vegas: 90 years of conservation and study of plant genetic resources. *Plant Genetic Resources Newsletter* 99, 6–14.

Estrada, J. and López, M.T. (1997) Los bioplaguicidas en la agricultura sostenible cubana. *Proceedings of Conferencias III Encuentro Nacional de Agricultura Orgánica. Villa Clara, Cuba*, p. 7.

Fernández, C.J.G. (1992) Effective selection criteria for assessing plant stress tolerance. In: Kuo, C.G. (ed.) *Adaptation of Food Crops to Temperature and Water Stress. Proceedings of an International Symposium, 13–18 August, Taiwan*, pp. 257–270.

Fernández-Larrea, O. (1997) Microorganismos en el control fitosanitario en Cuba. Tecnologías de producción. *Proceedings of Conferencias III Encuentro Nacional de Agricultura Orgánica, Villa Clara, Cuba*, p. 19.

Funes, F. (1997) Experiencias cubanas en Agroecologia. *Revista Agricultura Orgánica*, August–December, 3–7.

Galvez, G. (1985) Parámetros genéticos y estadísticos en genética cuantitativa. Disenos geneticos-estadisticos para la descomposicion de la varianza genetica. In: Cornide, M.T., Lima, G., Sigarroa, A. and Galvez, G. (eds) *Genética Vegetal y Fitomejoramiento*. Científico Técnica, La Habana, Cuba.

Grupo Nacional de Agricultura Urbana (1999) Lineamientos para los subprogramas de la agricultura urbana. Ministerio de la Agricultura, La Habana.

Guenkov, G. (1969) *Fundamentos de Horticultura Cubana*. Ciencia y Técnica, La Habana, Cuba.

Hammer, K., Esquivel, M. and Knüffer, H. (1992) *Evolution and Diversity of Cuban Genetic Resources*, Vols I and II. Institut für Planzengenetik und Kulturplazenforschung, Gatersleben, Germany.

Instituto de Investigaciones del Arroz (1999) *Informe Técnico del Cultivo del Arroz*. Ministerio de la Agricultura, Cuba.

Lira, R. (1992) Cucurbitas. In: *Cultivos Marginados. Otra perspectiva de 1942*. FAO, Rome, pp. 37–42.

Louette, D. and Smale, M. (2000) Farmers' seed selection practices and traditional maize varieties in Cuazalapa, Mexico. *Euphytica* 113, 25–41.

Masera, O. and Astier, M. (1993) Energía y sistema alimentario en Mexico: aportaciones de la agricultura alternativa. In: Masera, O. and Astier, M. (eds) *Agroecologia y Desarrollo en Mexico*. Universidad Autonoma Metropolitana de Xochimilco, Mexico.

Ministerio de la Agricultura (1987) *Boletin Informativo*. Agricultura no cañera. Ministerio de la Agricultura, Cuba.

Ministerio de la Agricultura (1988) *Boletin Informativo*. Agricultura no cañera. Ministerio de la Agricultura, Cuba.

Ministerio de la Agricultura (1989) *Boletin Informativo*. Agricultura no cañera. Ministerio de la Agricultura, Cuba.

Ministerio de la Agricultura (1992) *Boletin Informativo*. Agricultura no cañera. Ministerio de la Agricultura, Cuba.

Ministerio de la Agricultura (1993) *Boletin Informativo*. Agricultura no cañera. Ministerio de la Agricultura, Cuba.

Ministerio de la Agricultura (n.d.) *Normas técnicas para el cultivo de melon calabaza y pepino*. Ministerio de la Agricultura, Cuba.

Moya, C. (1987) Estudio de la variabilidad agromorfológica en tomate (*Lycopersicon esculentum* Mill.) y sus uso en la selección de líneas adaptadas a las condiciones tropicales. PhD thesis, Universidad Agraria de La Habana, Cuba.

Oficina Nacional de Estadísticas (1997) *Anuario Estadistico de Cuba*. Habana, Cuba.

Paris, H.S., Nerson, H. and Burger, Y. (1987) Leaf silvering of *Cucurbita*. *Canadian Journal of Plant Sciences* 67, 593–598.

Pastor, J. (1992) *External Shocks and Adjustment in Contemporary Cuba.* Working Paper of the International & Public Affairs Centre, Occidental College, California, USA.

Pérez Marín, E. and Muñoz Baños, E. (1992) Agricultura y alimentación en Cuba. *Agrociencia, serie Socioeconómica* May–Agosto, 2.

Pimentel, D. (1984) Energy low in the food systems. In: Pimentel, D. and Hall, C.W. (eds) *Food and Energy Resources.* Academic Press, New York.

Ríos, H. (1997) Cuban pumpkin genetic variability under low input conditions. *Cucurbits Genetics Cooperative Report* 20, 48–49.

Ríos, H. (1999) Selección de cultivares de calabaza (*Cucurbita moschata.* Duch. ex Lam.) para condiciones de bajos insumos. PhD thesis, Universidad Agraria de La Habana, Cuba.

Ríos, H. and Almekinders, C. (1999) El sector informal perserva la variabilidad y el rendimiento del maíz en Cuba. In: *Proceeding Fitomejoramiento Participativo en America Latina y el Caribe.* Programa de Investigación Participativa y Género del GCIA. http://www.prgaprogram.org/prga

Ríos, H. and Wright, J. (1999) Early attempts at stimulating seed flows in Cuba. *ILEA Newsletter* 15, 38–39.

Ríos, H., Batista, O. and Ramírez, F. (1994) Comportamiento de genotipos de calabaza cultivados en la localidad de Batabanó. *Cultivos tropicales* 15, 84–88.

Ríos, H., Fernández, A. and Casanova, E. (1996) Caracteristicas y potenciali-dades del germoplasma cubano de calabaza (*Curcubita moschata.* Duch). *Cultivos Tropicales* 17, 88–91.

Ríos, H., Fernández, A. and Casanova, E. (1998a) Tropical pumpkin (*Curcurbita moschata*) for marginal conditions: breeding for stress interactions. *Plant Genetic Resources Newsletter* 118, 4–7.

Ríos, H., Perera, Y. and Fernández, A. (1998b) Effectiveness of the informal seed sector for increasing yield of pumpkins developed under low input conditions. *Cucurbits Genetics Cooperative Report* 21, 23–24.

Rosset, P. and Benjamin, M. (1993) *The Greening of the Revolution, Cuba's Experiment with Organic Farming.* Ocean Press, Melbourne.

Seidel, P. (1996) Tolerance responses of plant to stress: the unused reserve in plant protection? *Plant Research and Development* 4, 81–100.

Simmonds, N.W. (1979) *Principles of Crop Improvement.* Longman, London.

Smale, M., Soleri, D., Cleveland, D.A., Louette, D., Rice, E.B. and Aguirre, A. (1998) Farmers' collaborative plant breeding as an incentive for on farm conservation of genetic resources: economic issues from studies in Mexico. In: Smale, M. (ed.) *Gene Banks and Crop Breeding. Economic Analyses of Diversity in Wheat, Maize, and Rice.* Kluwer Academic Publisher, Norwell, Massachusetts, pp. 239–257.

Trinks, M. and Miedema, J. (1999) Cuban experiences with alternative agri-culture. MSc thesis, Communication and Innovation Studies Group, Wageningen Agricultural University, The Netherlands.

Wessel-Beaver, L. (1995) Broadening the genetic base of *Cucurbita* spp.: strategies for evaluation and incorporation of germplasm. *Cucurbitaceae* 94, 69–74.

Withaker, T.W. (1962) *Cucurbits: Botany, Cultivation and Utilization.* Davis, Leonard Hill, New York.

# Participatory Plant Breeding in Rice in Nepal

# 10

## Krishna D. Joshi,[1]* Bhuwon Sthapit,[2] Madhu Subedi[1] and John R. Witcombe[3]

[1]Local Initiatives for Biodiversity, Research and Development (LI-BIRD), PO Box 324, Pokhara, Nepal; [2]International Plant Genetic Resources Institute (IPGRI), Asia, Pacific-Oceania, Buddha Marg, Pokhara, Nepal; [3]Centre for Arid Zone Studies, University of Wales, Bangor, Gwynedd LL57 2UW, UK

## Abstract

We describe here work over many years both in high-altitude villages of Nepal and in the lowland area of the Nepal *terai* on participatory crop improvement in rice. We focus on the interactions of farmers and scientists and the contributions each partner makes in participatory research. This has encompassed contractual, consultative, collaborative and collegial modes of participation. Since the goal of the programme was functional, that is, to improve efficiency of the breeding process, rather than the empowerment of farmers, the first three of these four modes of participation were the most important. Farmers contributed in goal setting, in identifying traits and in providing a testing system that was multi-farmer, multi-locational and allowed the trade-off between many traits. Breeders contributed their more formal scientific knowledge to this process and assisted in the scaling up of products identified from the participatory varietal selection (PVS) and participatory plant breeding (PPB) programmes. Breeding methods were adapted to accommodate the opportunities and constraints of PPB, and used a low-cross-number, high population-size breeding strategy. Within this strategy, modified bulk population breeding has been used extensively. The PVS and PPB programmes have identified or produced many varieties that farmers prefer. Their adoption is spreading through farmer-to-farmer networks. Governmental and non-governmental organizations have helped this process by leading more formal approaches to seed supply. The government extension service is now scaling up the PVS approach in several districts of Nepal.

---

* Present address: CIMMYT – South Asia Regional Office, PO Box 5186, Kathmandu, Nepal.

## Why the Need for Participatory Crop Improvement?

In conventional plant breeding, the development of varieties is the responsibility of plant breeders; farmers are recipients of varieties only when they have been released by scientists and included into the official recommendation list of the agricultural extension system. Our participatory research questions the assumptions that underlie this conventional approach:

- Farmer participation is unnecessary, because farmer involvement is costly, cumbersome and farmers are less knowledgeable than university-trained plant breeders (e.g. Bhandari, 1997).
- Varieties with high local adaptation are less desirable than varieties with wider adaptation; breeders should select from multi-locational trials only those varieties that perform well across locations.
- Selection must be done under optimal conditions where heritabilities, and hence responses to selection, are assumed to be higher (discussed by Ceccarelli *et al.*, 1996).
- Any low adoption of modern crop varieties (MVs) is due to ineffective extension, the insufficient supply of quality seed, or to the inefficient agronomic practices employed by farmers, and not due to deficiencies in the released crop varieties.
- Seed supply should only be in the domain of the formal sector, which is more efficient than farmer-led systems and the only possible source of quality seed; the informal seed supply system involves farmers in unnecessary risks as they have limited capacities and expertise.

As a result of these common assumptions, varietal testing and release systems have often involved little participation by farmers. It is, hence, unsurprising that the adoption of modern cultivars has remained low in many marginal environments. In response to these deficiencies, Chambers (1989) and Maurya (1989) argued the case for involving farmers more in variety selection and extension, particularly in more marginal environments defined as being complex, diverse and risk-prone. Witcombe (1999) later extended the arguments to high potential production systems (HPPSs).

The research we describe here on participatory crop improvement (PCI) in Nepal questioned the common assumptions underlying varietal release. The theory behind our main reason for involving farmers was simple: farmers know best what they need and they make the ultimate decision on what is adopted. To ignore local knowledge and farmers' preferences is very inefficient and modern cultivars are often rejected by farmers because of traits that have not been considered in the breeding process (Chemjong *et al.*, 1995; LARC, 1995).

## Participatory Varietal Selection and Participatory Plant Breeding

In the literature, participatory varietal selection (PVS) was the first method used for varietal improvement. This was an amalgamation of the terminology of Sperling *et al.* (1993) who used 'varietal selection' (a common term in the literature, almost as common as the more popular 'variety selection') for the process of selecting varieties, and 'participatory selection' for the participatory approach to varietal selection. We adopted the term 'participatory varietal selection' since it is a clearly defined activity that can:

* be included within the breeding process as a whole (when participatory plant breeding (PPB) is used in its broad sense by, e.g. Weltzien *et al.*, 2000); or
* be separated from the breeding process when PPB is used in its narrower sense (e.g. Witcombe *et al.*, 1996). Here, PVS is an activity that takes place after the breeding process – the generation of new variation and the selection of varieties from it – is complete.

We prefer to use PVS and PPB to describe separate activities for many reasons (Witcombe *et al.*, 1996). PVS and PPB differ dramatically in their resource requirements and the time they require. Moreover, as can be seen below from our experiences in Nepal and elsewhere, PVS is not an exclusive activity of plant breeders but is, indeed, a very appropriate activity for extension workers and development organizations that have no expertise in plant breeding *per se*. The term PVS is now widely used and many find the term understandable and a useful distinction from PPB. In practice there is rarely, if ever, any difficulty in making a division between PVS and PPB as this involves the basic distinction between choice and selection in plant breeding (see Chapter 1, this volume).

## The Research on PPB and PVS

The work in Nepal was carried out by the Lumle Agricultural Research Centre (LARC) and by the non-governmental organization, the Local Initiatives for Biodiversity, Research and Development (LI-BIRD). LARC, along with other research efforts in Nepal supported by DFID (Department for International Development, UK), had a long history of using participatory methods and was instrumental in developing new methods for on-farm research (Pound *et al.*, 1988). One example was the development of informal research and development (IRD) techniques (Joshi and Sthapit, 1990).

The lessons on farmer participation that we describe are from two separate breeding programmes: one for high-altitude, partially irrigated rice, the other for low-altitude, irrigated rice grown in the *terai* of Nepal.

## High-altitude rice

The adoption of rice varieties released for high altitudes had been disappointingly low. In search of a more effective approach, plant breeders of LARC decided in 1985 to work directly with farmers and involve them at an earlier stage of selection. At this time, LARC was collaborating with the Centre for Arid Zone Studies (CAZS), University of Wales, which was also concerned with developing methods of participatory research (Joshi and Witcombe, 1996). High-altitude rice breeding was initiated on-farm in the two high-altitude villages of Chhomrong and Ghandruk. In Chhomrong existing rapport with farmers strengthened the case for participatory research. When the PPB programme started, LARC had an employee based in the village who knew all of the villagers well. Farmers were already involved in the LARC farming system research programme (land was rented from farmers, and they were employed to cultivate the crop). Farmers collaborated in the breeding research because it gave them access to new germplasm and they wanted a better rice variety. Detailed materials and methods on the high-altitude rice breeding programme are reported in Sthapit *et al.* (1996).

## Rice in the *terai*

Several years after the high-altitude rice breeding programme, LI-BIRD and CAZS commenced research on PVS and PPB for the high potential production systems of the *terai*, a lowland region of Nepal that borders on India. Although the uptake of improved varieties was high, many of the varieties that farmers were growing were extremely old. We wished to test if PCI would be effective in a high potential production system, because the use of such methods in favourable environments had never been well documented and there were only rare mentions of this approach in less formal (grey) literature. Detailed material and methods on the research in the *terai* are reported in Witcombe *et al.* (2001).

This is a review of our work on PVS and PPB in Nepal focusing on the knowledge and practice of farmers and scientists. We first look at the types of participation we used and how they varied according to process and objective. We review the lessons we have learnt on how farmers' knowledge differs from that of plant breeders, and how it can contribute to improved efficiency of the breeding programme. We then review the theory underlying our participatory research and how our understanding of plant breeding has changed in theory and practice. We briefly review some of the important results of our work, and its future directions.

## Types of Participation

The terms used for what we call participatory crop improvement (PCI) vary in the literature. In PCI we include all aspects of crop improvement such as agronomic improvements and participatory approaches to aid local seed supply systems. Apart from this innovation, we avoided creating new terms by following published precedents.

Use of the term 'participatory' to describe the involvement of primary stakeholders in a process has a long, well-established history. The earliest references (e.g. Arnstein, 1969) in the late 1960s were based on citizen participation in urban development and renewal in the USA (Table 10.1). For agricultural research, participation became associated with farming systems research (FSR) in the 1980s (Farrington, 2000). It became an integral part of a sustainable development strategy and

**Table 10.1.** Types of participation.

| Author | Types | Notes |
|---|---|---|
| Arnstein (1969) | Ladder of citizen participation: (1) Manipulation, (2) Therapy, (3) Information, (4) Consultation, (5) Placation, (6) Partnership, (7) Delegated power and (8) Citizen control | Not devised for agricultural research |
| Biggs (1989) | Farmer participation in agricultural research was divided into four types: contractual, consultative, collaborative and collegial | Most widely used types in agricultural research |
| White (1996) | Four types: normal participation, consultative participation, action-oriented participation, decision-making/ design participation | Correspond precisely to those of Biggs (1989) but have different names |
| Pretty (1998) | Seven types: passive participation, participation resulting in information transfer, participation by consultation, participation for material incentives, functional participation, interactive participation and self-mobilization | Closest in agricultural research to that of Arnstein (1969). However, while some types match those of Arnstein (1969) others match those of Biggs (1989) |

is widely accepted within the United Nations (UN) and among international donor organizations (Friis-Hansen and Sthapit, 2000). The classification of Biggs (1989) was the first to be widely accepted in agricultural research, though other types have been proposed (Table 10.1).

## Types of participation in our work

Many workers in participatory research have found the definitions of Biggs (1989) the most useful. Biggs did not give different values to the participation types but pointed out that managers need to be clear about which mode is appropriate at a specific time and to create a working environment that promotes that mode. We have found these types very helpful to analyse which methods are the most cost-effective from a *functional* viewpoint; that is, to make conventional breeding more efficient and produce varieties that are more useful and which will be disseminated beyond the participating farmers. The appropriateness of any mode of participation depends on its goal. If it is the *empowerment* of farmers, then the greater the involvement of farmers the better; the ultimate goal is collegial participation where farmers run their own PPB programmes in a way that should be self-sustaining. However, if the purpose is functional, then contractual or consultative forms of participation may be the most effective (compare Ceccarelli and Grando, Chapter 12, this volume; Soleri *et al.*, Chapter 2, this volume).

### *Contractual participation*

We have used contractual participation to obtain land for many of the breeding procedures; for example, generation advance, small plot multiplication and progeny trials. Farmers were given compensation for the area used by the project. This was assumed to be equivalent to the yield the farmer would normally have obtained from that area, and yield per unit area was assessed by a crop cut in a farmer's field that was mutually agreed between the farmer and project staff. This contractual arrangement is essential when, as is the case with LI-BIRD, an NGO has no access to land of its own. A second example of contractual participation is when we have compensated farmers for labour involved in individual plant selection in very large plots. This occurs at the early stage of modified bulk breeding when a very large $F_4$ or $F_5$ population is divided into sub-bulks according to phenotype. This contractual participation creates the genetic material for subsequent collaborative participation.

### Consultative participation

Consultative participation in PPB has already proved its value in goal setting, in the selection among lines in the field and in the organoleptic testing of varieties. It is clear that successful PPB can be done entirely in a consultative mode, although collaborative PVS is required to test the genotypes produced by the PPB.

### Collaborative participation

The normal mode of participation for all PVS trials, including those that we have carried out, is collaborative, where farmers conduct experiments on their own fields. In PPB, participation is also collaborative when farmers grow the bulks they have been given in their own fields and select plants from them. Based on our own work, collaborative PPB appears to be very cost-effective since selection can be replicated across environments and across the individuals who carry out the selection.

### Collegial participation

Although empowerment was not a primary aim of our functional approach, collegial participation is evolving on its own. In the *terai*, farmers with whom we have been collaborating the most are forming a 'breeders club' and wish to put their collaboration with LI-BIRD on a more formal basis. They explain that, at present, agreements are on a personal level, but an institution (LI-BIRD) is a more assured long-term partner than individuals. The participation will be increasingly collegial because they intend the club to have a strong role in agenda setting.

## Farmers' Knowledge

### Quality traits

At the commencement of the PPB programme for high-altitude rice, scientists consulted with groups from the communities of the high-altitude villages (2000 m) of Chhomrong and Ghandruk in western Nepal. High grain yield, cold tolerance and resistance to sheath brown rot caused by *Pseudomonas fuscovaginae* had always been the predominant selection criteria in formal breeding programmes. To the surprise of the breeders, local women farmers considered breeding white-grained rice varieties as the most important goal. The villagers' reasoning was clear; white is the preferred colour for *bhat* (steamed rice) and is socially prestigious in Nepalese society, though it may have lower iron content than red-pericarped *bhat*. Red-pericarped rice imposes drudgery on women, as they need to spend considerable time removing the red bran with a manually operated rice pounder (Sthapit *et al.*, 1996). The participation of farmers was crucial in setting goals as they knew their needs for new varieties better. The breeders did not have pericarp colour as a breeding objective at the outset of the programme.

As the research progressed, other quality traits emerged for which farmers had a better knowledge of their importance (Sthapit *et al.*, 1996). These included:

- water-absorption capacity during cooking, because a high absorption capacity gives a greater volume of cooked rice;
- how long it took after eating the rice before hunger returned (appetite delay called *adilo*).

These characteristics were either not known, or at the least not selected for, in formal breeding, and no plant breeding textbook mentions this wisdom of farmers. Although, once discovered, these traits may appear simple, they have an important bearing on the overall ranking and final adoption of any variety targeted for subsistence farmers.

Although food-deficit farmers and agricultural labourers always look for varieties that delay appetite for longer, it does not mean that eating quality is not important. For example, *Chaite* rice (February sown) variety Radha 32 is very high yielding (producing nearly 25% more than the most widely grown variety CH 45) and has multiple disease and pest resistances including resistance to brown plant hopper, an increasingly important insect pest of rice in the *terai*. Breeders were confident it would be a highly successful variety, particularly as it initially spread rapidly from farmer to farmer (Witcombe *et al.*, 2001). However, farmers soon rejected it because it had very poor cooking and eating quality and low milling recovery; that is, a low ratio of grain to the other products of milling such as chaff. PNR 381 rice variety was widely adopted after the first year of PVS. In the next year, it was not grown in about 60% of the villages because of its poor postharvest traits (Joshi, 2001).

Quality is not a fixed attribute but varies with locality, socioeconomic class and use. The participation of farmers provides insights into their local preferences. Hence, in high-altitude villages expansion on cooking and appetite delay are important, while in the *terai* market price is paramount unless it is for the farmers' own consumption, when aroma and softness become important.

Market research is yet to be institutionalized in agricultural research. Gauchan *et al.* (2001) found that the market provides incentives and disincentives for the farmer's choice of variety through price signals, market margins and market channels. Despite this, hardly any effort is made in the formal varietal testing system to assess the potential market for new rice varieties. For example, in the high potential production systems in the *terai*, the market price of any rice variety is fixed against Masuli, the dominant rice variety that has medium grains. Mill owners, who play an important role in the market, pay less for rice varieties that do not meet the standard of Masuli. Farmers are well aware of this but breeders have paid little attention to it. This has resulted in the dominance of Masuli even 30 years after its introduction in Nepal. Millers rate all released rice varieties, except Basmati type rice, as inferior to Masuli, whatever their actual grain quality. Golden husk colour, shorter grains that aid high milling recovery and low gelatinizing temperature are some of the quality traits of Masuli that millers and consumers prefer. Hence, varieties that are superior in quality but have longer grains or do not have golden husks are rejected.

## Threshing and shattering

Easy threshing is a desirable trait in Nepal because threshing is not mechanized. Easy-to-thresh varieties require fewer beats of the rice bundle against the ground or a hard surface to remove the grain from the panicles. This increases labour efficiency and reduces drudgery. Difficult-to-thresh varieties obviously require more beatings to separate grains completely from the rice bundle. Less obviously, farmers consider plant height an important contributing factor to easy threshing. For simple mechanical reasons, if the straw is short, more beatings are required to remove the grains (Joshi *et al.*, 1997). Easy threshing is a particularly important selection criterion in *Chaite* rice production because threshing, which coincides with the monsoon rains, needs to be finished quickly to reduce the risk of mould. Farmers rejected IR13155 variety of *Chaite* rice in the lower hills and the *terai* because it was hard to thresh (Joshi, 2001).

Farmers can evaluate better than breeders both threshing and shattering (which is also a threshing trait; shattering varieties are too easy to thresh so grains are lost even before harvest or in handling before threshing). Farmers at Chhomrong village rejected Machhapuchhre-2 because it was extremely difficult to thresh (Sthapit *et al.*, 1996). Farmers of Marangche village detected that Machhapuchhre-3 was more prone to shattering, although breeders did not identify this before its release.

Screening requires a practical test that farmers apply as a matter of course but there is no easy way to reliably screen for shattering and threshing in formal breeding trials. Hence, in formal breeding programmes, ease of threshing is usually not assessed, and hard-to-thresh varieties would only be noticed if they could not be threshed by the mechanical thresher that is typically employed. We do not know of centralized breeding programmes that assess ease of threshing by the number of required beatings or consider its relationship to plant height.

## Other traits

Farmers have a more intimate knowledge of the crop (in part because they grow less entries than breeders do). They grow and manage the crop themselves from sowing through to harvesting and threshing, so they have ample opportunity to study all of the crop traits, including its adaptation to growing environments. Unlike breeders, farmers are involved in all postharvest processes. Farmers in the *terai* did not prefer Pusa Basmati-1 rice variety, although it fetched a premium price in the market, because they found that its long awns made milling more difficult (Joshi *et al.*, unpublished). Farmers evaluate complex traits that need close observation of the crop over time, such as proneness to shattering (see above). Chhomrong farmers considered the stay-green character of M-3 as an indicator of good straw quality but this is rarely, or never, used as a selection criterion by rice breeders (Joshi *et al.*, unpublished).

It is not doubted that farmers carefully observe the details of traits that concern them. In two instances, farmers noticed sprouted seeds, because of lack of seed dormancy in the panicles of test entries, when breeders and other project staff had not observed this deficiency.

## Choice of testing sites

Farmers' strategies for selecting trial sites differ dramatically from those of scientists. Breeders often optimize the growing environment by applying purchased inputs, irrespective of the farmers' practices, since this tends to even out within-field environmental heterogeneity and allows a better expression of the genetic potential for yield. Farmers adopt what appears to be a very careful strategy for choosing the testing site of new crop varieties by exploiting their detailed knowledge of the niches within their village and fields. In high-altitude rice growing areas, farmers tend to test new material in more stressed environments: near an inlet of cold water; in fields near to cattle sheds that are over-fertile and cause lodging and disease; in marginal patches of the field with shallower soil; or areas under shade (Sthapit *et al.*, 1996). If the varieties perform well under these conditions then they grow them in better environments, and indeed will apply more inputs to them than their original varieties.

We assume that farmers are simply adopting a risk-avoidance strategy; if the new variety fails in the hostile environment, it does not matter because in the same area the local variety also produces little. Perhaps it is more sophisticated; if the farmer is going to adopt a new variety, it should perform well in poor, as well as in more favourable, environments. The farmers may feel that a good performance in the poor environment is an indication of excellent performance in a favourable one.

In the *terai* we found from focus group discussions with participating farmers that once a rice variety is accepted following PVS, farmers improve agronomic management by increasing the use of inputs, particularly farmyard manure and chemical fertilizers. They did this with varieties that respond well to inputs, such as Swarna, BG 1442 and NDR 97, to maximize yields. The use of insecticides to control stem borer, rice bug and leaf folder was also common with more productive varieties even though they were no more susceptible; farmers believed they got a greater benefit from applying pesticides when the varieties had a higher yield potential.

## Identifying ecological niches

In our research we found that farmers discover subtle ecological preferences. Farmers of Chhomrong village detected that the cold tolerance of Machhapuchhre-3 is not as good as one of its parents, Chhomrong Dhan. To reduce the level of cold injury they altered its growing domain to the lower elevations of Chhomrong village from the higher elevations where it was originally tested. Machhapuchhre-3 was not specifically adapted to very cold environments, which was why it performed well and was adopted at lower altitudes than where it was originally selected. In the *terai*, in our PVS programme, Swarna rice variety was initially tested in varying water regimes. Farmers found it to be best suited to long-standing water areas where Masuli usually yields much less because it lodges. The area under Swarna is rapidly increasing now that farmers are convinced of its advantages over Masuli in this particular niche. Rice variety BG 1442 was identified by breeders for the *Chaite* season and farmers selected it in PVS trials for this season. However, farmers also multiplied it during the main season and because it performed well it is now becoming popular in both seasons. Kalinga III rice variety is very early maturing. In areas where farmers grow two rice crops a year, farmers grow Kalinga III in areas where they will raise the nursery beds for the main season rice. This is possible because Kalinga III matures nearly 15 days earlier than the widely grown *Chaite* rice variety CH 45.

# Farmers' and Breeders' Knowledge in PPB

## In the process of plant breeding

In the participatory breeding of high-altitude rice varieties, scientists shared the entire process of variety development. All the generation advancement and selection activities conducted at Chhomrong and Ghandruk were done in the farmers' fields with community participation. Farmers could observe and understand the different steps in the breeding process and scientists assisted farmers in taking care to maintain varietal purity by avoiding the mixing of different genetic materials.

Scientists were able to explain to farmers the concepts of genetic segregation in the progeny of crosses and (indirectly) the concept of heritability. This latter concept was explained by discussing how the amount of variation that is inherited differs from trait to trait; selection for grain colour, an important target trait, was likely to be very effective while selection for tiller number, for example, was likely to be less successful (compare Soleri *et al.*, Chapter 2, this volume).

Farmers were shown segregating rice lines from the $F_2$ to the $F_5$ generations from different crosses to illustrate practically how segregation takes place and how lines become more uniform and stabilized over generations. Scientists were able to share with them the need for selection and selfing over several generations in order to develop a population of plants that were more uniform and stable. This helped them to make distinctions between advanced generation lines that need little time and effort to make uniform, and $F_2$ or $F_3$ generations that require many years and considerable effort. They also learnt, when they saw the progeny of several crosses involving Chhomrong Dhan, the way that siblings from the same parents can differ. For example, they observed that a few lines had a grain type and colour very similar to Chhomrong Dhan, but were very different from it in other morphological traits. In contrast, some varieties that were morphologically very similar to Chhomrong Dhan had white grains.

In general, scientists have a greater knowledge of new germplasm. For example, breeders from the Nepal Agricultural Research Council were able to select Fuji 102 as an exotic parent for hybridizing with Chhomrong Dhan and obtain seed. Breeders also have more experience in hybridization so they made the crosses and the segregating progenies were given to farmers. Hence we did not train farmers to emasculate florets and to pollinate, which of course we would have done if the goal was empowerment rather than simply increasing efficiency.

Breeders and pathologists have a better knowledge of plant pathology, and understand the symptoms of micro-nutrient deficiency. They assisted the farmers in diagnosing the causes of symptoms and advised on those that could be safely ignored and those that were important. For example, sheath brown rot is an important disease at higher altitudes and farmers were told:

- how to recognize it;
- that selection against the disease would be effective; and
- that taller plants with better panicle exsertion will give both improved disease resistance and better chilling tolerance (Sthapit *et al.*, 1995).

### Selection strategies

In the PPB programme in the *terai* we compared the results of selection by farmers and breeders among more than 160 advanced lines derived from two crosses of *Chaite* rice (Kalinga III × IR64 and Kalinga III × Radha 32) (compare Ceccarelli and Grando, Chapter 12, this volume). First, farmers and breeders jointly evaluated the lines, but the farmers alone made the final decision on the ones that were selected. Five days later, two breeders (K.D. Joshi and J.R. Witcombe) independently evaluated the plots. The agreement in selections between the two breeders was reasonably high, varying from 50% to 78%, depending on the cross. The agreement between the breeders and the farmers was lower, and varied from 30% for both breeders in one cross, and 41% or 46% in the other.

The lower agreement between the breeders and farmers was further analysed. The number of lines selected by farmers but by neither of the two breeders was 36% or 41% depending on the cross. These lines were inspected and all were found to be early maturing. In all cases, breeders had not selected them because they had a high incidence of disease. Presumably farmers were prepared to tolerate these levels of disease in their fields -- they were not sufficient to have a considerable impact on yield – but breeders are more stringent because they realize that a susceptible variety, when widely grown, can create epidemics (see Smale, Chapter 3, this volume). None the less, it underlines the higher emphasis that farmers place on earliness compared with breeders. Farmers rarely selected entries that matured even at the same time as CH 45, the most widely grown variety in the *Chaite* season, let alone later entries, indicating their dissatisfaction with the long duration of this variety.

### After varieties have been produced

Breeders entered M-3 into the formal trials and tested it across agroclimatic zones. This was possible for formal-sector scientists but, because of a lack of knowledge and access, it would be unlikely that a farmer would enter his or her varieties in multi-locational trials. A knowledge of the theory of multi-locational testing that accounts for variations in climate, pathogens and pests across locations allows scientists to select for widely adapted varieties that perform well across locations (compare Ceccarelli and Grando, Chapter 12, this volume; Bänziger and de Meyer, Chapter 11, this volume). Multi-locational testing was combined with PPB to not only provide varieties that were specifically adapted to the needs of the participating farmers but also to identify those that met the needs of farmers far from the original area of the PPB programme.

The importance of the informal seed supply system (farmer-to-farmer networks) is very great in Nepal because formal extension and seed supply systems are poorly developed. In part this is because of the difficulty of transporting seed over the mountainous terrain. It often has to be carried on foot. Baniya *et al.* (1999) reported that more than 97% of rice area is planted with farmers' own seed (of both landraces and modern varieties) and 85% of farmers regularly obtain rice seed in social seed networks by exchange, gift or purchase. However, despite its importance, the speed of farmer-to-farmer spread is relatively slow and it tends to be limited within geographical areas. The formal sector, including both NGOs and government organizations, particularly Lumle Agricultural Research Centre, increased the speed and extent of the spread of M-3 (Joshi *et al.*, 2001).

## Novel Interpretations of Plant Breeding Theory

Much of plant breeding theory is uncontroversial and not open to very different interpretations. The theories of heritability, selection differential and selection response are straightforward and do not differ theoretically between conventional and participatory methods (see also Soleri *et al.*, Chapter 2, this volume). In practice, if heritabilities are expected to differ between farmers' fields and the research station it might present a case for or against PPB. In Nepal, at least in the target regions in which we have worked – high-altitude rice and low-altitude irrigated rice in the *terai* – there is no significant difference between the heritabilities found in farmers' fields and on research stations (see also Ceccarelli and Grando, Chapter 12, this volume). Sthapit *et al.* (1995) report on the heritabilities of field resistance to sheath brown rot disease in these crosses of high-altitude rice in Nepal. Heritabilities in farmers' fields in Chhomrong and Ghandruk villages and at the research station differed little, ranging from 72 to 84%, although the highest heritability for the trait was at the research station. Sthapit

(1994) reported on heritabilities for many characters in three crosses and in four locations, only one of which, Lumle, was a research station. There was no significant difference in the heritabilities between the research station and the other locations where the crop was grown on farmers' fields (Table 10.2). In the high potential production system of the *terai*, farmers' fields tend to be extremely uniform and there is no reason to suppose that the uniform farmers' fields will be more heterogeneous than research stations' fields.

Selection theory, that is, what selection units to use, and the intensities of selection that can be used in each generation are open to greater interpretation. For example, in outbreeding crops there is much empirical evidence on whether half-sib, full-sib, $S_1$ or $S_2$ families should be employed as the selection units or, indeed, whether mass selection is more cost-effective (e.g. for maize, Hallauer and Miranda, 1988). In inbreeding crops there is an equally voluminous literature, but no comprehensive review, on the efficiency of various breeding methods such as pedigree, bulk-pedigree, bulk, modified bulk or single seed descent. None the less, in practice, most methods, if executed in reasonable accordance with theory, will result in a genetic gain, and plant breeders have adopted methods that best fit the programme's human resources and infrastructure and its level and duration of funding (compare Duvick, Chapter 8, this volume).

It is clear from the economics of plant breeding theory that genetic improvement should be for traits of economic importance (e.g. Simmonds, 1979). However, formal trial systems have emphasized yield too much, have measured a limited number of traits and rarely, if ever, have employed a system of trade-offs between those traits (Witcombe *et al.*, 1998). Hence, a major impetus for participatory research was multi-trait evaluation by those that make the final decision on whether a variety is adopted or not, the farmers. Farmers evaluated many traits in focus group discussions, and traded them off, for example, by accepting lower yields for earlier maturity or higher quality. The results have somewhat different implications for PVS and PPB:

- In PPB, once the important traits are identified and the trade-offs among them are known, collaborative participation can be avoided. All the farmer-identified traits can be evaluated by scientists in their on-station breeding trials or in laboratory tests of harvested grain. A selection index can be created using the relative weights that farmers give to traits. Even if the selection index was less than optimal it increases the general appropriateness of the material that is bred by the programme to meet farmers' needs.
- In PVS, the case for the continuing involvement of farmers is stronger. The ultimate purpose of any multi-trait evaluation is to predict

**Table 10.2.** Heritabilities for several traits in three crosses across four locations, main season, 1993. Only in Yampaphant was there no chilling stress.

| Trait | Ch × Ma | | Ra × Ma | | IR36 × Ch | |
|---|---|---|---|---|---|---|
| | Lumle (1450 m) | Yampaphant (475 m) | Ghandruk (2000 m) | Yampaphant (475 m) | Chhomrong (2000 m) | Yampaphant (475 m) |
| *(a) Vegetative phase:* | | | | | | |
| Seedling height at GS1 (score) | 0.78*** | 0.83*** | 0.76*** | 0.59*** | 0.81*** | 0.82*** |
| Plant height at GS9 (cm) | 0.95*** | 0.93*** | 0.84*** | 0.65*** | 0.83*** | 0.87*** |
| Seedling vigour at GS1 (score) | 0.51*** | 0.28** | 0.50*** | 0.52** | 0.58*** | 0.48** |
| Chilling tolerance at GS1 (score) | 0.69*** | 0.67*** | 0.39*** | 0.58*** | 0.66*** | 0.34*** |
| Tillers at GS2 (no. per plant) | 0.48*** | 0.69*** | 0.72*** | 0.69*** | 0.77*** | 0.77*** |
| *(b) Phenology:* | | | | | | |
| Days to 50% flowering (days) | 0.85*** | 0.95*** | 0.89*** | 0.78*** | 0.92*** | 0.90*** |
| *(c) Reproductive phase:* | | | | | | |
| Panicle (no. per plant) | 0.51*** | 0.72*** | 0.70*** | 0.73*** | 0.76*** | 0.84*** |
| Panicle weight (g per plant) | 0.83*** | 0.56*** | 0.66*** | 0.79*** | 0.77*** | 0.86*** |
| Total florets per plant | 0.55*** | 0.75*** | 0.35*** | 0.78*** | 0.71*** | 0.90*** |
| Panicle exsertion (score) | 0.82*** | 0.64*** | 0.75*** | NR | 0.81*** | 0.66*** |
| Panicle length (cm) | 0.80*** | 0.85*** | 0.75*** | 0.76*** | 0.78*** | 0.91*** |
| 100 grain mass (g) | 0.81*** | 0.89*** | 0.63*** | 0.78*** | 0.79*** | 0.79*** |
| Sterile floret mass (g per plant) | 0.56*** | 0.85*** | 0.71*** | 0.74*** | 0.72*** | 0.85*** |
| Spikelet sterility (%) | 0.89*** | 0.78*** | 0.73*** | 0.73*** | 0.75*** | 0.59*** |
| Seed number (no. per plant) | 0.92*** | 0.59*** | 0.32*** | 0.78*** | 0.79*** | 0.85*** |
| Seed yield (g per plant) | 0.92*** | 0.51*** | 0.27*** | 0.78*** | 0.77*** | 0.84*** |

Ch = Chhomrong Dhan; Ma = Makwanpur; Ra = Raksali.

***$P \leq 0.001$; **$P \leq 0.01$ indicates probability of significance of the heritability indicated.

NR = not recorded.

whether farmers will adopt a variety, and PVS provides a simpler and more direct approach than using a selection index in multi-locational trials. PVS also perfectly accounts for any changes in farmers' preferences over time with, for example, changes in market price and fodder availability and for differences between socio-economic groups and localities. Using selection indices in multi-locational trials is more complex than using PVS. Since PVS is required to verify the selection indices employed, it is a reasonable strategy to also use PVS to test varieties.

Although our initial motivation for involving farmers was multi-trait evaluation, it also involved a move towards decentralization to farmers' fields. Hence, testing was in the targeted environments rather than multi-locational testing for wide adaptation. Selection in the target environment should be more effective and lead to better adapted, albeit more specifically adapted, varieties (e.g. Simmonds, 1984; Atlin and Frey, 1989; Ceccarelli and Grando, 1989). The existence of genotype × environment (G × E) interaction, as found in the case of high-altitude rice (Fig. 10.1), was thus involved in this decentralization. However, the theory of G × E and the detailed methods of analyses based on this theory were not used in this decision. Most methods of analysis that have been proposed to account for G × E (e.g. reviewed by DeLacey *et al.*, 1996) concentrate on analysis of a single trait, yield, and

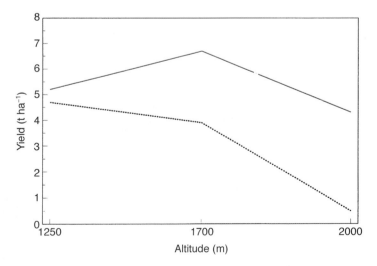

**Fig. 10.1.** The yield across three altitudes of Machhapuchhre-3 (solid line) compared with the mean of the three best entries selected at 1250 m from a nursery of the International Rice Research Institute in the National Rice Cold Tolerance Nursery (NRCTN) (dotted line), Nepal, 1995.

they have become increasingly complex, increasingly less intuitive and, with very few exceptions, are completely non-participatory (compare Soleri *et al.*, Chapter 2, this volume). Only the methods of Binswanger and Barah (1980) have attempted to take farmers' perceptions, in this case of risk aversion, into account.

Unfortunately, no matter how sophisticated the method of analysis, the G × E that occurs in the analysis is dependent on the environments of the test sites and these are often unrepresentative of target environments. Frequently the test sites fail to represent the diversity and the mean of the environments in farmers' fields (Packwood *et al.*, 1998). Increasing testing with farmers, and reducing the emphasis on formal multi-locational trials, is one of the best ways of improving the appropriateness of test sites. There is also a gain in exploiting specific adaptation to a target environment, although this advantage may be outweighed by the reduction in the scale of adaptation. Unfortunately, theory cannot predict an expected degree of specific adaptation for the genotypes that emerge from a breeding programme; that is, for the new genotypes, what will be the slopes of the genotype regressions against an environmental index (Finlay and Wilkinson, 1963) compared with those of control varieties. What little empirical evidence exists is insufficient in quality or quantity to give a reliable prediction of the outcome. In our work on high-altitude rice breeding, the best genotypes were not specifically adapted and performed even better at mid-altitudes lower than in the high-altitude ones in which they were bred (Fig. 10.2).

Once the decision has been made to decentralize to farmers' fields, to have the benefits of multi-trait evaluation and to exploit specific adaptation, practical issues intervene. Moving a breeding programme to farmers' fields meant that conventional plant breeding practices, which rely on the generation of a very large number of test units from many crosses, would only be possible if we were to devote resources to help farmers to avoid errors when sowing many entries. Given our limited human resources, the number of participating farmers and the short time period over which farmers sow, this was impracticable (compare Ceccarelli and Grando, Chapter 12, this volume). Hence, we used many fewer crosses and a much larger population size for each cross than is common in conventional centralized plant breeding. Even though it was contrary to common practice this was supported strongly by theory:

> Choice of germplasm to be used in a practical breeding programme may be the most critical decision facing the breeder, although the choice of breeding method and system of evaluation can be equally important. Goodman (1985) further pointed out that choice of materials is perhaps the most critical to the success of the programme. When one reviews the

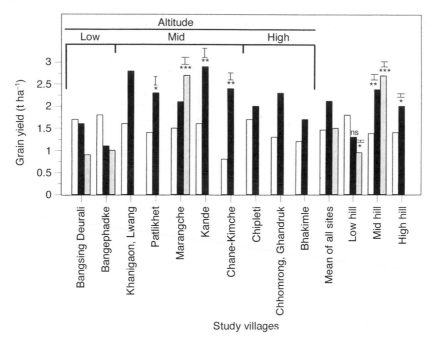

**Fig. 10.2.** Grain yield of Machhapuchhre-3 (solid bars) and Machhapuchhre-9 (hatched bars) compared with the local landrace (open bars) across low-, mid- and high-altitude villages of Kaski, Myagdi and Syangja districts, Nepal, 1998. The significance of yield difference is shown along with the standard error of the mean difference. \*\*\**P* < 0.001; \*\**P* < 0.01; \**P* < 0.05.

literature or reflects on plant breeding courses, however, far more time is devoted to breeding methods than to the selection of useful germplasm.

(Maunder, 1992)

This [the very high number of possible recombinants] suggests that populations should be as large as practicable in early generations, and that their management should be directed towards obtaining, in later generations, manageable numbers of promising pure lines for intensive evaluation of agricultural worth. Such evaluations, it is hoped, will reveal at least one pure line deemed good enough to be released to farmers. If not, at least some pure lines may be found that the breeder considers to be worthy of use as parents in additional cycles of hybridisation.

(Allard, 1999)

We have found that participatory methods assist greatly in reducing the number of crosses by reliably identifying superior parents: in all our recent PPB programmes at least one parent has been selected using PVS or farmers' current varietal adoption. Witcombe and Virk (2001) have examined in more detail the strategy of low-cross-number, high-population-size breeding.

# How Has Our Understanding of Plant Breeding Theory and Practice Changed?

## Theory: low-cross-number, high-population-size breeding

Following our adoption of using fewer crosses and larger populations in our PPB programmes, we reviewed the underlying theory (Witcombe and Virk, 2001). Our work has provided empirical evidence, to support theory, that a low-cross-number strategy in an inbreeding crop is effective. From a single cross, Kalinga III × IR64 in a PPB programme, variety Ashoka 200F has been released in eastern India. Another variety from this cross, Ashoka 228 is in the pre-release stage in eastern India and has been accepted by farmers in PVS trials in the Nepal *terai*. From the same cross, farmers in Nepal are accepting varieties for partial and well-watered conditions in the *Chaite* season and we will soon test promising varieties from this cross that are adapted to main season environments. Only two other crosses have been examined in Nepal to any extent in our PPB programme and both have produced promising genotypes that are currently entering PVS trials.

## Why most breeders do not do this

The question arises as to why most breeders do not adopt a low-cross-number strategy. One reason is that they are not in the situation of starting a breeding programme where the predominant variety has been cultivated by farmers for many years and where there is no ongoing breeding programme. In this situation the choice of parents is simple. When there is an ongoing breeding programme, and farmers are constantly adopting new varieties, then there are many more parents to choose from.

Another reason is that many breeders work in a competitive environment, where to be commercially successful it is not enough to breed a variety better than the one farmers are growing. In commercial environments breeders have to produce the highest-yielding cultivar in official trials, one that is better than those of all competitors. To do this breeders attempt to use the best possible parents. Given a genetic gain of 1–2% per annum by plant breeding (e.g. Byerlee and Heisey, 1990), the most recently produced varieties are most likely to be highest yielding and, hence, on performance *per se* are the best possible parents. Unfortunately, these are the varieties about which least is known; there is a trade-off between the increased probability that a new variety is the highest yielding genotype and the quantity of information on performance *per se*. Hence, to ensure that the best possible parents are included, many crosses are made among the newest material; for example, a diallel cross or partial diallel cross among the entries of an advanced yield trial. Of course, there is a cost to attempting to include the best parent, as more crosses mean that individual population sizes of each cross have to be smaller. Also it is certain that many of the parents would not have been used if more were known about their performance.

## Bulk-population breeding methods

We have also adopted bulk population methods of plant breeding since these are more suited to participatory approaches (Witcombe *et al.*, 2001). The theory supports this approach despite many breeding programmes relying on pedigree breeding. Bulk methods preserve variability in the early generations when plant-to-plant heritability is lower and selection begins when a high level of homozygosity has been achieved. The literature on such methods, particularly the adoption of single seed descent methods, indicates that these are efficient and successful (e.g. in rice, Fahim *et al.*, 1998), and our experience has strongly supported this view.

## Practice: manifold changes in approach

Plant breeding practice has changed dramatically with the adoption of PPB approaches. These have resulted in:

- the selection of parents for PPB on the basis of performance and acceptability in PVS trials;
- the selection of target traits on the basis of consultation with farmers (this has changed the selection criteria used in the breeding programme);
- a reduction in the number of crosses and an increase in the population size of the crosses;
- a reduction in the use of pedigree breeding and an increased emphasis on bulk population methods;
- an increased emphasis on consultative screening for postharvest quality traits with farmers (this cost-effectively reduces the number of entries to be tested by PVS, and ensures that more widely tested entries are not rejected because of poor grain quality);
- the testing of varieties first with farmers rather than in formal multi-locational trials; the most farmer-accepted lines are later entered in these trials.

# Some Impacts of our Work

## Participatory plant breeding

PPB for high-altitude areas of Nepal produced varieties that farmers adopted. One of the most popular, Machhapuchhre-3 (M-3), performed much better in the formal trials system than the products from centralized breeding (Fig. 10.1). In 1996 it was released by the Variety Registration and Release Committee of the National Seed Board, Nepal. From 1996 to 1999, the spread of M-3 and another variety, Machhapuchhre-9 (M-9), was monitored in high-altitude villages by interviewing individual households and groups, and by field verification (Joshi *et al.*, 2001) (Fig. 10.3). M-3 and M-9 were derived from the same cross and the latter variety was identified in a PVS programme in Marangche village. M-3 and M-9 spread from farmer to farmer and through interventions by NGOs and government organizations. Their adoption has steadily increased (Fig. 10.4) and commenced 5–6 years earlier than would have been the case in a conventional varietal release system. The PPB programme was decentralized – all selection was in only two villages in the same valley – but this did not result in specific adaptation. The varieties were adopted in distant villages situated at much lower altitudes than the original PPB sites and the greatest yield advantage of the varieties over the local landraces was also at these lower altitudes (Fig. 10.2).

## Participatory varietal selection

In the work on PVS in the *terai*, we tested two participatory approaches to varietal selection that differed in their resource requirement in both *Chaite* and main season rice (Joshi and Witcombe, 2002). Six new *Chaite* rice varieties were tested in over 300 trials, and 16 main season varieties in over 1100 trials. Surveys were done to record the extent of adoption and spread of the new rice varieties in the study villages. In many cases, farmers tested varieties for a second year before deciding whether to adopt or drop them. Several varieties were quite widely accepted, others were adopted for niches in a few villages, or they were rejected. The two participatory approaches used, called farmer managed participatory research (FAMPAR) and informal research and development (IRD), identified the same varieties, but IRD was much cheaper and, therefore, more cost effective.

We also studied the impact of the PVS programme in the *Chaite* season rice on on-farm varietal biodiversity through household surveys in two high potential production systems in Chitwan and Nawalparasi

**Fig. 10.3.** Adoption of Machhapuchhre-3 and -9 in Marangche village at 1500 m. The terraces in the middle (lighter shade) are earlier maturing Machhapuchhre varieties.

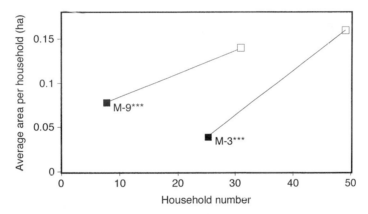

**Fig. 10.4.** The change in the area and number of adopting households for Machhapuchhre-3 (M-3) and Machhapuchhre-9 (M-9) from 1996 (■) to 1999 (□) in seven study villages. Both cultivars increased significantly in area over this period (***$P \leq 0.001$).

districts of Nepal (Witcombe *et al.*, 2001). Varietal diversity was extremely low in *Chaite* rice (weighted diversity 0.04) because one cultivar dominated, CH 45 (Witcombe *et al.*, 2001). Participatory varietal selection identified new varieties that farmers preferred such as Kalinga III, NDR97 and BG 1442. Their adoption by farmers increased on-farm varietal biodiversity within the period of the three cropping seasons that were studied. Despite the commonly assumed uniformity of high potential production systems, the new varieties occupied specific niches in the farming system. Farmers' preferences for different varieties – there were large differences in quality traits and maturity period among the new varieties – should help to maintain varietal diversity.

## On-farm conservation

The lesson learned from our work has been extensively used in IPGRI's global project, 'Strengthening the scientific basis of *in situ* conservation of agrobiodiversity on-farm' in Nepal and Vietnam (Joshi *et al.*, 1999; Sthapit and Jarvis, 1999). PPB is a strategy to strengthen the process of on-farm conservation by encouraging farmers to continue to select and manage local crop populations, and manage seed supply systems through informal networks. Diverse farmer preferences, agroecological niches and farming systems will conserve a reservoir of genetic diversity on-farm (compare Smale, Chapter 3, this volume). This reservoir can be considered as valuable pre-breeding germplasm.

Both PVS and PPB may impact negatively on the diversity of landraces, because both methods are intended to change the local crop population structure to make it more productive and useful to farmers. Studies are under progress in the villages of Begnas (600–1400 m altitude) and Bara (54–100 m), Nepal, to test if PPB will meet conservation goals. This is likely because the variable, segregating materials used in PPB are derived from landraces already in the local farming system.

## Conclusions and Future Directions

We conclude by discussing how the results of our work so far will affect its future directions. One of the major results of the work has been a realization of how important it is to involve farmers in setting breeding goals. Initially, the setting of the goal to breed white-pericarped, rather than red-pericarped, varieties for high altitudes was a consultative exercise. In more recent research, goal setting has been the result of collaborative research as part of PVS, where farmers have evaluated new varieties in their own fields. One example is variety Pusa Basmati-1 for which farmers identified deficiencies. We established a breeding programme with the objective of removing the two they considered most important: the presence of apical awns which made milling difficult and dwarf plant height that reduced straw yield. Farmers consider straw an important product as they feed it to their livestock. The progeny of natural outcrossing were selected from a very large irradiated population of Pusa Basmati-1 that were not only awnless and taller but had improved grain quality. Another example was the initial widespread acceptance of Radha 32 for its high yield and its equally widespread, subsequent rejection because of its very poor cooking quality. We used Radha 32 as a parent and crossed it to Kalinga III, which has better grain quality, with the breeding objective of producing lines that yield at least as much as Radha 32 but that do not have its poor grain quality.

In our high-altitude breeding programme we had a small number of $F_4$ bulk progenies to test with farmers. This bulk pedigree method was not perfect, because when farmers found some faults with the lines there were too few of them left in the breeding programme to have a realistic possibility of finding lines without these faults. Hence, in our PPB programme in the *terai*, which is targeted at more favourable environments, we have used methods that preserve a greater genetic diversity, such as bulk population breeding and equal seed descent. Equal seed descent can, using few resources, preserve a reservoir of genetic diversity in a heterogeneous, but homozygous, population of plants. It is simple to derive pure lines from such a population and we then evaluate them jointly with farmers in consultative participation.

Although we have had undoubted success with PVS in the *terai*, it is limited by the genetic diversity among released cultivars. Many of the varieties adopted by farmers have deficiencies that would have been selected against in a PPB programme. However, PVS was able to give quick results, whereas the first PPB products have taken more than 4 years to produce, and even then their testing is constrained by seed quantities. What is becoming clear, and leading the direction of much of our current PPB programme, is that PPB is the filling in a sandwich between two slices of PVS. The first slice identifies both potential parents and breeding goals, and the second immediately identifies those varieties produced by PPB that are acceptable to farmers.

PVS in the *terai* gave very promising results and the local government extension service, the District Agricultural Development Office (DADO), said that they wished to collaborate with LI-BIRD by testing varieties identified in the PVS programme in many of the villages where DADO was working. PVS is clearly an extension, as well as a research tool. DADO screened all of the PVS-identified varieties with farmers using the same participatory approaches that we used and are actively extending the most preferred varieties over the whole of Chitwan district. Now the programme is being extended to DADO offices in four far-flung districts of Nepal.

## Acknowledgements

This chapter is an output from project R7122 funded by the Plant Sciences Research Programme of the UK Department for International Development for the benefit of developing countries. The views expressed are not necessarily those of the DFID. The contribution of the Nepal Agricultural Research Council and Lumle Agricultural Research Centre in breeding Machhapuchhre-3 and Machhapuchhre-9 are gratefully acknowledged. We are grateful to Mr C.K. Devkota, District Agricultural Development Office, Chitwan, for his support in scaling up the participatory approaches in Chitwan district. Without the enthusiastic collaboration of farmers of many villages in the hills and the *terai* of Nepal this research would not have been possible.

## References

Allard, R.W. (1999) *Principles of Plant Breeding*, 2nd edn. John Wiley & Sons, New York, p. 91.

Arnstein, S.R. (1969) Ladder of citizen participation. *Journal of the American Institute of Planners* XXXV, 216–224.

Atlin, G.N. and Frey, K.J. (1989) Predicting the relative effectiveness of direct versus indirect selection for oat yields in three types of stress environments. *Euphytica* 44, 137–142.

Baniya, B.K., Subedi, A., Rana, R.B., Paudel, C.L., Khatiwada, S.P., Rijal, D.K. and Sthapit, B.R. (1999) Informal rice seed supply system and storage systems in mid-hills of Nepal. In: Sthapit, B., Upadhaya, M. and Subedi, A. (eds) *A Scientific Basis of* in situ *Conservation of Agrobiodiversity On-Farm: Nepal's Contribution to the Global Project*. IPGRI, Rome, Italy, pp. 79–91.

Bhandari, H.S. (1997) Participatory plant breeding in Nepal: an analytical review. LARC Discussion Paper No. 98/1. Lumle Agricultural Centre, Pokhara, Nepal, pp. 1–15.

Biggs, S.D. (1989) Resource-poor farmer participation in research: a synthesis of experience from nine National Agricultural Research Systems. *OFCOR Comparative Study Paper* No. 3. International Service for National Agricultural Research (ISNAR), The Hague.

Binswanger, H.P. and Barah, B.C. (1980) Yield risk, risk aversion, and genotype selection: conceptual issues and approaches. *Research Bulletin* No. 3, ICRISAT, Andhra Pradesh, India.

Byerlee, D. and Heisey, P.W. (1990) Wheat varietal diversification over time and space as factors in yield gains and rust resistance in the Punjab. In: Heisey P.W. (ed.) *Accelerating the Transfer of Wheat Breeding Gains in Farmers: a Study of the Dynamics of Varietal Replacement in Pakistan*. CIMMYT Research Report No. 1. International Centre for Wheat and Maize Improvement, Mexico, D.F., pp. 5–24.

Ceccarelli, S. and Grando, S. (1989) Efficiency of empirical selection under stress conditions in barley. *Journal of Genetics and Breeding* 43, 25–31.

Ceccarelli, S., Grando, S. and Booth, R.H. (1996) International breeding programmes and resources and resource-poor farmers: crop Improvement in difficult environments. In: Eyzaguirre, P. and Iwanaga, M. (eds) *Participatory Plant Breeding Proceedings of a Workshop on Participatory Plant Breeding, 26–29 July 1995, Wageningen, The Netherlands.* IPGRI, Rome, Italy, pp. 99–116.

Chambers, R. (1989) Institutions and practical change. Reversals, institutions and change. In: Chambers, R., Pacey, A. and Thrupp, L.A. (eds) *Farmer First: Farmer Innovation and Agricultural Research.* Intermediate Technology Publications, London, pp. 181–195.

Chemjong, P.B., Baral, B.H., Thakuri, K.C., Neupane, P.R., Neupane, R.K. and Upadhaya, M.P. (1995) *The Impact of Pakhribas Agricultural Centre Research in the Eastern Hills of Nepal: Farmer Adoption of Nine Agricultural Technologies.* Pakhribas Agricultural Centre, Dhankuta, Nepal.

DeLacey, I.H., Ratnasiri, W.G.A. and Mirzawan, P.D.N. (1996) Retrospective analysis of historical data sets from multi-environmental trials – case studies. In: Cooper, M. and Hammer, G.L. (eds) *Plant Adaptation and Crop Improvement.* CAB International, Wallingford, UK; IRRI, Los Baños, Philippines; ICRISAT, Andhra Pradesh, pp. 269–290.

Fahim, M., Dhanapala, M.P., Senadhira, D. and Lawrence, M.J. (1998) Quantitative genetics of rice. II. A comparison of the efficiency of four breeding methods. *Field Crops Research* 55, 257–266.

Farrington, J. (2000) The development of diagnostic methods in FSR. In: Collinson, M. (ed.) *A History of Farming Systems Research.* CAB International, Wallingford, UK; FAO, Rome, pp. 59–66.

Friis-Hansen, E. and Sthapit, B.R. (eds) (2000) Concepts and rationale of participatory approaches to conservation and use of plant genetic resources. In: *Participatory Approaches to the Conservation and Use of Plant Genetic Resources.* International Plant Genetic Resources Institute, Rome; Centre for Development Research, Copenhagen, pp. 16–21.

Finlay, K.W. and Wilkinson, G.N. (1963) The analysis of adaptation in a plant breeding programme. *Australian Journal of Agricultural Science* 14, 742–754.

Gauchan, D., Chaudhary, P., Sthapit, B.R., Upadhaya, M.P., Smale, M. and Jarvis, D. (2001) A participatory marketing system research approach to analysing market based incentives and disincentives: a case study of rice, central terai, Nepal. Paper presented at the *National Workshop of* in situ *Conservation of Agrobiodiversity On-farm: Nepal Project. April 24–26 Lumle, Pokhara, Nepal.*

Goodman, M.M. (1985) Exotic maize germplasm: status, prospects, and remedies. *Iowa State Journal of Research* 59, 497–527.

Hallauer, A.R. and Miranda, J.B. (1988) *Quantitative Genetics in Maize Breeding,* 2nd edn. Iowa State University Press, Ames, Iowa.

Joshi, A. and Witcombe, J.R. (1996) Farmer participatory crop improvement. II. Participatory varietal selection, a case study in India. *Experimental Agriculture* 32, 461–477.

Joshi, K.D. (2001) Rice varietal diversity and participatory crop improvement in Nepal. PhD thesis, The University of Wales, Bangor, UK.

Joshi, K.D. and Sthapit, B.R. (1990) Informal research and development (IRD): a new approach to research and extension. *LARC Discussion Paper* No. 90/4. Lumle Agricultural Research Centre, Pokhara, Nepal.

Joshi, K.D. and Witcombe, J.R. (2002) A comparison of two participatory methods for varietal selection of rice in Nepal. *Euphytica* (in press).

Joshi, K.D., Subedi, M., Rana, R.B., Kadayat, K.B. and Sthapit, B.R. (1997) Enhancing on-farm varietal diversity through participatory varietal selection: a case study for *Chaite* rice in Nepal. *Experimental Agriculture* 33, 335–344.

Joshi, K.D., Rijal, D.K., Rana, R.B., Khatiwada, S.P., Chaudhary, P., Shrestha, K.P., Subedi, A. and Sthapit, B.R. (1999) Adding benefits. I: through PPB, seed networks and grassroots strengthening. In: Jarvis, D., Sthapit, B. and Sears, L. (eds) *Conserving Agricultural Biodiversity* in situ: *a Scientific Basis for Sustainable Agriculture. Proceedings of a workshop, 5–12 July 1999, Pokhara, Nepal*, pp. 206–209.

Joshi, K.D., Sthapit, B.R. and Witcombe, J.R. (2001) How narrowly adapted are the products of decentralised breeding? The spread of rice varieties from a participatory plant breeding programme in Nepal. *Euphytica* 122, 589–597.

LARC (1995) *The Adoption and Diffusion and Incremental Benefits of Fifteen Technologies for Crops, Horticulture, Livestock and Forestry in the Western Hills of Nepal.* LARC Occasional Paper 95/1. Lumle Agricultural Research Centre, Pokhara, Nepal.

Maunder, A.B. (1992) Identification of useful germplasm for practical plant breeding programs. In: Stalker, H.T. and Murphy, J.P. (eds) *Plant Breeding in the 1990s. Proceedings of the Symposium on Plant Breeding in the 1990s.* CAB International, Wallingford, UK, pp. 147–169.

Maurya, D.M. (1989) The innovative approach of Indian farmers. In: Chambers, R., Pacey, A. and Thrupp, L.A. (eds) *Farmer First: Farmer Innovation and Agricultural Research.* Intermediate Technology Publications, London, pp. 9–13.

Packwood, A.J., Virk, D.S. and Witcombe, J.R. (1998) Trial testing sites in all India co-ordinated projects – how well do they represent agro-ecological zones and farmers' fields? In: Witcombe, J.R., Virk, D.S. and Farrington, J. (eds) *Seeds of Choice: Making the Most of New Varieties for Small Farmers.* Oxford and IBH Publishing Co., New Delhi; Intermediate Technology Publications, London, pp. 7–26.

Pound, B.P., Budhathoki, K. and Joshi, B.R. (1988) Mountain agricultural technology development and diffusion: the Lumle model, Nepal. In: Jodha, N.S., Baskota, M. and Tej Pratap (eds) *Sustainable Mountain Agriculture: Perspectives and Issues.* ICIMOD, Oxford; IBH Publishing Company, New Delhi, pp. 711–736.

Pretty, J. (1998) *The Living Land: Agriculture, Food and Community Regeneration in Rural Europe.* Earthscan Publications, London.

Simmonds, N.W. (1979) *Principles of Crop Improvement.* Longman Group, Harlow, UK.

Simmonds, N.W. (1984) Decentralised selection. *Sugar Cane* 6, 8–10.

Sperling, L., Loevinsohn, M.E. and Ntabomvura, B. (1993) Rethinking the farmer's role in plant breeding: local bean experts and on-station selection in Rwanda. *Experimental Agriculture* 29, 509–519.

Sthapit, B.R. (1994) Genetics and physiology of chilling tolerance in Nepalese rice. PhD thesis, The University of Wales, Bangor, UK.

Sthapit, B.R. and Jarvis, D. (1999) Participatory plant breeding for on-farm conservation. *Ileia Newsletter* 15, 40–41.

Sthapit, B.R., Pradhanang, P.M. and Witcombe, J.R. (1995) Inheritance and selection of field resistance to sheath brown rot disease in rice. *Plant Disease* 79, 1140–1144.

Sthapit, B.R., Joshi, K.D. and Witcombe, J.R. (1996) Farmer participatory crop improvement. III. Participatory plant breeding, a case study for rice in Nepal. *Experimental Agriculture* 32, 479–496.

Weltzien, E., Smith, M., Meitzner, L.S. and Sperling, L. (2000) Technical issues in participatory plant breeding from the perspective of formal plant breeding. A global analysis of issues, results and current experiences. Working Document No. 3. CGIAR, Systemwide Programme on Participatory Research and Gender Analysis for Technology Development and Institutional Innovation. CIAT, Cali, Colombia.

White, S.C. (1996) Depoliticising development: the uses and abuses of participation. *Development in Practice* 6, 6–15.

Witcombe, J.R. (1999) Do farmer-participatory methods apply more to high potential areas than to marginal ones? *Outlook on Agriculture* 28, 43–49.

Witcombe, J.R. and Virk, D.S. (2001) Number of crosses and population size for participatory and classical plant breeding. *Euphytica* 122, 451–462.

Witcombe, J.R., Joshi, A., Joshi, K.D. and Sthapit, B.R. (1996) Farmer participatory crop improvement. I. Varietal selection and breeding methods and their impact on biodiversity. *Experimental Agriculture* 32, 445–460.

Witcombe, J.R., Virk, D.S. and Raj, A.G.B. (1998) Resource allocation and efficiency of the varietal testing system. In: Witcombe, J.R., Virk, D.S. and Farrington, J. (eds) *Seeds of Choice: Making the Most of New Varieties for Small Farmers*. Oxford and IBH Publishing, New Delhi; Intermediate Technology Publications, London, pp. 135–142.

Witcombe, J.R., Joshi, K.D., Rana, R.B. and Virk, D.S. (2001) Increasing genetic diversity by participatory varietal selection in high potential production systems in Nepal and India. *Euphytica* 122, 575–588.

# Collaborative Maize Variety Development for Stress-prone Environments in Southern Africa

# 11

MARIANNE BÄNZIGER AND JULIEN DE MEYER

*International Maize and Wheat Improvement Center (CIMMYT), PO Box MP163, Harare, Zimbabwe*

## Abstract

Highly variable and stress-prone crop growing conditions, and financial, structural and human resource constraints have impeded the realization and impact of traditional plant breeding approaches in southern Africa to the extent that the food security of farming communities, countries and the region as a whole continues to be threatened. This chapter describes a collaborative maize breeding approach that seeks to improve the relevance of current breeding approaches to resource-poor farmers in this region. The interventions chosen consider the fundamental elements of breeding progress. Selection of maize breeding germplasm under managed abiotic and biotic stress factors that are most relevant in farmers' fields, and collaborative trials that evaluate promising experimental varieties for a wide range of growing conditions and characteristics important to smallholder farmers resulted in a better representation of farmers' real environments and preferences in the selection process. Experimental procedures and a wider involvement of partners permitted the evaluation of a larger number of germplasm accessions while keeping heritability and selection intensity high. Currently this approach involves collaboration with more than 40 organizations and 600 farmers in ten countries, and gives them and the wider farming community the opportunity to make informed and potentially diverse selection decisions and design strategies to access seed of more appropriate maize varieties. We conclude that collaborative plant breeding approaches have an advantage over traditional approaches in selecting and deploying more appropriate varieties for a target environment that is highly variable, stress-prone and hampered by resource constraints.

# Introduction

More than 100 million people in southern Africa (including Angola, Botswana, Lesotho, Malawi, Mozambique, Namibia, South Africa, Swaziland, Tanzania, Zambia and Zimbabwe) get their food and income from small-scale farming. Maize is the preferred staple food and is grown on 71% of the cereal area planted, or about 11 Mha (Aquino *et al.*, 2001). Many smallholder farmers sell surpluses from home maize production and maize is one of the most important crops traded.

Formal maize breeding research in this region spans more than 90 years (Eicher and Kupfuma, 1997) and large-scale commercial farmers have successfully used improved agronomic practices for a similar time period. However, maize yields in southern Africa (excluding South Africa) still average 1.2 t ha$^{-1}$ (Aquino *et al.*, 2001). Only about 35% of all farmers in any given year use purchased seed of modern maize varieties (Morris, 2001). Average inorganic fertilizer application on cereals is less than 30 kg ha$^{-1}$ (deduced from FAO statistics for 1998). This is less than would be needed to maintain a balanced soil nutrient budget (Smaling and Braun, 1997), and nutrient depletion and soil fertility decline are widespread in smallholder farming systems (Kumwenda *et al.*, 1997). Apparently, the results of formal plant breeding and improved crop management techniques have not reached many farming communities in southern Africa (compare Ceccarelli and Grando, Chapter 12, this volume; Joshi *et al.*, Chapter 10, this volume; McGuire, Chapter 5, this volume; Zimmerer, Chapter 4, this volume).

Growing environments in southern Africa are highly stress-prone, and farmers' access to and use of external inputs are limited. Variation in annual rainfall and crop production are the highest in the world (Heisey and Edmeades, 1999). Annual maize production averaged 16.2 Mt over the past 20 years which barely results in self-sufficiency of this region. However, during the same period, production varied between 7.3 and 22.4 Mt (FAO statistics for 1981 to 2000), implying that crop failure is common. Farmers investing in agricultural inputs bear a considerable risk of obtaining little or negative economic returns and – given the extreme poverty of the majority of smallholder farmers – they often choose not to accept this risk, avoiding it instead by minimizing cash investment in crop production (McCown *et al.*, 1992; see Soleri *et al.*, Chapter 2, this volume). The agricultural input and credit sector is little developed and, together with poor transport infrastructure, this increases input prices and reduces the incentive and possibility for their use. With the demand for food rising at 3.0–3.5% per year, improving food security is a vital issue for farming families, countries and the region as a whole, and increasing the productivity of maize assumes a central role in this (Eicher and Byerlee, 1997).

The International Maize and Wheat Improvement Centre (CIMMYT) and the Southern African Centre for Cooperation in Agricultural and Natural Resources Research and Training (SACCAR) initiated in 1996 a new breeding approach aimed at increasing and stabilizing maize production at the low agronomic input level existing on most smallholder farms in the Southern African Development Community (SADC). Maize varieties that were performing better under conditions typical for resource-poor farmers in southern Africa should be developed and disseminated. What started as an approach to improve food security developed into a collaborative plant breeding (CPB) approach that currently involves over 40 institutions and receives systematic feedback from more than 600 farmers in six countries. We describe this breeding approach as a case study in this book as it interweaves the comparative advantages of formal plant breeding and seed dissemination with those of farmers (and their environments) in assessing and selecting varieties and designing follow-up strategies to increase food security.

## Developing Maize Varieties for Smallholder Farmers

The unsatisfactory impact of formal plant breeding on productivity in southern Africa may have several origins, among them: (i) farmers have no access to seed of modern varieties (MVs); (ii) farmers do not adopt MVs when they are available; and (iii) when farmers do adopt MVs, they may demonstrate little breeding progress under conditions where low inputs are the norm and abiotic stresses are common ('stress environments').

### Developing stress tolerant varieties

Plant breeders have generally been hesitant to try to increase productivity in stress environments because genetic variance and breeding progress are less than under high-yielding conditions (Rosielle and Hamblin, 1981; Simmonds, 1991). The exploitation of genetic progress has therefore often been connected to the use of inputs and breeders have focused their efforts on favourable and high-input conditions (Bänziger and Cooper, 2001). Likewise, most MVs in sub-Saharan Africa have been selected for high yields under agronomically optimal conditions. However, when grown by smallholder farmers under their own management practices, they frequently failed to provide a sufficiently large advantage to be adopted (e.g. Kamara et al., 1996). Spillover effects from selection gains achieved under optimal conditions have been shown to decrease along with the yield level of the environment where the variety is deployed (Castleberry *et al.*, 1984; Duvick, 1984; Bänziger *et al.*, 1997; Ceccarelli and Grando, Chapter 12, this volume). It may therefore be of little surprise that relatively few varieties with superior performance under the very low yield levels of resource-poor farmers in southern Africa have been identified, by both researchers and farmers.

There have been many (mostly informal and little published) assessments of the type and relative importance of stresses that continue to prevail in farmers' fields in southern Africa, and of farmers' preferences when choosing among different maize varieties. Drought and nitrogen stress are generally recognized as the most prominent stress factors (Kumwenda *et al.*, 1997; Pixley *et al.*, 1997; Heisey and Edmeades, 1999). Their importance is often amplified by insufficient weed control and late planting caused by labour constraints (Kumwenda *et al.*, 1997). Other stress factors include leaf diseases (mainly grey leaf spot, northern leaf blight, maize streak virus and common rust), low pH soils with associated nutrient deficiencies and toxicities, ear rots, stem borers, and infestation with parasitic weeds (*Striga* sp.) (e.g. Pixley *et al.*, 1997). Postharvest damage by storage pests may assume as important a role as drought and nitrogen stress (SADC, unpublished). Apart from yield-related traits, smallholder farmers in southern Africa frequently mention their preference for: (i) early maturing varieties, as they escape late season drought, provide food when home stores become depleted and command a higher market price when sold as green maize; (ii) hard endosperm kernel types, as they are easier to pound, give more flour and are perceived to resist storage pests better than soft endosperm types; and (iii) good husk cover to protect the grain against storage pests and ear rots (e.g. Gebrekidan and Gelaw, 1990; Smale *et al.*, 1995; Rubey *et al.*, 1997; de Meyer and Bänziger, 2000).

Most formal breeding programmes in the region select maize for better disease resistance, and the consideration of smallholder farmers' preferences, such as earliness and hard endosperm grain type, has become more important in these breeding programmes. Because of lack of cost-effective breeding methodologies, little was done in the past to address abiotic constraints. Several studies have concluded that selection under abiotic stress conditions is more effective than selection under non-stress conditions for improving grain yield in environments where that specific abiotic stress occurs (Pederson and Rathjen, 1981; Atlin and Frey, 1990; Ceccarelli *et al.*, 1992; Bänziger *et al.*, 1997; Edmeades *et al.*, 1999). However, the difficulty of choosing appropriate selection environments that would lead to consistent gains in a target environment where type, intensity and occurrence of stress factors vary greatly over space and time has impeded the application of this concept. Multi-environment trials with a sufficiently large number of entries (to ensure significant breeding progress) and sites (to adequately sample the target environment), as conducted by formal breeding programmes in developed countries (e.g. Jensen, 1995; Cooper *et al.*, 1996), required more resources than most breeding programmes in southern Africa could afford. Formal breeding programmes therefore opted for exploiting the spillover effects to stress environments from selection under optimal conditions, though rarely questioning the cost-effectiveness of this approach.

Several national programmes and CIMMYT-Zimbabwe started in 1996 to employ a new breeding approach for improving the tolerance of maize to the most important abiotic stresses present in farmers' fields in southern Africa. The approach was based on a breeding methodology developed by CIMMYT during the 1980s and 1990s (Bänziger *et al.*, 2000a), for which substantial selection gains have been documented (Bolaños and Edmeades, 1993; Lafitte and Edmeades, 1994; Pandey *et al.*, 1994; Edmeades *et al.*, 1999). Sites were established on research stations in Angola, Botswana, Malawi, Mozambique, South Africa, Tanzania, Zambia and Zimbabwe where maize breeding materials can be screened for tolerance to drought, nitrogen stress and, more recently, low soil pH. In contrast to farmers' fields, it is possible to screen large numbers of breeding materials at these sites while managing the target stress uniformly. Results from these sites give information about the tolerance of the genetic material being evaluated to a specific abiotic stress factor. This information is combined with other selection criteria, such as responsiveness to good conditions, disease resistance and known preferences of farmers, for selecting those varieties that are likely to be more suitable to farmers' real conditions.

The rationale for this stress breeding approach has recently been reviewed and compared with other approaches targeted at stress environments (Bänziger and Cooper, 2001). In essence, it considers the fundamental elements of breeding progress by ensuring: (i) adequate genetic variation and a high selection pressure for priority traits; and (ii) an evaluation strategy that permits retention of high heritability while ensuring better genetic correlations between selection and target environment. The latter was achieved by making selection decisions based on results combined across managed stress and non-stress environments. There is increasing evidence that this selection strategy results in materials with improved yield and yield stability across a wide range of environments, including environments and stress levels other than where the germplasm was selected (Granados *et al.*, 1993; Byrne *et al.*, 1995; Bänziger *et al.*, 1999, 2000b; de Meyer and Bänziger, 2000).

The strategy of using managed stress and non-stress selection environments and how it contrasts with other approaches can be more easily understood by considering the following. When looking at sufficiently large numbers (i.e. a *random* set) of genotypes for two characteristics: (i) performance under a certain abiotic or biotic stress; and (ii) performance under non-stress, those genotypes can be sorted into the following four categories. A, genotypes poor for both traits; B, genotypes good for both traits; C, genotypes poor under stress and good under non-stress; and D, genotypes good under stress and poor under non-stress (Fig. 11.1 is an idealized representation of categories). B-type genotypes are desirable in areas such as southern Africa because they are more widely adapted, performing well under the most important stresses, and yet responsive to more favourable conditions. Thus, they are appropriate for small-scale farmers in marginal areas who cannot normally afford inputs, yet respond well when they can afford those inputs, and at the same time are appropriate for farmers who regularly use inputs or farm in more optimal environments. Indeed, many smallholder farmers in southern Africa grow maize under a range of conditions. They apply fertilizer or manure to fields with better soils or that are located near the homestead whereas other fields remain unfertilized or are on poorer soils. Likewise, the same variety may be exposed over the years to very different weather conditions. An appropriate maize variety thus must have adaptation to a range of stress levels.

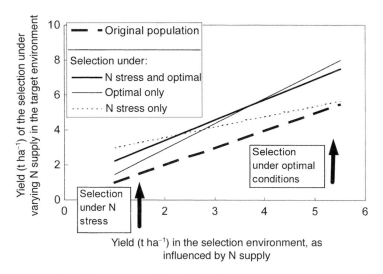

**Fig. 11.1.** Idealized description of the effect of selection under N stress and/or optimal conditions on the performance of selected materials under a range of N stress levels in the target environment.

Several studies have shown that genetic correlations between performance under stress and non-stress conditions are quite low (Pederson and Rathjen, 1981; Atlin and Frey, 1990; Ceccarelli et al., 1992; Bänziger et al., 1997; Ceccarelli and Grando, Chapter 12, this volume). As a result, there is very little prediction and improvement of performance under one type of condition possible (e.g. stress) if genotypes are evaluated only under the other type of condition (e.g. non-stress). To distinguish B-types from among the other genotypes (A, C, D), a breeding programme must therefore evaluate/select them under both stress and non-stress conditions. If genotypes are selected only under favourable conditions, performance under stress will not improve much, and C-types will be favoured. Likewise, if genotypes are selected only under stress conditions, performance under favourable conditions will not improve much, and mainly D-types will be selected.

Breeding theory says that a breeding programme seeking to improve two traits (performance under stress *and* non-stress) will make less progress for each individual trait compared with a breeding programme that pursues only one trait (performance under stress *or* performance under non-stress). Given the same amount of resources invested, C-types (selected for one trait only) are typically better than B-types (selected for two traits) in the favourable environment, and D-types are typically better than B-types in the stress environment. A breeding programme must therefore carefully assess the relative importance of stress and non-stress conditions in the target environment to appropriately apply selection pressure to each of the two conditions; overemphasized selection pressure for one type of conditions implies reduced breeding progress for the other type of conditions.

CIMMYT's Maize Program in southern Africa is currently pursuing the development of B-type germplasm in order to increase yields across a range of stress levels. As described above, this is done by evaluating the performance of breeding materials under severe (but carefully managed) stress and non-stress conditions. As selection pressure is applied for both traits, both traits are improved through inclusion of B-genotypes and exclusion of others and, as evidence shows (Bolaños and Edmeades, 1993; Bänziger et al., 1997), the selection response seems to be effective across a range of stress levels for that particular stress, and may also have spillover effects to other stress factors (Bänziger et al., 1999). By evaluating breeding materials under a range of important stress factors, each individually managed, *and* optimal conditions, it seems that a fairly small number of experiments can simulate to some extent a target environment that comprises a wide range of combinations of those stress factors. Interestingly, plant breeders have used the concept of managed stress environments widely for biotic stresses while many fewer attempts were made to assess the usefulness of this approach for abiotic stresses. Limited experience is therefore available with this approach for other crops.

## Characterizing new varieties at the regional level

Several public and private formal maize breeding programmes are developing new maize varieties in southern Africa. Among them are eight publicly funded national breeding programmes of varying strength, and 4–6 regionally operating private breeding programmes. National agricultural research and extension programmes, seed services, private seed companies and non-governmental organizations conduct trials and demonstrations of new (genetically fixed) varieties, in the past with little coordination among the different organizations. Funding constraints dictated that many of these variety trials were conducted at relatively few sites, less than would be required to adequately represent the highly variable growing conditions present in the area where the variety will be deployed. In addition, and with few exceptions, researchers and extension staff applied optimal crop husbandry to these trials, and release decisions and recommendations were made to smallholder farmers based on those results.

The practice of using optimal agronomic management in variety trials and demonstrations had two main justifications. Because trials were conducted at a few sites only, the use of farmer-managed (low-yielding) conditions would have increased the risk of obtaining results that are obscured by non-repeatable genotype × environment interactions and large error variances (Bänziger and Cooper, 2001), delaying the identification of varieties that seem superior across several trials. In addition, the 'seed-fertilizer paradigm' supported the promotion of MVs together with improved agronomic management practices (Heisey and Mwangi, 1997). Evaluating new crop varieties exclusively under optimal conditions probably promoted different crop varieties than if the same varieties had been evaluated under abiotic stress conditions. In the case of nitrogen stress, there is a decreasing relationship (genetic correlation) between the varieties that perform best under well-fertilized conditions and those that perform best under nitrogen stress, with the yield level decreasing as nitrogen stress becomes more severe (Fig. 11.2). Given the frequency of smallholder farmers growing MVs in southern Africa under less than optimal N supply (see e.g. Byerlee and Jewell, 1997; Kumwenda et al., 1997; Heisey and Mwangi, 1997), these results suggest that many inappropriate release decisions and recommendations have been made for maize in the past.

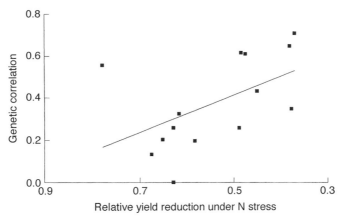

**Fig. 11.2.** Genetic correlation between grain yield under N stress and well-fertilized conditions plotted against the relative yield reduction under N stress [1-(GYLow N/GYHigh N)] for 14 maize trials comprising a total of 2978 different genotypes (from Bänziger et al., 1997).

CIMMYT has traditionally evaluated its germplasm with national agricultural research programmes and private seed companies, a practice similar to that of other International Agricultural Research Centres (IARCs). This testing approach was revised in 1998 by: (i) including the evaluation of trials at managed stress screening sites in different countries; and (ii) inviting other germplasm developers in the region (private companies, national programmes) to submit promising experimental varieties for testing. The objective was to evaluate new varieties from *any* breeding programme (i.e. not only those that adopted the new stress breeding approach) in the major agroecologies of southern Africa and under the stresses that are most common under resource-poor farmers' conditions. Through the collaboration of more than 40 scientists in over 25 institutions, data on performance under drought, nitrogen stress and low pH stress, resistance to grey leaf spot, northern leaf blight, maize streak virus and common rust, ear rots, grain weevils and *Striga* infestation are now routinely being collected, and characteristics that may affect farmers' acceptance, such as grain texture, husk cover and earliness, are being established (e.g. Bänziger *et al.*, 2000b; Vivek *et al.*, 2001).

Interest in these kinds of trials, as evidenced by the number of trial sets distributed to collaborators in southern Africa, grew rapidly, from 152 trial sets distributed in 1998 to 317 in 2001. Every year, approximately 150 varieties and 50 inbreds are characterized in five or six different trials and, given the current return of data, data from 35–50 sites are reported for each variety (Vivek *et al.*, 2001). Results are summarized based on a regional GIS-based classification of maize growing environments (Hartkamp *et al.*, 2000). The results of the trials are widely distributed to any interested party, including ministries of agriculture, seed companies, non-governmental organizations and seed relief programmes.

## Variety evaluation and selection with farmers

Selection and evaluation under environmental conditions similar to those found in farmers' growing environments increased the frequency and number of varieties that supposedly better met farmers' needs. At this stage, collaboration with farmers was essential for assessing the performance and acceptance of such pre-selected varieties under their conditions, and for a wide range of stakeholders (including farmers) being able to select those that are specifically adapted to a given target area. Researchers and NGOs have been using farmer-participatory variety evaluation in southern African countries. Inadequate human and financial resources, high transport costs and poor road infrastructure often limited such efforts to a relatively small sample of sites and farmers in each country. And, due to reasons explained above, farmer-participatory variety evaluation was usually restricted to assessing farmers' preferences while variety performance under farmers' management practices was rarely determined.

In 1999, we initiated a pilot project in Zimbabwe to develop a new variety testing system that evaluates the performance and acceptance of new maize varieties under farmers' conditions, and creates a cost-effective and simple flow of information between breeders, extension staff and farmers. The resulting testing scheme was based on two concepts: (i) an innovative experimental design – the 'Mother–Baby Trial Design' (Snapp, 1999) – that was adapted for variety evaluation and linked to regional trials; and (ii) wide-ranging and organized partnerships with farmers and farmer organizations interested in variety testing.

Snapp (1999) first used the Mother–Baby Trial Design to evaluate soil fertility management practices with farmers in Malawi: a replicated 'conventional' on-farm trial including a range of treatments was grown in the centre of a farming community. Several farmers of that community evaluated the four most promising treatments in an unreplicated design on their own farm. The design proved useful for farmer-participatory research because the Mother Trial permits farmers to choose from a range of new technologies those that they want to evaluate in Baby Trials on their own farm during the following season (Snapp, 1999).

## Adapting Mother–Baby Trials for variety evaluation

We adapted the Mother–Baby Trial Design for a farmer-participatory variety evaluation scheme in the following manner. The Mother Trial was a replicated researcher-managed trial, planted in the centre of a farming community, typically with a school or a progressive farmer. It evaluated 12 cultivars under two input levels, using two-row plots and three replications. Input levels were determined individually for the sites where the trials were evaluated. One level represented recommended input application as advised by extension services or the local partner organization for the area where the trial was grown. The other level was grown using the amount of fertilizer and/or lime that was representative for farmers' practice in the area, often none at all.

Farmers' practice was established in discussion with local partners and was based either on a survey or, in the case of partners that have been working for several years in the area, on their expert opinion. The two input levels of one trial were planted side-by-side on the same date and were weeded as recommended for the site. Cob weight, shelling percentage and grain moisture were measured at harvest and data were analysed using a spatial analysis program (Gilmour *et al.*, 1998). From the breeding perspective, the two input levels created two environments per site, each representing relevant growing conditions. In our case, inorganic fertilizer application was the main difference between the two input levels. However, other factors such as land preparation, planting date or application of organic fertilizers could also be included in the two input levels, particularly if local partners are interested in demonstrating and promoting such practices to farmers.

Baby Trials were grown by at least six farmers in a community that hosted a Mother Trial. The local partner organization decided how to choose farmers for hosting Baby Trials and we describe the implications of this approach below. Each Baby Trial contained four of the varieties evaluated in the Mother Trial and all entries in the Mother Trial were represented among the Baby Trials. Cultivars were allocated to Baby Trials using an incomplete lattice design (Patterson and Williams, 1976); each Baby Trial (or farm) was an incomplete block and there were two replications in each community. Plot size in the Baby Trial was determined by the amount of seed: 650 seeds per variety. Farmers were asked to plant the seed using a plot length of about 15 m, but choosing their own planting distance. The seed was packed in coloured plastic bags with the name of the variety on it, and four stones painted with the same colours (blue, red, yellow and green) were included as plot markers. Farmers were requested to treat the four cultivars uniformly but follow their own management practices. Preferably the Baby Trial was grown next to the farmer's own maize to allow for comparisons between the farmer's choice of variety and trial entries. Cob weight was taken from a subplot at harvest, leaving some border rows for farmers to harvest as green maize. Using a simple questionnaire that was filled out together with a representative from the local partner organization, farmers provided information on their socioeconomic circumstances, type of characteristics they value in maize varieties and an assessment of the four varieties in the Baby Trial for a common set of criteria.

Trial entries came from several public and private breeding programmes. Trial entries were chosen by the breeders supplying the seed 'as being the best-bets for smallholder farmers' conditions'. This modification of Snapp's (1999) original Mother–Baby Trial method was made because of the project's goal of exposing farmers to *new* varieties. As the system generates information on farmers' criteria for varietal choice, those criteria are expected to influence the choice of the varieties (selected, for example, from regional trials) to be evaluated with farmers in future years. In the pilot project, trial entries consisted of newly released hybrids that were little known to farmers and experimental hybrids and open-pollinated varieties. The entries were allocated to Baby Trials in a manner so that each farmer evaluated no more than two experimental entries. Farmers were informed that not all of the seed they were evaluating was available on the market but that their feedback would help to decide whether they will become available on the market.

## Involving partners and farmers

The new variety testing scheme took into consideration that many farmers, farmer organizations, rural development projects and non-governmental organizations (NGOs) have an interest in variety testing. It exposes farmers to new genetic variability and enables them to make more informed choices about crop varieties that best meet their preferences. The Mother–Baby Variety Testing Scheme was first tried in Zimbabwe during the 1999/2000 season. Partners interested in variety testing were identified by word of mouth. Within the first season, 43 sites (Mother–Baby Trial sets) were being planted with a total of 11 partners and 249 farmers (Table 11.1, Fig. 11.3). Participation in the scheme increased to 14 partners planting 49 sets with 316 farmers in the second season, and more communities and farmers would have participated if additional trial sets had been available.

Types of partners included extension services, research stations, NGOs, rural development projects, farmer associations and secondary schools. Many of them had previously attempted to evaluate maize

**Table 11.1.** Partners involved in the Mother–Baby Trial Scheme in Zimbabwe in 1999/2000 and 2000/2001.

| | | Number of sites | |
|---|---|---|---|
| Name | Type of partner | 1999 | 2000 |
| Partner supervising one or several Mother–Baby Trial sets | | | |
| Agritex | Public extension services | 12 | 15 |
| CARE International | NGO | 9 | 9 |
| Farmers' groups | Informal farmer association | 7 | 5 |
| CIMMYT | International agricultural research centre | 4 | 2 |
| Department for Research and Specialist Services (DR&SS) | National agricultural research programme (public) | 3 | 4 |
| University of Zimbabwe | Public university | 2 | 2 |
| Horse Shoe Farmer Association | Formal farmer association | 2 | 1 |
| Intermediate Technology Development Group (ITDG) | NGO | 1 | 2 |
| Scientific and Industrial Research and Development Center (SIRDC) | National agricultural research programme (public) | 1 | 1 |
| Smallholder Dry Area Resource Management Project (SDARMP) | Rural development project | 1 | 1 |
| Southern African Unit for Land Resource Development (SALRED) | NGO | 1 | 1 |
| Development Aid From People to People (DAPP) | NGO | – | 3 |
| Rural Unity Development Organization (RUDO) | NGO | – | 2 |
| Chiweshe Secondary School | Secondary school | – | 1 |
| Total number of Mother–Baby Trial sites | | 46 | 49 |
| Host of the Mother Trial | | | |
| Secondary school | | 16 | 20 |
| Farmer | | 20 | 22 |
| Research station | | 7 | 7 |
| Total number of Mother–Baby Trial sites | | 43 | 49 |

varieties with farmers, based on requests by the farming communities they were working with. Several of them had failed because: (i) they did not have the technical know-how to conduct and analyse a trial; (ii) they lacked access to seed of different varieties; or (iii) they did not know what varieties to grow. Partners appreciated that they were given a trial set with all the seed prepared, technical backstopping from the coordination unit (in this case CIMMYT) and access to the data

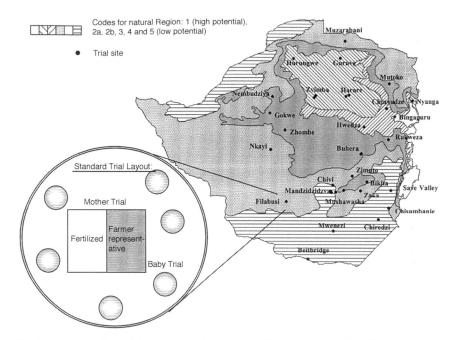

**Fig. 11.3.** Location of the sites hosting Mother–Baby Trials in Zimbabwe during the 1999/2000 season, with detail of the standard layout of trials. GIS information is from Corbett *et al.* (2001).

gathered country-wide. Partners were ready to contribute to trial costs in terms of land, labour, transport and trial supervision. This resulted in considerable cost-savings for this system compared with conventional on-farm testing approaches.

The Mother–Baby Trial set gave each partner a range of opportunities to interact with the community. The Mother Trial was the venue for field days organized by partners and demonstrated both different varieties and the effect of different input levels to the local farming community. On field days, farmers hosting Baby Trials reported on the performance of the varieties in their own fields. Baby Trials were planted within walking distance to the Mother Trial to foster interaction and discussions among farmers. Some farmers reported that up to 50 neighbours and community members visited their Baby Trial, demonstrating some of the informal interaction potential of the Mother–Baby Trial concept. Several partners planted the Mother Trial at a secondary school with the help of the agricultural teacher and students. This further fostered information exchange with the community.

Farmers hosting Baby Trials appreciated that they could try out new varieties under their own conditions and management practices, and hence make an informed decision about what variety to grow in the coming season. The local partner discussed trial results with the participating farmers and the wider community. This typically resulted in farmers agreeing on a preference for one or two varieties. Where a community selected an open-pollinated variety (OPV), foundation seed and simple seed production instructions were made available to initiate community-based seed production, usually with the help of the local partner. Where farmers selected hybrids, they started to look for the seed in the shops or organized bulk purchases through the local partner. Approximately half of the hybrid entries were experimental and the seed were not yet available on the market. Farmers seemed to accept that the information on these hybrids would be used to decide on future releases, and they were satisfied with being able to make a choice among the already released hybrids or choose an OPV for community-based seed production.

Among the many elements incorporated in the Mother–Baby Variety Evaluation Scheme, training of partners and the use of a colour scheme for keeping track of individual varieties in farmers' fields proved of particular value. Both were key to timely and decentralized planting of Mother and Baby Trials by partners and farmers, respectively. Farmers might have been unable to recall or read the name of the variety on the seed bag, but they were always aware of the colour of a certain variety and its location in their own field. Because of the limited number of colours available, it was not possible to assign a unique colour to each variety (i.e. across all trials). Farmers therefore relied on a literate person to make that type of comparison. The first year of establishing the evaluation scheme involved a considerable effort to train partners, but it was reduced considerably in the second year because partners and farmers were familiar with the procedures. Moreover it became apparent that the coordination unit of such trials can easily cope with a certain proportion of partners and farmers entering and leaving the system every year as their interest changes.

Within a year of the initial Mother–Baby pilot project in Zimbabwe, the national agricultural research programmes in five other countries (Angola, Malawi, Mozambique, South Africa, Tanzania) initiated similar schemes, with the other southern African countries planning to follow in coming years. Several NGOs have initiated community-based seed production schemes and demonstration plots with the OPVs selected by farming communities from the Mother–Baby Trials. Through the involvement of extension staff in these trials, they were more confident about making variety recommendations to the wider community of farmers, and adoption of trial entries that had previously not been known to farmers increased.

Some of the OPVs that farmers (and breeders) selected after the 2000 season for community-based (and commercial) seed production originated from the stress breeding approach initiated in 1996. The components of these OPVs were selected from among more than 2000 maize breeding materials evaluated during 1997, using results combined across managed stress and non-stress screening sites. The OPVs were composed in 1998, evaluated in regional trials during 1999 and 2000, and in Mother–Baby Trials during 2000 and 2001. By the beginning of the 2001 season, these OPVs had been evaluated in 86 regional trials and 54 Mother and Baby Trials, probably a much more thorough, appropriate and faster evaluation than had been conducted in the past for new varieties in this region.

## Farmer Participation in Breeding and the Need for Collaborative Plant Breeding Approaches

There has been, in recent years, an increasing interest in participatory research in general and towards farmers' involvement in plant breeding in particular. Following the early work of Rhoades and Booth (1982), scientists have become increasingly aware that user participation in technology development may increase considerably the probability of success for the technology (Ceccarelli and Grando, 2000). The multitude of benefits arising from involving farmers in the formal breeding process have been well documented: enhanced genetic diversity on farm, reduced research costs, encouraged dialogue between farmers and scientists, better understanding of farmers' criteria, concentration of efforts and resources towards further improving preferred varieties, increased appreciation of indigenous knowledge by scientists, and increased rates of technology transfer and variety adoption. In many examples, farmers benefited from new genetic materials 5–6 years in advance of the conventional formal system and with less effort (Sperling et al., 1993; Otim Nape et al., 1994; Ipinge et al., 1996; Kamara et al., 1996; Witcombe et al., 1996).

A range of approaches have been used that involve farmers more closely in crop development and seed supply. They may be differentiated into farmer-led approaches that support farmers' own systems of crop development and seed exchange, and approaches where farmers join in experiments that have been initiated by the formal sector (Smith et al., 2000). Some of the most common approaches are as follows:

1.  A group of farmers is brought to the research station to give their input in the selection process. The selected varieties are tested on farm subsequently (Sperling et al., 1993; Ipinge et al., 1996).
2.  Evaluation of pre-release varieties on farm for assessing acceptability (Butler et al., 1995; Joshi and Witcombe, 1996; Thiele et al., 1997).

**3.** Farmers grow and select segregating materials in their own field (Ceccarelli *et al.*, 1996; Sthapit *et al.*, 1996).
**4.** Use of a parallel breeding scheme on farm (for selection) and on station (to fulfil the demands of the formal release system) (Witcombe *et al.*, 1996).

Each of these approaches has its challenges and benefits, and different approaches have been used in different environments or for different goals and crops. In most cases, farmers participated during the testing of genetically fixed materials in the latest stage of selection, or even in the evaluation of released varieties: a group of representative farmers select a limited number of varieties on station for growing in their fields; or after involving farmers in priority setting and defining breeding targets, a group of farmers was given a number of varieties for testing in their own fields. However, participation in this last stage of the breeding process lacks the cyclic nature of plant breeding, and it is not clear whether, how and when a farmer will have another chance in participating in variety selection. Therefore many examples are part of a linear process and could be described as sporadic, episodic or occasional participation (Ceccarelli and Grando, 2000).

There is little doubt that farmers' skills in selecting appropriate materials can match those of breeders and the process of selection alone does not justify a formal plant breeding approach (Witcombe *et al.*, 1996; Ceccarelli and Grando, 2000). Establishing organized but collaborative structures in a plant breeding approach is desirable and needed because:

**1.** Logistical constraints prohibit a large number of farmers from being exposed to a reasonably large number of genetic combinations;
**2.** Environmental variation across sites may represent some of the variation encountered at one site (or by one farmer) over time, and users therefore benefit from information gathered for a variety at other sites;
**3.** As the trust in information compiled by others increases, users may change more rapidly to other varieties as their preferences and socio-economic circumstances change;
**4.** Significant selection progress is often the result of a cyclical process with a relatively large cost–benefit ratio when viewed from the individual farm perspective, with the result that farm-based breeding rarely attracts the continued attention of a large number of farmers.

The most intrinsic advantage for involving farmers in the breeding process is that they contribute a sample of the 'real' growing conditions, they express their preferences, and these preferences are based on the socio-economic constraints they encounter. Breeders' intrinsic

advantage is that they have the essential understanding of the underlying genetics of parental selection and subsequent genetic segregation
(Witcombe *et al.*, 1996). The main objective of a CPB approach, like the
one described here, is therefore to expose representative farmers to
relevant genetic variation and establish a sustainable collaboration
that enables breeders, as well as other stakeholders involved in the seed
sector, and the wider population of farmers to make better decisions on
what variety to promote or use. It is natural that different parties will
make different selection decisions in this system. This fosters the use of
genetically diverse germplasm and allows the exploitation of specific
adaptation of varieties.

## Collaborative Elements and their Benefits in the Maize Breeding Approach for Southern Africa

Scaling-up has been a major issue and constraint when involving farmers in plant breeding. There are more than 100 million men, women
and children dependent on agriculture in southern Africa. Giving them
the opportunity to benefit from useful genetic variation, as a result of a
formal or informal selection process, is the ultimate challenge to plant
breeding in this region. The maize breeding approach described in the
first part of this chapter has as a premise that, because of the intricate
and highly variable biophysical and socio-economic environment,
no single breeding approach can reach all potential beneficiaries. It
therefore invests in a multi-level approach that attempts to consider the
diverse strength of stakeholders, their diverse interest in the breeding
process, and the diverse type of products needed to match the demands
of different environments, management practices, preferences and
socio-economic circumstances.

Plant breeding – whether done by farmers or scientists – is about
the generation of new genetic combinations and the selection of those
combinations that seemed the most useful under a particular set of
biophysical and socio-economic conditions. Similarly, new genetic
combinations generated by different parties and selection done by the
same and other parties and farmers in a region such as southern Africa
can be viewed as one large breeding effort for which the fundamental
concepts of plant breeding theory apply. Given that resource-poor
farmers have insufficiently benefited from the results of modern
plant breeding, creating collaborative approaches that lead to more
appropriate selection of varieties within this effort is the challenge.

Considering the fairly recent introduction of maize into sub-
Saharan Africa (Byerlee and Heisey, 1997), the genetic diversity of
maize varieties used by farmers is likely to be less than that existing

globally and particularly in Latin America. We estimate that several hundred thousand new gene combinations are created every year by formal maize breeding programmes in southern Africa, and to the extent that they include introductions from Latin America, the USA and Europe, it can be assumed that the genetic base is widened. Formal breeding programmes therefore have an advantage over farmers in accessing genetic variability (as it originates from recombining different and introducing new breeding materials), which is a premise for finding trait combinations that better suit the needs of farmers and their environments.

Because the existence of useful genetic variance was probably not a limiting factor in making breeding progress in southern Africa, the emphasis on improving the relevance of breeding for poorer farmers has focused on influencing formal maize breeding to appreciate small-holder farmers' needs and on giving a wider range of stakeholders the opportunity to make informed and potentially diverse selection decisions. Given the existing breeding practices in the region, the following three interventions were chosen:

**1.** Selection of maize breeding materials for tolerance to the most important abiotic stress factors prevalent in farmers' fields.
**2.** Evaluation of promising new varieties from any germplasm developing programme through common regional trials for a wide range of growing conditions and characteristics important to small-holder farmers.
**3.** Farmer-participatory evaluation of pre-selected varieties that match the perceived conditions and preferences in a country through Mother–Baby Trials with direct feedback to the criteria chosen to select entries for such trials.

## Collaborative trials

Evaluating new breeding materials through collaborative trials is not a new approach. It has been used under resource-constrained conditions where individual breeding programmes cannot sample the entire target environment, or under circumstances where germplasm users – often farmer organizations – want more information on the relative merit of varieties. Unique to the approach in southern Africa, it exposes a relatively large number of breeding materials from different germplasm developers to very diverse conditions, less in the search for the 'best' variety but for rapidly and reliably assessing the relative merit of new varieties for a wide range of growing conditions and characteristics that are important to farmers in southern Africa. Both the information created and the linkages established among institutions enabled users with diverse interests (farmers, breeders, seed companies, government services, NGOs, agencies involved in seed relief, etc.) to design individual follow-up strategies.

Regional trials and Mother–Baby Trials influenced formal plant breeding programmes in several aspects. In the case of selection for abiotic stress tolerance, regional trials and Mother–Baby Trials established a platform where differences in the performance of maize varieties under optimal conditions and abiotic stress conditions or farmers' conditions became quickly transparent. This has increased the credibility and acceptance of this new breeding approach. Several public and private maize breeding programmes in the region are now using managed stress screening sites, yet more experience with the methodology is needed for breeders to increase the selection pressure for these traits to a level that is likely to be more appropriate given their importance in the target environment. In the case of farmers' preferences, formal breeding programmes have been directly and immediately exposed through Mother–Baby Trials to farmers' rejection of varieties that breeders may have perceived as being appropriate given their performance on station and breeders' knowledge of farmers' preferences. Finally, there is the potential for formal plant breeding programmes to use the information coming from regional trials and Mother–Baby Trials for making strategic decisions on seed production and geographical deployment.

To the wider range of stakeholders involved in the seed sector, regional trials and Mother–Baby Trials created a virtual market place where seed disseminators (private companies, NGOs, government agencies, seed importers) can approach germplasm developers (CIMMYT, national programmes, private seed companies) and seed producers (private seed companies, community-based seed production schemes) for seed of varieties with desirable trait combinations for further testing, commercialization, sale or seed relief activities. As conditions and preferences vary with the target environment where the varieties should be deployed, different varieties are being selected. Several national programmes and NGOs have started to use regional trial results for selecting maize varieties for further testing in a country, more recently by using the Mother–Baby Trial approach.

Local partners and farmers involved in Mother–Baby Trials benefited from information about new germplasm, the possibility of making their own choice on what variety to grow next year, and the possibility of designing their own follow-up strategies (e.g. bulk purchases, community-based seed production schemes) to get access to seed of desired OPVs or hybrids. Local farmers and partners suggested that information from Mother–Baby Trials should be made available to retailers to increase the availability of appropriate varieties.


## Partnerships

Given the financial, structural and personnel constraints in regions like southern Africa, attempts by individual researchers and even organizations to evaluate new crop varieties in a reasonable time period and in a manner that does justice to the highly variable growing conditions, diverse crop management practices and user preferences are prone to failure. Lack of access to communication and transport, poor road networks and high cost of travel combined with financial and personnel constraints result in few trials being planted, delayed planting, poor supervision of trials and poor contact with farmers. The same constraints render it difficult to make use of the benefits of farmer-participation in breeding.

Collaboration between different actors – governmental and non-governmental – interested in variety testing and seed production systems is an effective way to avoid these constraints. Some NGOs treat research as one of various means of seeking viable production alternatives for improving food and income security, others are more interested in the aspect of farmer empowerment. As variety testing seems to fall into the domain of researchers, extension services have often been ignored even though extension staff are better positioned for frequent contacts with farmers and are often queried by farmers about varieties. Agriculture is taught at secondary schools and many agricultural teachers show a lively interest in research. The idea of evaluating a Mother–Baby Trial together with their students and the community is more appealing to them than simply growing a maize crop at the school plot. Farmer organizations typically show enthusiasm and keen interest in the information created through collaborative research, and the benefit to their community organization is apparent in the follow-up to the information created by the trials.

In our approach, local partners provided established links to the community and knowledge of the problems faced by farmers in the area. Because the trials were located nearby, it was easy for a partner to visit and monitor them and provide feedback to the coordinating unit. Collaboration with an interested partner was only established if the farming community had identified improved knowledge of maize varieties as a priority topic and we think that this was an important reason for the success of the trials. Local partners decided how to select the farmers that participated in Mother–Baby Trials. It emerged that different partners were addressing different groups of farmers. Some of our partners were working with more wealthy influential farmers or more informed farmers, others were focusing on working with the poorest farmers. Some partners were providing credits to farmers, others were working in very remote stress-prone areas with little or no access to inputs. Thus, different partners helped to achieve better representation of different farmers and farming communities.

As established by many farmer-participatory breeding approaches, farmers took a unique part in this approach. Farmers contributed the representative growing environments, feedback on preferences and insights into their socio-economic conditions. They transmitted an urgent demand to breeders and seed disseminators for developing and making seed of more appropriate varieties available, and contributed innovative ideas for desirable follow-up strategies. As users and ultimate beneficiaries they took part in the technology development and transfer process, and through their interaction with the wider farming community, they obviously facilitated the adoption and dissemination of new varieties.

## Conclusions and Challenges

The challenge of CPB is larger than merely establishing better collaboration between farmers and individual plant breeding programmes. It is in the poorest countries where the population is the most dependent on agriculture, where the environments are the most challenging and where there is the most urgent need to give farmers widespread access to opportunities that will improve their livelihood. It is in these same countries where financial, structural and human resource constraints persist and impede the realization and impact of traditional breeding approaches.

The breeding approach described in this chapter (re-)applies the theoretical basis of plant breeding to a target environment that is highly variable, stress-prone and hampered by resource constraints. Theory says that selection response in the target environment is the largest when each of the four factors, genetic variance, selection intensity, heritability in the selection environment and genetic correlation between selection and target environment, is high (adapted from Falconer, 1989). Managed stress screening sites allowed the exposure of more breeding germplasm to stress factors that are relevant in farmers' fields while keeping heritability at the screening sites high and applying a high selection intensity. Collaborative trials and partnerships among stakeholders involved in the formal and non-formal seed sector were then powerful approaches to better expose varieties pre-selected by a range of formal breeding programmes to a wider range of environments, thus better representing growing conditions, management practices and farmers' preferences in the potential target environment of those varieties.

Collaborative breeding approaches have been successfully used under resource-constraining conditions, and the example of maize in southern Africa is only one of them. Establishing a collaborative breeding approach is still non-conventional, and requires plant breeders (and seed release authorities) to rethink their premises and accept that they will achieve their objectives less effectively by simply copying a traditional 'first-world approach'. In temperate areas, farmers' environments are more similar to optimal experiment station environments and resources are available for individual breeding programmes to sample the target environment adequately during the testing phase. Given the experience gained with maize in southern Africa, we think that private and public partners can pursue their basic goals through a collaborative approach. For example, the multitude of partners involved in the evaluation of collaborative trials gives an independent assessment of the merit of new varieties. This information can be used for traditional 'release' decisions, yet the closer collaboration between germplasm developers and clients in a CPB approach is likely to achieve the same goal. Some germplasm developers may be uncomfortable with the transparency established in common trials, yet cost-effectiveness, the exposure to a wider range of environments and potential users, and the marketing effect convinced virtually all breeders in our region of the merit of collaborative trials, and they supplied what they considered to be their best materials (compare Duvick, Chapter 8, this volume). Collaboration may have been facilitated in our case by the fact that competitors cannot easily 'steal' hybrid varieties and that OPVs have a relatively small market value. We acknowledge that trust is an important element in collaborative approaches and cannot be simply duplicated under different circumstances.

Making informed choices about crop varieties and having access to seed of those varieties are priorities to farmers and these interests are represented by development organizations/projects, grassroots level organizations and extension services. A CPB approach gives these stakeholders and farmers the possibility of influencing breeding priorities of the formal sector, to apply diverse selection criteria for diverse purposes and environments, and design appropriate and diverse follow-up strategies (such as community-based seed production schemes; bulk purchases of seed through farmer organizations, local partner organizations or dealers; further promotion of farmer-selected varieties through demonstration plots, etc.) that stimulate the dissemination of more appropriate and genetically diverse varieties. As more appropriate varieties are being deployed through different channels, the benefits of useful genetic variation are expected to improve productivity and stimulate an increased demand for seed and other agricultural inputs. Thus, in contrast to the perception of some seed producers, the approach we used in southern Africa shows that increased transparency and diverse deployment strategies are less in competition with the formal seed system but rather stimulate the demand for seed of appropriate varieties.

Given the level of collaboration achieved in southern Africa, the challenges ahead lie in the further exploration of the merits of collaboration particularly in the area of seed production and dissemination, the institutionalization of the approach, and the adjustment of policies in support of this system.

## Acknowledgements

Credit should go to all our colleagues, collaborators and farmers whose commitment and critical and innovative input shaped the breeding approach described in this chapter.

## References

Aquino, P., Carrión, F., Calvo, R. and Flores, D. (2001) Selected maize statistics. In: Pingali, P. (ed.) *CIMMYT 1999–2000 World Maize Facts and Trends. Meeting World Maize Needs: Technological Opportunities and Priorities for the Public Sector.* CIMMYT, Mexico D.F., pp. 45–60.

Atlin, G.N. and Frey, K.J. (1990) Selecting oat lines for yield in low-productivity environments. *Crop Science* 30, 556–561.

Bänziger, M. and Cooper, M.E. (2001) Breeding for low-input conditions and consequences for participatory plant breeding – examples from tropical maize and wheat. *Euphytica* (in press).

Bänziger, M., Betrán, F.J. and Lafitte, H.R. (1997) Efficiency of high-nitrogen selection environments for improving maize for low-nitrogen target environments. *Crop Science* 37, 1103–1109.

Bänziger, M., Edmeades, G.O. and Lafitte, H.R. (1999) Selection for drought tolerance increases maize yields over a range of N levels. *Crop Science* 39, 1035–1040.

Bänziger, M., Edmeades, G.O., Beck, D. and Bellon, M. (2000a) *Breeding for Drought and Nitrogen Stress Tolerance in Maize.* CIMMYT, Mexico D.F.

Bänziger, M., Pixley, K.V., Vivek, B. and Zambezi, B.T. (2000b) *Characterization of Elite Maize Germplasm Grown in Eastern and Southern Africa: Results of the 1999 Regional Trials Conducted by CIMMYT and the Maize and Wheat Improvement Research Network for SADC (MWIRNET).* CIMMYT, Harare, Zimbabwe.

Bolaños, J. and Edmeades, G.O. (1993) Eight cycles of selection for drought tolerance in lowland tropical maize. I. Responses in grain yield, biomass and radiation utilization. *Field Crops Research* 31, 233–252.

Butler, L.M., Myers, J., Nchimbi-Msolla, S., Massangye, E., Mduruma, Z.O., Mollel, N. and Dimosa, P. (1995) Farmer evaluation of early generation bean lines in Tanzania: comparisons of farmers' and scientists' trait preferences. In: *Southern Africa Development Community (SADC) Regional Bean Research Workshop, 2–4 October 1995.* Oil and Protein Seed Center, Potchefstroom, South Africa.

Byerlee, D. and Heisey, P.W. (1997) Evolution of the African maize economy. In: Byerlee, D. and Eicher, C.K. (eds) *Africa's Emerging Maize Revolution.* Lynne Rienner Publishers, Boulder, Colorado, pp. 9–22.

Byerlee, D. and Jewell, D.C. (1997) The technology foundation of the revolution. In: Byerlee, D. and Eicher, C.K. (eds) *Africa's Emerging Maize Revolution.* Lynne Rienner Publishers, Boulder, Colorado, pp. 127–143.

Byrne, P.F., Bolaños, J., Edmeades, G.O. and Eaton, D.L. (1995) Gains from selection under drought *versus* multilocation testing in related tropical maize populations. *Crop Science* 35, 63–69.

Castleberry, R.M., Crum, C.W. and Krull, C.F. (1984) Genetic yield improvement of U.S. maize cultivars under varying fertility and climatic environments. *Crop Science* 24, 33–36.

Ceccarelli, S. and Grando, S. (2000) Quality of science in participatory plant breeding. In: PRGA Program (ed.) *Uniting Science and Participation in Research. Proceedings of the Third International Seminar of the CGIAR Program on Participatory Research and Gender Analysis for Technology Development and Institutional Innovation, 6–9 November 2000, Nairobi, Kenya.* CIAT, Cali, Colombia.

Ceccarelli, S., Grando, S. and Hamblin, J. (1992) Relationship between barley grain yield measured in low- and high-yielding environments. *Euphytica* 64, 49–58.

Ceccarelli, S., Bailey, E., Grando, S. and Tutwiler, R.N. (1996) Decentralized participatory plant breeding: a link between formal plant breeding and small farmers. In: PRGA Program (ed.) *New Frontiers in Participatory Research and Gender Analysis. Proceedings of the First International Seminar on Participatory Research and Gender Analysis for Technology Development, 9–14 September 1996.* CIAT, Cali, Colombia, pp. 65–75.

Cooper, M., Brennan, P.S. and Sheppard, J.A. (1996) A strategy for yield improvement of wheat which accommodates large genotype by environment interactions. In: Cooper, M.E. and Hammer, G.L. (eds) *Plant Adaptation and Crop Improvement.* CAB International, Wallingford, UK, pp. 487–511.

Corbett, J.D., Collins, S.N., Bush, B.R., Jeske, R.Q., Martinez, R.E., Zermoglio, M.F., Lu, Q., Burton, R.A., Muchugu, E.I., White, J.W. and Hodson, D.P. (2001) *Almanac Characterization Tool. A Resource Base for Characterizing the Agricultural, Natural and Human Environments for Selected African Countries.* Texas Agricultural Experiment Station, Texas A&M University System, Blackland Research Center Report 01–08, March 2001.

de Meyer, J. and Bänziger, M. (2000) *Evaluation of Maize Varieties in Farmers' Fields in Zimbabwe Using the Mother-Baby Trial Scheme. Results of the 1999/2000 Season.* CIMMYT, Harare, Zimbabwe.

Duvick, D.N. (1984) Genetic contribution to yield gains of U.S. hybrid maize, 1930 to 1980. In: Fehr, W.R. (ed.) *Genetic Contribution to Yield Gains of Five Major Crop Plants,* CSSA Special Publication 7. CSSA and ASA, Madison, Wisconsin, pp. 15–47.

Edmeades, G.O., Bolaños, J., Chapman, S.C., Lafitte, H.R. and Bänziger, M. (1999) Selection improves tolerance to mid/late season drought in tropical maize populations. I. Gains in biomass, grain yield and harvest index. *Crop Science* 39, 1306–1315.

Eicher, C.K. and Byerlee, D. (1997) Accelerating maize production: synthesis. In: Byerlee, D. and Eicher, C.K. (eds) *Africa's Emerging Maize Revolution.* Lynne Rienner Publishers, Boulder, Colorado, pp. 247–262.

Eicher, C.K. and Kupfuma, B. (1997) Zimbabwe's emerging maize revolution. In: Byerlee, D. and Eicher, C.K. (eds) *Africa's Emerging Maize Revolution.* Lynne Rienner Publishers, Boulder, Colorado, pp. 25–43.

Falconer, D.S. (1989) *Introduction to Quantitative Genetics,* 3rd edn. Longman Scientific and Technical, Burnt Mill, Harlow, UK.

Gebrekidan, B. and Gelaw, B. (1990) The maize mega-environments of eastern and southern Africa and germplasm development. In: Gebrekidan, B. (ed.) *Maize Improvement, Production and Protection in Eastern and Southern Africa: Proceedings of the Third Eastern and Southern Africa Regional Maize Workshop.* CIMMYT, Nairobi, Kenya, pp. 197–211.

Gilmour, A.R., Cullis, B.R., Welham, S.J. and Thompson, R. (1998) *ASREML.* NSW Agriculture, Orange, NSW, Australia.

Granados, G., Pandey, S. and Ceballos, H. (1993) Response to selection for tolerance to acid soils in a tropical maize population. *Crop Science* 33, 936–940.

Hartkamp, A.D., White, J.W., Rodriguez Aguilar, A., Bänziger, M., Srinivasan, G., Granados, G. and Crossa, J. (2000) *Maize Production Environments Revisited: a GIS-based Approach.* CIMMYT, Mexico D.F.

Heisey, P.W. and Edmeades, G.O. (1999) Maize production in drought-stressed environments: technical options and research resource allocation. Part 1. *CIMMYT 1997/98 World Maize Facts and Trends.* CIMMYT, Mexico D.F., pp. 1–36.

Heisey, P.W. and Mwangi, W. (1997) Fertilizer use and maize production. In: Byerlee, D. and Eicher, C.K. (eds) *Africa's Emerging Maize Revolution.* Lynne Rienner Publishers, Boulder, Colorado, pp. 193–211.

Ipinge, S.A., Lechner, W.R. and Monyo, E.S. (1996) Farmer participation in on-station evaluation of plant and grain traits: the case of pearl millet in Namibia. In: Leuschner, K. and Manthe, C.S. (eds) *Drought-tolerant Crops for Southern Africa. Proceedings of the SADC/ICRISAT Regional Sorghum and Pearl Millet Workshop, 25–29 July 1994, Gaborone, Botswana.* ICRISAT, Patancheru, Andhra Pradesh, India, pp. 35–42.

Jensen, S. (1995) Genetic improvement of maize for drought tolerance. In: Jewell, D.C., Waddington, S.R., Ransom, J.K. and Pixley, K.V. (eds) *Maize Research for Stress Environments. Proceedings of the Fourth Eastern and*

*Southern Africa Regional Maize Conference, 28 March – 1 April 1994, Harare, Zimbabwe.* CIMMYT, Mexico D.F., pp. 67–75.

Joshi, A. and Witcombe, J.R. (1996) Farmer participatory crop improvement. II. Participatory varietal selection, a case study in India. *Experimental Agriculture* 32, 461–477.

Kamara, A., Defoer, T. and de Groote, H. (1996) Selection of new varieties through participatory research, the case of corn in southern Mali. *Tropicultura* 14, 100–105.

Kumwenda, J.D.T., Waddington, S.R., Snapp, S.S., Jones, R.B. and Blackie, M.J. (1997) Soil fertility management in southern Africa. In: Byerlee, D. and Eicher, C.K. (eds) *Africa's Emerging Maize Revolution.* Lynne Rienner Publishers, Boulder, Colorado, pp. 157–172.

Lafitte, H.R. and Edmeades, G.O. (1994) Improvement for tolerance to low soil nitrogen in tropical maize. II. Grain yield, biomass production, and N accumulation. *Field Crops Research* 39, 15–25.

McCown, R.L., Keating, B.A., Probert, M.E. and Jones, R.K. (1992) Strategies for sustainable crop production in semi-arid Africa. *Outlook on Agriculture* 21, 21–31.

Morris, M.L. (2001) Assessing the benefits of international maize breeding research: an overview of the global maize impact study. In: Pingali, P. (ed.) *CIMMYT 1999–2000 World Maize Facts and Trends. Meeting World Maize Needs: Technological Opportunities and Priorities for the Public Sector.* CIMMYT, Mexico D.F., pp. 25–34.

Otim-Nape, G.W., Bua, A. and Baguma, Y. (1994) Accelerating the transfer of improved production technologies: controlling African Cassava Mosaic Virus disease epidemics in Uganda. *African Crop Science Journal* 2, 479–495.

Pandey, S., Ceballos, H., Magnavaca, R., Bahia Filho, A.F.C., Duque-Vargas, J. and Vinasco, L.E. (1994) Genetics of tolerance to soil acidity in tropical maize. *Crop Science* 34, 1511–1514.

Patterson, H.D. and Williams, E.R. (1976) A new class of resolvable incomplete block designs. *Biometrika* 63, 83–89.

Pederson, D.G. and Rathjen, A.J. (1981) Choosing trial sites to maximize selection response for grain yield in spring wheat. *Australian Journal of Agricultural Research* 32, 411–424.

Pixley, K.V., Harrington, L. and Ransom, J.K. (1997) Regional priorities for maize research in eastern and southern Africa. In: Ransom, J.K., Palmer, A.F.E., Zambezi, B.T., Mduruma, Z.O., Waddington, S.R., Pixley, K.V. and Jewell, D.C. (eds) *Maize Productivity Gains through Research and Technology Dissemination. Proceedings of the Fifth Eastern and Southern African Regional Maize Conference, 3–7 June 1996, Arusha, Tanzania.* CIMMYT, Addis-Ababa, Ethiopia, pp. 169–172.

Rhoades, R. and Booth, R. (1982) Farmer-back-to-farmer: a model for generating acceptable agricultural technology. *Agricultural Administration* 11, 127–137.

Rosielle, A.A. and Hamblin, J. (1981) Theoretical aspects of selection for yield in stress and non-stress environments. *Crop Science* 21, 943–946.

Rubey, L., Ward, R.W. and Tschirley, D. (1997) Maize research priorities: the role of consumer preferences. In: Byerlee, D. and Eicher, C.K. (eds) *Africa's Emerging Maize Revolution.* Lynne Rienner Publishers, Boulder, Colorado, pp. 145–155.

Simmonds, N.W. (1991) Selection for local adaptation in a plant breeding programme. *Theoretical and Applied Genetics* 82, 363–367.

Smale, M., Heisey, P.W. and Leathers, H.D. (1995) Maize of the ancestors and modern varieties: the microeconomics of HYV adoption in Malawi. *Economic Development and Cultural Change* 43, 351–368.

Smaling, E.M.A. and Braun, A.R. (1997) Soil fertility research in sub-Saharan Africa: new dimensions, new challenges. In: Hood, T.M. and Benton Jones, J. Jr (eds) *Soil and Plant Analysis in Sustainable Agriculture and Environment.* Marcel Dekker, New York, pp. 89–110.

Smith, M.E., Weltzien, E., Meitzner, L.S. and Sperling, L. (2000) *Technical and Institutional Issues in Participatory Plant Breeding from the Perspective of Formal Plant Breeding. A Global Analysis of Issues, Results and Current Experience.* Working Document 3. PRGA Program. CIAT, Cali, Colombia.

Snapp, S. (1999) Mother and baby trials: a novel trial design being tried out in Malawi. *Target – Newsletter of the Soil Fertility Research Network for Maize-based Cropping Systems in Malawi and Zimbabwe.* 17, 8.

Sperling, L., Loevinsohn, M.E. and Ntabomvura, B. (1993) Rethinking the farmer's role in plant breeding: local bean experts and on-station selection in Rwanda. *Experimental Agriculture* 29, 509–519.

Sthapit, B.R., Joshi, K.D. and Witcombe, J.R. (1996) Farmer participatory crop improvement. III. Participatory plant breeding, a case study for rice in Nepal. *Experimental Agriculture* 32, 479–496.

Thiele, G., Gardner, G., Torrez, R. and Gabriel, J. (1997) Farmer involvement in selecting new varieties: potatoes in Bolivia. *Experimental Agriculture* 33, 275–290.

Vivek, B., Bänziger, M. and Pixley, K.V. (2001) *Characterization of Maize Germplasm Grown in Eastern and Southern Africa: Results of the 2000 Regional Trials Coordinated by CIMMYT.* CIMMYT, Harare, Zimbabwe.

Witcombe, J.R., Joshi, A., Joshi, K.D. and Sthapit, B.R. (1996) Farmer participatory crop improvement. I. Varietal selection and breeding methods and their impact on biodiversity. *Experimental Agriculture* 32, 445–460.

# Plant Breeding with Farmers Requires Testing the Assumptions of Conventional Plant Breeding: Lessons from the ICARDA Barley Program

**12**

SALVATORE CECCARELLI AND STEFANIA GRANDO

*The International Center for Agricultural Research in the Dry Areas (ICARDA), PO Box 5466, Aleppo, Syria*

## Abstract

It is widely recognized that modern agriculture and plant breeding have been beneficial for better-off farmers in more optimal environments, but not so beneficial for the poorest farmers in marginal environments. Whether addressing optimal or marginal environments, plant breeding practice is often based on a number of assumptions which in turn are based on widespread conventional interpretations of theory regarding the biological or the social and institutional basis of plant breeding. This chapter documents tests of these assumptions using mostly data from the barley breeding programme at ICARDA, in collaboration with farmers and national breeding programmes in a number of countries in West Asia and in Africa. Results of these tests have led to greater understanding of the application of basic biological theory of plant breeding to marginal environments, about the potential of farmers to work with plant breeders and to contribute to professional plant breeding and to innovations in plant breeding methods and practice. We conclude that conventional assumptions about the application of basic plant breeding theory are often unfounded and that, therefore, small-scale, poor farmers in marginal environments could be served by a different way of doing plant breeding.

## Introduction

It is widely recognized that modern agriculture and plant breeding have been beneficial to farmers in high potential environments or those who could profitably modify their environment to suit new cultivars.

They have not been so beneficial to the poorest farmers who could not afford to modify their environment through the application of additional inputs (Byerlee and Husain, 1993) and who continue to suffer from chronically low yields, crop failures and, in the worse situations, malnutrition and famine (Bänziger and de Meyer, Chapter 11, this volume). In addition, the successes in favourable environments have not been without negative consequences, particularly in relation to the adverse environmental effects of the indiscriminate use of high inputs and the loss of genetic diversity, and it is not clear which are greater, the successes or the shortcomings (Tilman, 1998). Whether addressing high potential or marginal environments, plant breeding practice is often based on a number of assumptions that can be summarized as follows:

- Selection has to be conducted under optimum or near-optimum conditions, because these selections will be superior also in marginal environments due to the 'spillover effect'.
- Cultivars must be genetically uniform.
- Cultivars must be widely adapted over large geographical areas.
- Locally adapted landraces must be replaced.
- The dissemination of improved cultivars must take place through mechanisms and institutions such as variety release committees, seed certification schemes and governmental or private seed production organizations.
- The end users of new varieties are not involved in selection; they are only involved at the end of the consolidated breeding routine (generation of variability, selection between and within segregating populations, researcher managed trials, verification trials) during the final testing, to verify if the choices made for them by others are appropriate or not.

Each of the assumptions affecting plant breeding practice listed above is based on one or more predominant interpretations of theory regarding the biological or the social and institutional basis of plant breeding. For example, selection has to be conducted under optimum or near-optimum conditions because these are generally (but not always) associated with larger genetic variances and heritabilities; genetic uniformity is required mainly on the assumption that the seed of improved cultivars will be produced by seed companies and because of plant breeders' rights; local germplasm is not adequately used in 'modern' breeding programmes because it is assumed that locally adapted landraces are low yielding, lodging and disease susceptible, and hence are not responsive to the addition of external inputs; eventually the end users are not involved in the development of new cultivars because it is assumed that they do not have the necessary 'knowledge' or that their knowledge is not relevant to the efficiency and effectiveness of plant breeding.

This chapter documents tests of these assumptions using mostly data from the barley breeding programme at the International Center for Agricultural Research in the Dry Areas (ICARDA), but also from other breeding programmes on different crops. It will be shown that these assumptions are often unfounded and that therefore marginal environments and small-scale and poor farmers could indeed be served by a different way of doing plant breeding.

## Conventional Breeding Programmes

The assumptions described above have led to centralized and non-participatory breeding programmes, which, with few exceptions, are conducted in the strictly controlled conditions of research stations. By *centralized* we mean that all the operations and decisions are the responsibility of a team of scientists and are taken in locations (usually research stations) that may be more or less similar to some target environment(s), but are not the target environment(s). By *non-participatory* we mean that users are not involved in those operations and decisions except in those cases in which they are providing the land and except in the very final stages of plant breeding.

Although plant breeding programmes differ from each other depending on the crop and on the breeder, they all have in common some major stages that Schnell (1982) has defined as 'generation of variability', 'selection', and 'testing of experimental cultivars' (Fig. 12.1). The generation of variability is the shortest stage, consisting often of the process of making crosses and producing a number of segregating populations, and takes place on research stations. The second stage is longer, and consists firstly of the evaluation of the breeding value of the different segregating populations ('cross-evaluation' or 'selection between crosses'), and then the selection of the best plants within the superior populations, or various combinations of the two. Also the second stage usually takes place on research stations (although there are exceptions), and in some crops it can be shortened by the use of techniques such as single seed descent (SSD) and double haploids (DH). During the second stage the breeding material is exposed to relevant biotic and abiotic stresses, often in different research stations. The end product of the second stage is usually a population of up to several thousand uniform breeding entities (clones, hybrids, pure lines) even in those situations where uniformity is not a farmer's necessity.

The third stage is also long, consisting of the comparison of yield (usually of grain in those crops where the grain is the main commercial product) between the breeding entities produced during the second stage. This phase is usually subdivided into two sub-stages. The first takes place in (sometimes several) research stations and the trials are known as multi-environment trials (MET). The second (when the breeding entities have been reduced to between 10 and 20) takes place in farmers' fields. In some exceptional cases, such as most of the Australian breeding programmes, the third phase takes place entirely in farmers' fields, and therefore is fully decentralized.

As shown in Fig. 12.1, plant breeding is a continuous process: each year (or cropping season) a new cycle begins with new crosses made largely using material derived from previous cycles as parents. There- fore, each year, breeding materials belonging to the three stages, and to different steps within each stage, are grown together, and this implies not only a considerable investment in land to grow the parental material, the various generations of segregating populations and the various levels of yield testing, each representing a different breeding cycle (often amounting to several tens of thousands of plots), but also in people and in facilities to handle the considerable amount of seed and of data that the process generates.

In the majority of plant breeding programmes, only a small fraction of the entire process takes place in farmers' fields (Fig. 12.1). One of the main consequences is that a large amount of breeding material is discarded without knowing whether it could have been useful in the real conditions of farmers' fields, and it is surprising how few breeding programmes even ask the question of how large this loss of useful

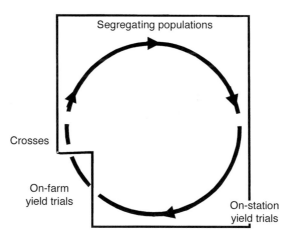

**Fig. 12.1.**   The stages of a centralized non-participatory plant breeding programme, namely generation of variability (crosses), selection (segregating populations) and testing of breeding material (on-station and on-farm yield trials). The line represents the fence of a research station.

breeding material is. We have shown (Ceccarelli, 1989) that for crops grown in environments poorly represented by the research stations, this often results in discarding useful breeding material. A recent example of this danger is offered by the results of a trial with 100 pure lines extracted from local barley landraces conducted in four locations in Eritrea (Tekle *et al.*, 2000), namely Halhale (a research station) and three villages (Embaderho, Tera-Emni and Senafe). As shown in Fig. 12.2 the ten lines with the lowest grain yield in the research station included two of the highest yielding lines in Embaderho, and one of the highest yielding lines in Senafe. Similarly, in the case of total biomass yield (data not shown), the ten lines with the lowest biomass yield in the research station included three of the highest yielding lines in Senafe and one each of the highest yielding lines in Tera-Emni and Embaderho.

## Theoretical Issues and Assumptions Underlying Breeding for Wide vs. Specific Adaptation

### Genotype × environment interactions

A fundamental problem in plant breeding is the relationship between selection and target environment. As Falconer (1952) pointed out, direct selection (i.e. selection in the target environment) is always the most efficient. In the case of indirect selection, selection efficiency decreases as the selection environment becomes increasingly different from the target environment, and genotype × environment $(G \times E)$

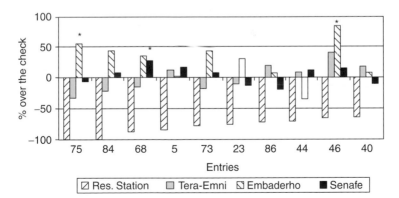

**Fig. 12.2.** Performance (as yield in % over the check) of the bottom ten barley lines in the research station when evaluated in three villages in Eritrea in farmers' fields (the entries with * are among the best 5% in the given farmer's field) (based on Tekle *et al.*, 2000).

interactions limit the efficiency of breeding programmes. Plant breeders can either avoid G × E interactions by selecting material that is broadly adapted to a range of target environments, or exploit them by selecting a range of material, each adapted to a specific target environment (Ceccarelli, 1989).

The issue of selection for broad and specific adaptation has been debated since the early 1920s (Hayes, 1923; Engledow, 1925) and is still highly controversial. One of the causes of controversy is the definition of stress environment, which is well illustrated by the distribution of the crops in different agroclimatic environments. For example, in a country like Syria, with a large spatial variability of rainfall within short distances, bread wheat, durum wheat and barley among the cereals, and faba bean, chickpea and lentil among the food legumes, are grown in progressively drier environments with some overlapping. Therefore, a stress environment for faba bean or bread wheat is moderately favourable for durum wheat and chickpea, and a stress environment for durum wheat and chickpea is moderately favourable for barley and lentil. At the drier end, barley and lentil are the only rainfed crops, and the other cereals or legumes are only grown under supplementary or full irrigation.

Another cause of the controversy is the confusion between adaptation over time and adaptation over space, even though the distinction is of fundamental importance. Adaptation over time (also called stability or dependability) refers to the performance of a cultivar in a given location across several years; if the cultivar performs consistently better than a reference cultivar, it is said to be stable. Adaptation over space refers to the performance of a cultivar in several locations; if the cultivar performs consistently better than a reference cultivar, it is said to be widely adapted. One recurrent problem in the literature on this topic is the use of the term 'environments', to indicate locations, years, but more often combinations of locations and years. This makes it difficult to understand whether wide or narrow adaptation refers to time or space or both (see for example Cooper, 1999). An additional problem has been the widespread use of the term 'wide' adaptation for those varieties grown on large areas. The term has confused geographically wide adaptation, often associated with lack of vernalization requirement and of photoperiod sensitivity, with environmentally wide adaptation, i.e. the ability to perform consistently well in locations or years which differ for soil characteristics (type, depth, fertility, etc.) water availability, biotic stresses, etc. (Cleveland et al., 2000).

It can be argued that wide adaptation over time (also defined as stability) is much more important to farmers than wide adaptation over space. The latter is, for obvious reasons, the major concern of seed producers. Two other causes of the controversy are the range of environmental variation sampled and the type of genetic (or breeding) material being used. Most of the studies comparing the two strategies are biased either because they fail to note the different ranges of environments considered in each (Cleveland and Soleri, 2002), or because breeding materials selected for specific adaptation to stress environments are not included.

One of the most recent examples of using a narrow range of environments is on two-row barley in Canada (Atlin *et al.*, 2000) where the ratio between the highest yielding environment and the lowest (nearly 3 t ha⁻¹) is 1.8, and where, not surprisingly, no crossover interaction was found. On the other hand, most of the studies supporting the concept of breeding for wide adaptation, counting on the so-called 'spillover effect' in marginal environments from selection conducted in optimal or sub-optimal environments, are based on comparisons between modern varieties (MV) and farmer varieties (FV) and are not correct. The right comparison should be between varieties selected under optimum conditions and varieties selected under stress. Since few breeding programmes have conducted enough breeding cycles under marginal conditions to achieve measurable gains, studies comparing MVs selected for favourable environments and MVs selected for unfavourable environments are few. The few studies that do exist invariably show repeatable crossover interactions (Calhoun *et al.*, 1994; Ceccarelli, 1996a). Many of the arguments relating repeatability of crossover interactions with breeding for specific or wide adaptation or for a combination of the two (Cooper, 1999; Weikai *et al.*, 2000) do not take into account that in plant breeding programmes material has to be selected and material has to be discarded at each cycle.

Several international breeding programmes implicitly recognize the existence of repeatable crossover interactions by targeting the development of germplasm for discrete target areas. For example, CIMMYT divides the global target environments for bread wheat into 12 megaenvironments (Braun *et al.*, 1996), which are further subdivided. For example, megaenvironments 4, 5, 8 and 11 are divided into 4a, 4b, 4c, 5a, 5b, 8a, 8b and 11a and 11b based on other biotic and/or abiotic stresses. Breeding for megaenvironment 4a, which includes West Asia and North Africa, has been decentralized by posting a breeder in that region (CIMMYT, 2000). Similarly, the International Rice Research Institute (IRRI) targets nurseries to seven irrigated, three rainfed, one deepwater and two upland ecosystems (Chaudhary and Ahn, 1996). The basic philosophy of these programmes is to generate and distribute widely adapted germplasm and to add traits that confer specific adaptation at a later stage. In both programmes the subdivision between target environments is an implicit recognition of specific adaptation, and in the case of rice, in breeding for individual target environments with independent breeding programmes. In the case of wheat, all the breeding material, regardless of the megaenvironment for which it is targeted, undergoes a heavy selection pressure for yield under optimum conditions and for disease resistance, and therefore is specifically adapted to all those locations and countries sharing reliable rainfall (or where the crop is irrigated), and affected by biotic stresses.

Breeding for specific adaptation is particularly important in the case of crops predominantly grown in unfavourable conditions, because unfavourable environments tend to be more different from each other than favourable environments (Ceccarelli and Grando, 1997). Furthermore, unfavourable environments tend to produce repeatable crossover interactions (Ceccarelli, 1989, 1996a). Breeding for specific adaptation to unfavourable conditions is often considered an undesirable breeding objective because it is usually associated with a reduction of potential yield under favourable conditions. This issue has to be considered in its social dimension and in relation to the difference between adaptation over space and adaptation over time: for example Australian farmers prefer maximizing yield in favourable years, while for North African and Near Eastern farmers yield in very poor years is more important (farmers, personal communication).

## Uniformity and heterogeneity

The trend of modern breeding is towards uniformity, often as a response to the needs of seed companies and breeders without taking into consideration farmers' requirements and preferences. The seed systems associated with commercial agriculture favour a small number of geographically broadly adapted varieties; they do not tolerate intravarietal diversity (Berg, 1996). This is in sharp contrast with the attitude of farmers in stress environments who tend to maintain genetic diversity in the form of different crops, different cultivars within the same crop, and/or heterogeneous cultivars to retain adaptability, that is, to maximize adaptation over time, rather than adaptation over space (Martin and Adams, 1987). Diversity serves to disperse or buffer the risk of total crop failure due to environmental variation. Several examples of the presence, the value and the use of diversity in various countries and crops are given by Almekinders and de Boef (2000).

While the advantages of uniformity (both in the sense of few really different cultivars, and of homogeneity within cultivars) are obvious for seed companies and for plant breeders' rights, they are much less obvious for farmers. It is not clear why in a country where barley is used as a feed crop, modern cultivars must be uniform, when for several thousand years small ruminants have done reasonably well feeding on heterogeneous landraces. In market-oriented agriculture, cultivars must obviously be uniform for traits associated with mechanized commercial cultivation (for example height and maturity) and for specific traits (malting quality, sugar content, etc.). However, it is not clear why they should be uniform also for traits that have nothing to do with market requirements. This includes disease resistance, about which much has been written (see for example Akem *et al.*, 2000).

Finally, when discussing uniformity it is difficult to use the same standard everywhere and for every crop: different crops and even the same crop in different countries or within the same countries may have different requirements for homozygosity and homogeneity depending on who is growing them, and for what purpose.

## Use of landraces in breeding programmes

In many developing countries, and for crops grown in low-input conditions and in stress environments, landraces are still the backbone of agricultural production. The reasons why farmers prefer to grow only landraces or continue to grow landraces even after partial adoption of modern cultivars are not well documented, but include quality attributes such as food and feed quality, and seed storability. Another important reason is that landraces are often able to produce some yield even in difficult conditions where modern varieties are less reliable, for example where farmers have adopted modern cultivars but have kept the landraces in the most unfavourable areas of the farm (Cleveland *et al.*, 2000; McGuire, Chapter 5, this volume; Smale, Chapter 3, this volume; Zimmerer, Chapter 4, this volume).

The value of landraces is well documented in the case of barley in Syria (Grando *et al.*, 2001). The comparison between landraces and modern germplasm in a range of conditions from severe stress (low input and low rainfall) to moderately favourable conditions (use of inputs and high rainfall) (Table 12.1) has consistently indicated the following:

1. Landraces yield more than modern germplasm under low-input and stress conditions.
2. The superiority of landraces is not associated with mechanisms to escape drought stress, as shown by their heading date.
3. Within landraces there is considerable variation for grain yield under low-input and stress conditions, but all the landrace-derived lines yield something whereas some modern germplasm fail.
4. Landraces are responsive to both inputs and rainfall and the yield potential of some lines is high, though not as high as modern germplasm.

**Table 12.1.**   Grain yields (kg ha$^{-1}$) of pure lines derived from Syrian landraces and modern germplasm at three levels of stress[a] in northern Syria.

| Environment | Landraces $n = 44$ | Modern $n = 206$ | Difference | $P$[b] |
|---|---|---|---|---|
| Stress | 1038 | 591 | 447 | < 0.01 |
| Intermediate | 3105 | 3291 | −186 | n.s. |
| Non-stress | 4506 | 6153 | −1647 | < 0.01 |

[a]As defined by average precipitation and soil fertility (Ceccarelli, 1996b).
[b]Based on $t$-tests for groups of unequal size; n.s. = not significant.

**5.**   It is possible to find modern germplasm which under low-input and stress conditions yield almost as well as landraces, but their frequency is very low.

The data in Table 12.1 also suggest that selection conducted only in high-input conditions is likely to miss most of the entries that would have performed well under low-input conditions.

   The assumption of most breeding programmes that landraces are genetically inferior is based on work conducted in research stations. This assumption has rarely been challenged even by those breeding programmes addressing target environments that have low yield potential because of the combination of biotic and abiotic stresses.

## Heritability and genetic correlation

An argument commonly used in favour of selecting in good environments is that heritabilities are higher there than in poor environments (Blum, 1988). First, this argument is based on comparisons between estimates of heritabilities in contrasting environments with no indication of whether the differences are significant or not. Yet, it is well known that heritability estimates usually have large standard errors unless based on very large numbers (Falconer, 1981). Second, the argument neglects that when selection is conducted in one environment (usually, as mentioned above, a high yielding environment) and response is measured in another environment (for example a low-yielding, marginal environment), the response depends not only on the heritabilities, but also on the genetic correlation coefficient between them. Falconer (1981) showed that when a character (A) is measured in two different environments (X and Y) it should be treated as two different characters. Selection for A in environment X aimed at improving A in environment Y is a case of indirect selection, and the response in environment Y is a correlated response to selection. The relative efficiency of indirect versus direct selection can be predicted by the magnitude of the heritabilities and the genetic correlation coefficient. If A is the trait to be improved in environment X by selecting in environment Y, then (Falconer, 1981):

$$CR_X/R_X = r_G (h_Y/h_X)$$

where $CR_X$ is the correlated response in environment X when selection is done in environment Y, $R_X$ is the direct response when selection is done in environment X, $r_G$ is the genetic correlation coefficient between $A_X$ and $A_Y$, and $h_X$ and $h_Y$ are the square roots of heritabilities of A in the two environments.

It is obvious from the formula that when $h_X = h_Y$, the maximum value of $CR_X/R_X$ is 1 when $r_G = 1$. Therefore, when the heritability of the trait is the same in the two environments, indirect selection will never be more effective than direct selection because the genetic correlation coefficient can never be more than one. When $h_Y$ is twice as large as $h_X$, $r_G$ must be larger than 0.5 before $CR_X$ is greater than $R_X$. With low genetic correlation coefficients (0.1–0.2), $h_Y$ must be more than 5–10 times higher than $h_X$ for $CR_X$ to become higher than $R_X$. Thus, comparing heritabilities of the same trait measured in two environments is not sufficient to draw conclusions on the optimum environment for selection (Atlin and Frey, 1989; Ceccarelli, 1989). This is obvious when $r_G$ is negative as the ratio between the two heritabilities becomes irrelevant. When only two environments are considered, the departure of the genetic correlation coefficient from +1 is a measure of the genotype × environment interaction (Falconer, 1952; Itoh and Yamada, 1990).

That the magnitude of heritability alone is not sufficient to identify the best environment for selection is also shown by the relationship between the correlated response to selection ($CR_X$) and the correlated selection differential ($S'_X$), where $r_P$ is the phenotypic correlation between $A_X$ and $A_Y$ (Falconer, 1981: 288):

$$CR_X = r_G/r_P \, h_Y \, h_X \, S'_X$$

The theory has been further developed by Rosielle and Hamblin (1981) with specific reference to selection in stress and non-stress environments. They showed that selection for tolerance to stress will generally result in reduced yield in non-stress environments and in a reduced average mean yield in stress and non-stress environments. Their results imply that at each of the several loci indirectly controlling yield through other traits, the alleles controlling ultimately the expression of yield under stress are different from those controlling the expression of yield under non-stress conditions. Simmonds (1991), using numerical simulation, concluded that: (i) selection for low-yielding conditions has to be conducted under low-yielding conditions; (ii) using environments with intermediate yield levels is ineffective for either stress or non-stress conditions; and (iii) alternating selection cycles in low- and high-yielding environments (shuttle breeding) is equally ineffective. His results imply that genotypes carrying positive alleles for yield in low-yielding conditions cannot be identified in high-yielding conditions where the same alleles might have a negative effect on yield. The two contrasting environments used in the case of the CIMMYT's shuttle breeding for wheat, Ciudad Obregon and Toluca, have mean grain yield of 5475 and 2955 kg ha$^{-1}$, respectively (Braun *et al.*, 1992), both well above the yield levels of rainfed and stressful environments. The shuttle breeding between the two locations has been very effective in developing photoperiod insensitive germplasm (with geographically wide adaptation) with high yield potential and resistance to diseases, but cannot be the cause of adaptation to contrasting production environments because they both represent a high production environment.

Furthermore, the experimental evidence (reviewed by Ceccarelli *et al.*, 1994) shows that heritability in low-yielding conditions is not consistently lower than that in high-yielding conditions (see Joshi *et al.*, Chapter 10, this volume); the ratio between the two ranges from 0.54 to 2.89, and even when it is lower it is often not sufficiently lower to offset the effect of a low correlation coefficient. Our own experience with barley (Singh and Ceccarelli, 1995; van Eeuwijk *et al.*, 2001) also suggests that the positive correlation between yield levels and magnitude of heritability is not a firm rule. Furthermore there have been many reports that narrow sense heritability increases in response to increased environmental stresses (Ward, 1994). This agrees with a crossover type of genotype × environment interaction, where the lowest heritability is expected at the crossover point and explains some of the disagreement among breeders on the issue of the optimum selection environment.

When the target environments lie above the crossover point the best discrimination between breeding lines occurs in the highest yielding environments. Close to the crossover point, breeding lines tend to look alike, hence the low heritability which is often found. When the target environments lie below the crossover point, the largest differences between genotypes are found at the lowest yield levels, with the identification of the best genotypes becoming increasingly difficult as yield levels rise. An example is given in Table 12.1 where 44 pure lines from Syrian barley landraces and 206 modern barley cultivars are compared at three levels of stresses resulting from different amounts of inputs and rainfall. At the lowest and highest yield levels there were significant differences between the two groups of germplasm, while the difference was not significant at the intermediate level, near to the crossover point.

The assumptions underlying each of the conventional interpretations of theory described above were challenged by the environments, germplasm and farming communities encountered in ICARDA's barley improvement efforts. Our response to these challenges coalesced into a number of strategies including decentralization and, later, participation, the subjects of the next two sections.


## Decentralized Breeding in the Barley Programme at ICARDA

The use of the concept of decentralization in a plant breeding programme, which represents a dramatic departure from the basic assumptions of conventional breeding programmes, is discussed in detail in the next section largely derived from Ceccarelli *et al.* (2001). Among the International Agricultural Research Centres, ICARDA has the global responsibility for barley improvement. While the goal of the barley breeding programme has remained the same over the years, namely to contribute to a sustainable increase in barley productivity in developing countries, thus contributing to alleviation of poverty, the way in which the objective is being pursued has changed considerably. These changes have occurred largely as a consequence of testing some of the assumptions on which most of the breeding programmes are based.

One of the major changes associated with the recognition of the importance of specific adaptation, particularly but not only for plant improvement in marginal environments, and of the potential of locally adapted germplasm, was to move from a fundamentally centralized breeding programme to a largely decentralized breeding programme. It is important at this point to specify that the term decentralization has been used often to describe two fundamentally different processes, namely decentralized selection and decentralized testing.

The term decentralized selection was first used by Simmonds (1984) and defined as selection in the target environment(s). Decentralized selection has also been referred to as *in-situ* or on-site selection. In the case of self-pollinated crops it consists of selection among early segregating populations (such as $F_2$) in a number of locations representing the target environment(s) (climate, soil, farming system and management) the breeding programme aims to serve. Decentralized selection becomes selection for specific adaptation when the selection criterion is the performance in specific environments rather than the mean performance across environments, and thus focuses on spatial more than temporal environmental variation. Selection for mean performance across a number of environments (years and locations) tends to exclude breeding material that performs very well in the lowest yielding years and/or locations but not particularly well in the highest yielding years and/or locations, unless data are standardized. In contrast, selection for the highest yielding breeding material in specific locations or areas, will automatically include breeding material performing well across all locations. In other words, selection for specific spatial adaptation will not exclude breeding material with wide spatial adaptation, while selection for wide adaptation tends to eliminate breeding material with specific adaptation.

Decentralized selection is different from decentralized testing, which is a common feature of breeding programmes and takes place, usually in the form of multi-location trials and on-farm trials, after a number of cycles of selection in one or few environments (usually with high levels of inputs).

In 1991, ICARDA's barley breeding programme started a gradual process of decentralization of selection work to the four Maghreb countries, Morocco, Algeria, Tunisia and Libya (Ceccarelli *et al.*, 1994). This was extended to Iraq in 1992, to Egypt in 1995, and it is gradually being implemented in the Mediterranean highlands of Turkey and Iran and in the Central Asia and Caucasus countries.

The type of decentralization implemented in North Africa (Fig. 12.3) involves moving all the selection of germplasm to national programmes from the very initial stages. National programmes scientists

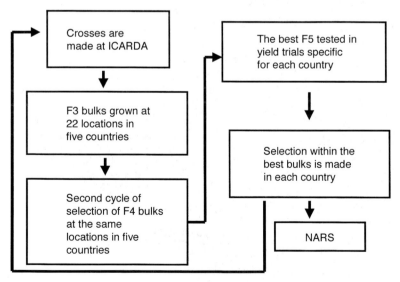

**Fig. 12.3.** Scheme of decentralized barley breeding between ICARDA and five national programmes in North Africa.

identify suitable parents, crosses are made at ICARDA, the material is advanced as bulks (without any selection) to the $F_3$ generation, and then distributed to the national programmes. The reason for advancing material to $F_3$ is to ensure the availability of sufficient seed for distribution. Selection between populations is made in the target environments in each country by national programme scientists.

Since barley is mostly used as animal feed in most of the countries in North Africa and the Near East, farmers do not require uniform cultivars, and successful bulks could, in theory, be released directly as cultivars after adequate testing. Unfortunately, however, a high degree of genetic uniformity is still legally required for variety release, even in those countries where farmers grow heterogeneous landraces. To meet these requirements, selection within the best populations is done within each country and at ICARDA through single seed descent, and the best lines are used for further cycles of crosses.

The decentralization in Iraq is similar, in principle, to the one described for North Africa, but much simpler in its implementation. Two types of targeted crosses are made at ICARDA with elite germplasm identified in Iraq, one for the irrigated areas with warm winters, high temperature and salinity stress, where barley is used for dual purposes (cut for forage 3–4 times and left to produce grain), and one for the rainfed areas characterized by low rainfall and continental climate with cold winters, where straw and grain are used as animal feed. The $F_2$ are sent to Iraq and all the successive stages of selection are conducted in Iraq without backup from ICARDA. Released cultivars and/or promising lines are sent back to ICARDA for crosses.

Another type of decentralization has been practised in Ethiopia where, in 1984, the barley breeding programme was almost entirely based on exotic germplasm and where, now, the major thrust is on Ethiopian landraces, tested in farmers' fields at low input levels. In this case it is the methodology (selection of landraces in the target environment) that has been decentralized rather than germplasm (Lakew et al., 1997; Semeane et al., 1998).

Breeding for resistance to pests and diseases is also increasingly conducted in a decentralized mode. For example, breeding for resistance to Russian wheat aphid (RWA), a devastating pest in some North African countries and in Ethiopia consists of: (i) screening sources of resistance in the target country; (ii) transferring the resistance to breeding material (including landraces) adapted to the target country and screening for resistance in the segregating populations at ICARDA headquarters; and (iii) screening for adaptation in the target country of the resistant $F_4$ or $F_5$ families. These examples show that decentralization can take several forms depending on the nature of the problem, and the strength and the expertise of the national programmes.

Since it started, the process of decentralizing the barley breeding programme has been gradually extended to other countries and regions. For each country and/or region where the two most important conditions for decentralized breeding exist: (i) large G × E interactions between ICARDA research station(s) and the target environment(s); and (ii) availability of local expertise in plant breeding, decentralization generally follows three steps. First, we send a special nursery to identify suitable parents; second, we start a specific crossing programme aimed at developing a specific germplasm pool for that country/region; and, third, we distribute the segregating populations. When fully implemented, the first step is replaced by the routine in-country screening of various germplasm sources. This, together with the decentralized screening for resistance to pests and diseases, assures that more and more parental material is supplied by the individual countries.

Decentralized selection is very powerful in maintaining genetic diversity. As an example we show the frequency of $F_4$ bulks and of $F_5$ bulks which were selected from the $F_3$ bulks distributed in 1994, 1995 and 1996 (Table 12.2) following the scheme shown in Fig. 12.3. The selection pressures used in decentralized selection appear much milder than those commonly applied in breeding programmes, so that much of the original diversity present in the original sets of $F_3$ bulks is still present in the subsequent cycles; as much as 80% of the bulks are still retained after two cycles of selection in the sets distributed in 1994 and 1995. In the third set the frequency of bulks after two cycles of selection drops to 56% (mostly because no data were returned from Algeria) and consequently those bulks that would have been selected only in Algeria are missing. As mentioned earlier, the bulks can reach the final stages of evaluation and could, in theory, be released directly as cultivars. These, being heterogeneous populations, are comparable in terms of population structure to landraces, with a different frequency of heterozygotes in the case of self-pollinated crops. Unfortunately, the legal requirement for uniformity makes it difficult to exploit this

**Table 12.2.** Number and frequency of bulks selected in two successive cycles of selection in four North African countries starting from $F_3$ bulks distributed in 1994, 1995 and 1996.

| $F_3$ bulks | | $F_4$ bulks | | | $F_5$ bulks | | |
|---|---|---|---|---|---|---|---|
| Year | No. | Year | No. | % | Year | No. | % |
| 1994 | 224 | 1995 | 197 | 87.9 | 1996 | 180 | 80.4 |
| 1995 | 208 | 1996 | 185 | 88.9 | 1997 | 169 | 81.3 |
| 1996 | 269 | 1997 | 257 | 95.5 | 1998 | 151 | 56.1 |

additional advantage of decentralized breeding within the framework of a formal seed system, while this is entirely possible with informal seed production.

There are some major implications for adopting decentralized breeding as a philosophy in international breeding programmes. Some are purely biological; for example, many varieties will be generated by national programmes, each adapted to specific conditions, and the superior performance of the varieties developed for low-input and less-favoured lands will not depend on agronomic practices that require large amounts of inputs. Therefore, a breeding programme based on this philosophy is less likely to endanger biodiversity and the environment.

Another important implication is the change in the role of national programme scientists who, with decentralized breeding, have the power to shape the breeding material according to the needs of their country, while in centralized breeding their role is only to test breeding materials developed by others. Eventually, one important implication is that the responsibility for developing and releasing varieties, and the reputation that goes with it, is given back to the national programmes. Consequently, the efficiency and the effectiveness of a decentralized breeding programme cannot be measured directly or solely by the number of varieties released and the extent of the area covered by each variety, as is done in conventional breeding. In fact, if success is measured by increases in farmers' net income, environmental sustainability and biodiversity, then it is likely that the success of international plant breeding programmes at the IARCs (International Agricultural Research Centres) may often be indicated by *low* numbers of varieties released by IARC breeders, a large and diverse basket of varieties released by NARS (National Agricultural Research System) breeders, and by *small* areas of adoption per variety.

Although the concept of decentralized breeding is usually well received by national programmes, decentralization *per se* will not necessarily respond to the needs of resource-poor farmers in less-favoured areas, if it is only a decentralization from the research station(s) of ICARDA to the research stations of the national programmes, and if the research stations of the national programmes do not represent, as is often the case, the difficult environments where the crop is predominantly grown. To exploit the potential gains from specific adaptation to low-input conditions, breeding must be decentralized from research stations to farmers' fields. Although decentralization and farmer participation are unrelated concepts, decentralization to farmers' fields almost inevitably leads to the participation of farmers in the selection process. Therefore, in the case of ICARDA's barley programme the idea of farmer participation originally arose as a type of decentralized selection to exploit G × E interactions and to make use, within a formal breeding programme, of the farmers' knowledge about the crop, its specific uses and its specific adaptation.

## Theoretical Issues and Assumptions Underlying Participation vs. Non-participation in Plant Breeding

The most serious limitation of decentralized selection for specific adaptation to unfavourable environments is the large number of target environments. Moreover, the number of target environments increases if we consider that environment is not only climate, soil, agronomic practices, farming system, etc., but also people living in that physical environment, their perception of risk associated with yield variation over time, their uses of the crop, and the consequent importance of quality traits even if neutral in terms of adaptation to the physical environment. Clearly, selection for specific adaptation to unfavourable conditions needs a larger sample of selection environments than selection for favourable environments.

The participation of farmers in the very early stages of selection offers a solution to the problem of fitting the crop to a multitude of both target environments and users' preferences (Ceccarelli et al., 1996, 2000). It is worth mentioning that, although farmer participation is often advocated on the basis of equity, there are sound scientific and practical reasons for farmer involvement to increase the efficiency and the effectiveness of the breeding programme. It is also expected that decentralized participatory plant breeding could be particularly effective in those situations where seed is supplied by the informal seed system as is the case for several crops in marginal environments.

The idea of farmers' participation in technology development, including plant breeding, is neither new nor revolutionary (Rhoades and Booth, 1982; Sperling et al., 1993; Farrington, 1996). For 10,000 years women and men have been consciously and unconsciously moulding the phenotypes (and so the genotypes) of hundreds of annual and perennial plant species, as one of their many routine activities in the normal course of making a living (Harlan, 1992). This traditional form of plant breeding by farmers produced hundreds of distinct varieties (Duvick, 1996), each adapted to the environmental and social conditions of particular farmers or communities.

The majority of the participatory plant breeding work published in refereed journals would be better defined as participatory variety selection (PVS) (Witcombe and Joshi, 1996). In PVS, farmers either choose between a limited (generally between 10 and 30) number of varieties on station, and then grow those chosen varieties in their fields, or are given a number of varieties (also between 10 and 30, but sometimes just one) to test in their own fields. PVS has been very successful both in facilitating adoption by poor farmers in marginal environments, not previously reached by formal plant breeding, and in understanding farmers' preferences (Maurya *et al.*, 1988; Sperling *et al.*, 1993; Joshi and Witcombe, 1996). However, PVS lacks the cyclical nature of plant breeding with a continuous flow of genetic material from one stage to next, and it is not clear from the literature on PVS whether, how and when a farmer or a farm community that has practised PVS will have another chance of participating in variety selection. Therefore, many examples of PVS are linear processes and could be described as sporadic, episodic or occasional participation. Also, to be successful, PVS has to assume that at least some of the varieties produced by a centralized non-participatory breeding programme are adapted to the target environment and meet farmers' requirements.

Nearly 50 examples of programmes defined as 'participatory plant breeding' (PPB) with a variable degree of involvement of formal breeding programmes and farmers have been recently described by Weltzien *et al.* (1999). They cover crops such as maize, chickpea, cowpea, beans, potatoes, rice, barley, pearl millet, sorghum and cassava, in Asia, Africa, Central and South America and in a variety of conditions, from the dry desert margins to high rainfall conditions, and from the lowlands to high altitudes. Many of these programmes are relatively new, having started within the last 10 years, and have been working on a small scale. The degree of participation varies from mere consultation to addressing issues or problems identified by farmers.

The best examples of participatory selection conducted for a number of cycles and starting from early segregating populations, are those reported by Sthapit *et al.* (1996) on rice, by Kornegay *et al.* (1996) on common bean, and by Ceccarelli *et al.* (2000) on barley. These are closer to PPB because farmers are exposed to breeding material at a much earlier stage than in PVS and for a number of cycles of selection, but they are not yet PPB. In fact, even in these cases, the difference from a participatory plant breeding programme is that farmers are not exposed to a continuous flow of germplasm, but only to an initial 'flush' of segregating populations from which to choose.

An important and obvious point to make is that PPB is based on the same genetic theories as non-participatory plant breeding (van Eeuwijk *et al.*, 2001), and therefore it is not a different type of plant breeding, at least in terms of biological theory. In fact, low heritability, unsuitability of germplasm, wrong choice of the selection environment(s), inappropriate selection methods, strong $G \times E$ interactions, and setting wrong objectives, all have a negative effect on selection efficiency and effectiveness no matter whether plant breeding is participatory or non-participatory. However, the belief that PPB is based on different fundamental theory from formal breeding is strongly rooted, and because most participatory work is sporadic, episodic or occasional, there is substantial discussion of the issue of breeding methods for PPB. Successful PPB does not need special breeding methods (see also the point made above), but it is certainly true that some breeding methods are more suitable than others to be used in PPB. It should also be remembered that farmers' skills almost inevitably improve in the course of truly participatory plant breeding, and therefore there may be a need to adjust the breeding method as the participatory breeding programme proceeds.

A final point is that plant breeding has been historically unable to reach small and poor farmers in marginal environments as efficiently and as effectively as better-off farmers in favourable environments (Zeigler, 1997). This is not necessarily due to lack of participation, but often to the use of the wrong selection environment(s) as a result of assumptions based on interpretations of theory that are inappropriate for the small-scale, poor farmers working in difficult environments. However, this lack of appreciation for the need for decentralization has discouraged consideration of farmer participation. As pointed out above, while decentralization is the major contributor to improved biological results of plant breeding, participation can significantly contribute to these outcomes as well.

## Comparing decentralization and participation

PPB trials are not structurally different from the MET described earlier when the latter are conducted in farmers' fields. However, there are two major differences between MET and PPB trials. The first is that MET are established with the primary objective to sample target physical environments, while PPB trials are meant to sample both physical and socio-economic environments and different types of users. The second is that MET data are usually analysed to estimate or predict the genotypic value across all locations, while in PPB trials the emphasis is on estimating or predicting the genotypic value over time in a given location.

Although the comparison between decentralization and participation is a crucial issue, only in a very few cases is this comparison possible because selection must be conducted both on station and in farmers' fields by both breeders and farmers. Results obtained in Syria and Yemen (Table 12.3) show that decentralization has the largest effect, as a consequence of large G × E interactions of a crossover type, but that in some cases participation adds a further and significant gain

**Table 12.3.** Effectiveness (measured by the % of high yielding entries included among the selections) of different selection strategies in barley and lentil.

| Type of selection | Barley in Syria[a] | Lentil in Yemen[b] |
|---|---|---|
| Decentralized participatory | 33.3a | 32.8a |
| Decentralized non-participatory | 17.2b | 26.6ab |
| Centralized participatory | 11.3bc | 19.6bc |
| Centralized non-participatory | 9.1c | 21.0bc |

[a]Ceccarelli *et al.*, 2000.
[b]Informal report to the Systemwide Program on Participatory Research and Gender Analysis for Technology Development and Institutional Innovation. Percentages followed by the same letter are not significantly different ($P < 0.05$).

in efficiency. It could be hypothesized that this gain may be most likely to occur in situations where the specific needs of a group of farmers are not well known or understood by the formal breeding system.

We have discussed many of the theoretical issues fundamental to any form of plant breeding, and how assumptions about their interpretation have contributed to formal plant breeding systems that do not always meet the needs of the poorest farmers working in the most difficult environments. However, there are questions specifically about the design of a participatory form of plant improvement that are significant for making decisions concerning practical application. Some of these questions are discussed in the next section, with a focus on the experience of ICARDA's barley programme.

## Participatory Breeding in the Barley Programme at ICARDA

ICARDA is involved in a number of participatory barley breeding programmes in Syria, Egypt, Jordan, Tunisia, Morocco and Eritrea and in one programme on both barley and lentil in Yemen. The results and information generated from ICARDA's ongoing work as well as other PVS and PPB experiments with various crops will be used to discuss some issues particularly relevant in plant breeding terms to decisions about whether and how to apply PPB, namely:

- Which type of breeding material and how much material farmers can handle.
- Is farmers' selection effective?
- Are farmers' and breeders' selection criteria different?
- Is decentralized participatory plant breeding effective in enhancing/conserving biodiversity?

- Is decentralized participatory plant breeding effective in increasing/speeding-up adoption?

## Quantity and type of breeding material

The amount of breeding material that can be evaluated in decentralized participatory breeding programmes is important for achieving some of the main objectives: particularly adaptation to the physical, production and social environment; an enhanced adoption rate; and the maintenance of biodiversity (compare Joshi *et al.*, Chapter 10, this volume). Too many so-called participatory programmes are based on a very small number of fixed or nearly fixed lines, and it is not clear what the difference is between them and the final stages of variety testing conducted by any private or public formal breeding programme.

There is a widespread and untested assumption that farmers are not able to examine, judge a large number of breeding lines, and translate that judgement into a quantitative score. As shown in Table 12.4, the assumption about the amount of material that farmers are able to handle needs to be ascertained in each project, and the programme should be designed accordingly. Similarly, plot size can be very different, and can also be very small, as in the case of the mountain-terraced agriculture in Yemen where the breeding trials have to be accommodated within the limited space of the terraces. This is important because it allows the participation of small farmers. The cases reported in Table 12.4 are all based on early segregating populations ($F_2$ or $F_3$ bulks) indicating that participation of farmers is feasible even at such an early stage of a breeding programme of a self-pollinated crop.

**Table 12.4.** Number of villages, number of entries, plot size used in different farmer selection projects in barley conducted by ICARDA and number of farmers involved.

| Country | No. of villages | No. of entries | Plot size (m²) | No. of farmers per village |
|---|---|---|---|---|
| Syria phase 1 | 9 | 208 | 12 | 5–9 |
| Syria phase 2 | 8 | 200–400 | 12 | 6–11 |
| Yemen | 3 | 100 | 3 | 15–20 |
| Morocco | 6 | 30–210 | 4.5 | 6–15 |
| Tunisia | 6 | 25–210 | 4.5 | 10–20 |
| Eritrea | 3 | 155 | 3 | 10–12 |
| Egypt | 8 | 60 | 6 | 5 |
| Jordan | 7 | 200 | 3 | 6–15 |

## Efficiency of farmer selection

One of the classic data sets on the efficiency of farmers' selection is the case of beans in Rwanda described by Sperling *et al.* (1993). The varieties selected by farmers on station out-yielded the locally grown mixtures 64–89% of the time with yield increases up to 38%. In contrast, the breeders' selections out-yielded the local mixtures about 50% of the time but with considerably smaller yield increases. Similarly, in the case of barley in Morocco, farmers seemed to be as able as the breeder to identify the highest yielding entries both in terms of grain and straw yield (Table 12.5).

One of the most remarkable results of the work in Syria was that farmers were effective in identifying high yielding entries in their fields, the entries they selected in six out of the nine locations had a significantly higher average grain yield than the population mean (Table 12.6). The breeder's selections were also significantly higher yielding than the population mean in six out of the nine farmers' fields. The entries selected by the farmers and by the breeder in the nine locations never differed significantly in grain yield with the exception of Al Bab, where farmers' selections yielded significantly more grain and more biomass than the breeder's selections.

**Table 12.5.** Farmers' and breeder's efficiency[a] in selecting early segregating populations of barley in Morocco.

| | | Farmers' field | | | | Research stations | |
|---|---|---|---|---|---|---|---|
| Selection done by | Attribute | S. Boumahdi | Chemaia | Oued Zem | Zhiliga | Merchouch | J. Shaim |
| **Experiment 1 (30 lines in two replications)** | | | | | | | |
| Farmer | Grain | 0.20 | 0.21 | 0.33 | 0.30 | 0.25 | 0.20 |
| Breeder | Grain | 0.17 | 0.18 | 0.31 | 0.29 | 0.18 | 0.38 |
| Farmer | Straw | 0.18 | 0.28 | 0.24 | 0.23 | 0.22 | 0.17 |
| Breeder | Straw | 0.42 | 0.27 | 0.25 | 0.24 | 0.18 | 0.38 |
| **Experiment 2 (50 lines in two replications)** | | | | | | | |
| Farmer | Grain | 0.16 | – | – | 0.29 | 0.50 | 0.14 |
| Breeder | Grain | 0.21 | – | – | 0.26 | 0.29 | 0.20 |
| Farmer | Straw | 0.34 | – | – | 0.30 | 0.07 | – |
| Breeder | Straw | 0.21 | – | – | 0.29 | 0.25 | – |

[a]Efficiency is expressed as the ratio between the number of high yielding lines for either grain or straw yield and the total number of lines selected.

**Table 12.6.**  Grain yield (kg ha$^{-1}$) and total biological yield (kg ha$^{-1}$) of the entries selected by the farmers and by the breeder in each of the nine farmers' fields (modified from Ceccarelli et al., 2000).

| Location | Grain yield | | | Biological yield | | |
|---|---|---|---|---|---|---|
| | Farmer | Breeder | Comparison[a] | Farmer | Breeder | Comparison[a] |
| Ibbin | 4615*** | 3971*** | n.s. | 10687** | 9686*** | n.s. |
| Ebla | 3498* | 3199** | n.s. | 8743 | 8233 | n.s. |
| Tel Brak | 4235 | 4020* | n.s. | 8729* | 8036 | n.s. |
| Jurn El-Aswad | 2049* | 1724** | n.s. | 10535** | 8429* | n.s. |
| Baylonan | 454* | 324 | n.s. | 3198 | 2816 | n.s. |
| Al Bab | 649*** | 488*** | *** | 2272*** | 1787*** | *** |
| Melabya | 915 | 920*** | n.s. | 4127** | 3246* | n.s. |
| Bari Sharki | 1366* | 1129 | n.s. | 5276 | 4708 | n.s. |
| Sauran | 2561 | 2654 | n.s. | 6796 | 7257 | n.s. |

$*P < 0.05$; $**P < 0.01$; $***P < 0.001$, in comparison with the population mean.
[a]Comparison between the breeder's and farmers' selections based on $t$-test for samples of unequal size.

### Farmers' selection criteria

A common inefficiency in formal breeding programmes is the selection and release of varieties that do not meet farmers' requirements and needs, and that therefore are not adopted – particularly but not only – in more marginal conditions. Information about farmers' selection criteria is one of the most common outputs of PPB programmes and some of the most common features emerging from studies of various crops in various countries are the following:

Usually farmers are interested in a wider range of traits or of combinations of traits than breeders expected. These are related to adaptation to various growing conditions but also to marketability. A typical example is provided by barley in the dry areas of Syria where farmers prefer genotypes which are tall even under severe drought (they can be harvested by combine even in dry years), with a soft straw (this is considered to be associated with palatability for sheep) and with black seed (the darker the seed, the higher the selling price).

Although farmers nearly always rank yield as their most important selection criterion, they in fact select for several other traits as long as yields are above an acceptable minimum.

Farmers' selection criteria vary according to the environment. In Syria, farmers in dry areas select barley entries taller than population mean and with soft straw and black seed on their farm, but entries

**Table 12.7.** Tall or short? Plant height (cm) of barley entries selected by the breeder and the farmer in a research station (favourable environment) and in a farmer's field in a dry area, compared with the population mean.

|  | Selected at | |
| --- | --- | --- |
| Selected by | Research station | Farmer's field |
| Farmer | 71.1* | 45.1*** |
| Breeder | 71.8* | 42.8* |
| Pop. mean | 77.5 | 39.6 |

*, *** differences significant at $P < 0.05$ and $P < 0.001$, respectively, in a *t*-test for samples of unequal size.

shorter than the population mean (lodging-resistant) on station (Table 12.7). Also, farmers' preference for modern germplasm and landraces (Fig. 12.4) both in the research station and in their field largely depends on the selection environment. The work on barley in Syria has shown that the environment affects the breeder's selection criteria much less as indicated by the similarity coefficients among the selections made by the same breeder in nine fields with mean environmental yields ranging from less than 300 kg ha$^{-1}$ to nearly 3700 kg ha$^{-1}$ (Fig. 12.5). The figure also confirms the large differences between a breeder's and farmers' selections in the same environment.

### Effects of PPB on biodiversity

One of the most common findings of PPB programmes is that different farmers in different communities select different varieties (as shown by the low similarity coefficients between farmers' selections in Fig. 12.5), and there are examples (rice in Nepal, bean in Rwanda and cassava in Colombia) of substantial increases in the number of different varieties grown by communities after one cycle of a PVS programme.

It is not easy to distinguish in the literature the effect of PPB and PVS on different levels of biodiversity, namely diversity within the same farm (more than one variety of the same crop in the same field), or diversity within a community or an area (different farmers growing different varieties), and it is even more difficult to understand whether PPB can affect the choice between uniform and heterogeneous breeding material.

In barley, we found that in individual locations decentralized participatory selection may lead to the same decrease of diversity as centralized non-participatory selection. In our project in Syria, the

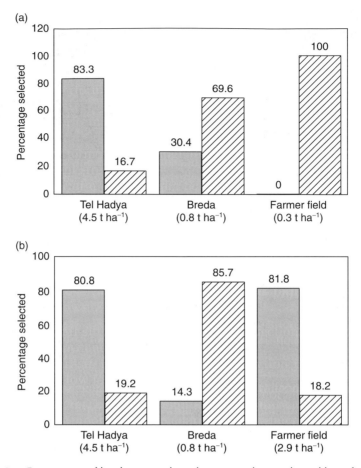

**Fig. 12.4.** Percentage of landraces and modern germplasm selected by a farmer from a low rainfall site (a) and by a farmer from a high rainfall site (b) in a research station located in a high rainfall site (Tel Hadya), in a research station located in a low rainfall site (Breda), and in their own field (the environmental mean yields are in parentheses). ▦ modern; ▨ landraces.

initial population of 208 entries included 48% modern germplasm and 52% landraces. Two cycles of selection on station led to the disappearance of the landraces (Fig. 12.6), but the same effect was observed in a farmer's field located in an environment similar to the research station. In contrast, two cycles of selection in a dry site led to the opposite result, namely the disappearance of the modern germplasm. These data seem to suggest that farmers' selection may have the same narrowing effect on the biodiversity available in the original breeding material. However, because different farmers select different material, the biodiversity over the total area is maintained or even increased.

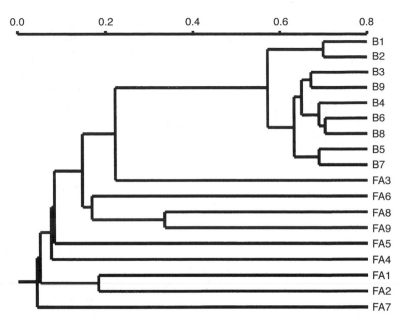

**Fig. 12.5.** Dendrogram based on cluster analysis of the selections of nine farmers and of a breeder in farmers' fields (FA = farmer, B = breeder). Individual farm locations are indicated with numbers from 1 to 9 (from: Ceccarelli *et al.*, 2000).

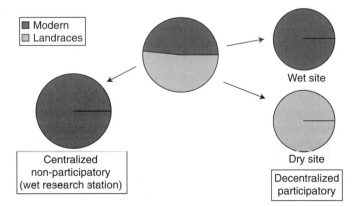

**Fig. 12.6.** Change in the frequency of modern germplasm and landraces after two cycles of centralized non-participatory selection in a research station (favourable conditions) and two cycles of decentralized participatory selection in a high rainfall (wet site) and in a low rainfall (dry site) location (modified from Ceccarelli, 2000).

The percentage of entries after one cycle of decentralized participatory selection is always among the highest when compared with selection strategies with other combinations of centralization and participation (Table 12.8). The major effect in the case of barley in Yemen

**Table 12.8.** Entries selected with four different strategies of selection in Syria on barley and in Yemen on barley and lentil in percentage of the original population size.

| Strategy of selection | Barley in Syria | Barley in Yemen | Lentil in Yemen |
|---|---|---|---|
| Centralized non-participatory | 0.34 | 0.20 | 0.22 |
| Centralized participatory | 0.61 | 0.24 | 0.40 |
| Decentralized non-participatory | 0.79 | 0.52 | 0.46 |
| Decentralized participatory | 0.74 | 0.50 | 0.56 |

is clearly the selection environment, with decentralized selection retaining more than twice the number of entries retained by centralized selection. However, in the other two cases, farmers' selection led to a larger percentage of entries selected even on station (centralized participatory selection), presumably as an effect of individual differences and of the different types required in the farmers' home areas. It is interesting to note that these effects were not due to the total number of selected entries, as the farmers always selected a lower number of entries than the breeder (Ceccarelli *et al.*, 2000). Most likely, this is because the breeders also select materials that do not meet the requirements of a finished cultivar, but that can be used as parents for crosses, while farmers are looking for material ready to be used for large-scale cultivation.

## Effects of PPB on adoption

Although PPB and PVS programmes are relatively recent, there are already some examples of impact. For example, there are cases where varieties preferred by farmers were identified in environments where no improved varieties have ever been available, such as the rice variety combining the frost tolerance of a landrace with the higher yield of a modern variety, which has been adopted in the mountains of Nepal (Sthapit *et al.*, 1996); and a bean variety combining disease resistance with a desirable coat colour which was adopted in north-eastern Brazil before the variety could be formally recommended (Zimmermann, 1996). Other examples are provided by rainfed rice varieties in India (Witcombe *et al.*, 1999), potatoes in Bolivia (Thiele *et al.*, 1997) and, surprisingly, irrigated wheat and rice in Gujarat (Witcombe, 1999), in India, demonstrating that even in the areas where formal plant breeding has been particularly successful, farmer participation can identify desirable varieties at an earlier stage than in conventional breeding.

One of the best examples of fast adoption through farmers' participation in variety testing (PVS), is the spreading of the rice variety Kalinga III in India (Witcombe *et al.*, 1999). Seed of this variety, which has neither been recommended nor released, was initially made available to farmers in three villages in Rajasthan in 1993 through a farmer-managed participatory research trial. By 1997, 65% of the area in one of the villages was planted with Kalinga III, while the village with the lowest adoption planted the variety in about 20% of its area. The variety also spread to other villages with very high rates of adoption, and the number of villages growing Kalinga III increased by a factor of 2.3–7.0.

There are other examples of very rapid adoption of new varieties by participating farmers. Obviously, much of the success depends on seed availability and also on the reaction of the formal system to having varieties (or breeding lines) bypass the official certification and release channels.

## Quality of Participation and Assumptions

As mentioned earlier, one of the assumptions most commonly made by formal breeding programmes is that farmers do not have the knowledge needed to contribute to plant breeding. When this assumption is tested by conducting participatory breeding in the way we have described, not only farmers' skills, such as the ability to identify superior entries, become obvious, but these skills are inevitably enriched and developed by the continuous exposure to a wide (in several cases much wider than ever experienced before) range of genetic diversity. Farmers' skills have been shown in a number of countries and with several crops, and it became clear that the specific type of knowledge is different in different countries (and possibly even within the same country) and among different users. In Syria we have monitored the process of skill building more closely than in other countries, even though informally. In the course of 5 years of participatory barley breeding, the level of participation has increased to the point that now farmers decide which type of germplasm they would like to use for selection (as well as which type they do not want to use), which type of additional variables to add to the trials (for example farmers in three communities wanted two sets of breeding material to test genotypic responses to different rotations, soil depths and soil types), they ask to see the results of the statistical analysis, and eventually suggest crosses between entries with desirable but different traits. One negative consequence of the skills development process is that visitors to the PPB trials in Syria are so impressed with the farmers' skills that they hardly believe that these were originally a random sample of farmers. Therefore, they leave with the impression that PPB is possible only with exceptionally gifted farmers.

One particularly interesting development in our PPB project in Syria was the interest to know more about the origin and pedigree of both good and bad looking entries. This interest was initially shown during visual selection in the field, and later during selection based on quantitative data. This has resulted in a gradually increasing awareness of the value of landraces in comparison with modern germplasm under the physical and management conditions of central and northern Syria. Even though landraces and modern cultivars have been compared several times in farmers' fields in the on-farm trials, farmers were never told about the genetic nature of the material being compared.

These examples show that participatory breeding (and probably all participatory research) is a dynamic process during which participants change. Farmers' increased awareness of what formal breeding can offer, changes the quantity and the quality of demands on the formal breeding programme. This consequently changes the power relationships because de facto most of the important decisions, not least the choice of parents for the following cycle of recombination, are gradually taken by farmers. The most important challenge for the formal breeders is to progressively adapt and respond to those changes.

One farmers' reaction that has been common during the first year of implementation of a participatory breeding programme in barley in several countries has been the request to extend this approach to other, often more important, crops. In some countries this has considerably speeded up the process of institutionalization of participatory plant breeding.

For these reasons we wonder whether there can be participation without empowerment, given that by definition participation is sharing of knowledge, responsibilities, resources and decision taking, and that in fact this is why there is so much resistance to changing to a participatory mode (compare Joshi et al., Chapter 10, this volume). Yet it should be only natural for agricultural scientists committed to increase food production and to eradicate poverty, to facilitate the process of skill building, and increase the capacity of farmers to actually redirect plant breeding and shape agricultural research towards their needs.

One specific advantage of decentralized PPB is to rapidly adapt the crops to changing climatic and agronomic conditions. Eventually, PPB could be the only possible type of breeding for crops grown in remote regions, for crops for which a high level of diversity is required within the same farm, or for those crops locally important but globally considered as minor crops and therefore neglected by formal breeding.

# Conclusions

This chapter has shown that decentralized participatory plant breeding is not an alternative type of plant breeding, somewhat opposed to formal plant breeding, but is rather a different way of applying the same genetic principles and breeding theories to specifically address situations such as marginal environments where G × E interactions are large and repeatable, precluding the adaptation of one or few varieties, or where there is a variety of different requirements (quality, crop duration, management, etc.). Therefore, decentralized participatory plant breeding recognizes that the same genetic principles and breeding theories translate into different breeding practices depending on the biological and social situations.

In the case of the barley breeding programme at ICARDA this has been the result of testing the assumptions on which centralized, non-participatory plant breeding is based. As a result of this process, it has been found that:

**1.** Selection in optimum or near-optimum conditions, such as those on research stations, tends to produce cultivars that are superior to local landraces only under improved management and not under the low-input conditions typical of the farming systems of stress environments. As a consequence, many new cultivars out-yield local landraces on a research station and some are released, but few (if any) actually replace landraces in difficult environments.

**2.** Farmers' selection criteria may be different from those of breeders (Hardon and de Boef, 1993; Sperling *et al.*, 1993). Typical examples are traits such as taste, colour, cooking properties and feed characteristics that are usually at least as important as yield to farmers, while breeders often use grain yield as the sole selection criterion.

**3.** Resource-poor farmers seldom use the formal seed-supply systems. They frequently rely on their own or on neighbours' seed (Almekinders *et al.*, 1994). Therefore, when the appropriate cultivar is selected, adoption is much faster through non-market methods of seed distribution (Grisley, 1993).

**4.** Farmers know their crop, their environment and the interactions between the two well. Often the knowledge of the crop dates back hundreds or thousands of years, yet modern plant breeding is systematically ignoring this knowledge and does not even test its nature and value. Participatory plant breeding merges farmers' and breeders' knowledge.

For thousands of years and until 100 years ago, farmers' agronomic and selection practices have fed the world. Modern agricultural science has taken these processes out of the hands of the farmers, has changed them to feed an ever-increasing population, without acknowledging the farmers' contribution and attempting to have them as partners in the development of new technologies. In fact, it appears that some of the assumptions discussed in this chapter are mostly correct in the case of seed companies rather than farmers. Participatory plant breeding puts farmers back as the focal point of the breeding processes.

## Acknowledgements

The authors thank BMZ, IDRC, the OPEC Fund for International Development, the Government of Italy, the Government of Denmark, the Systemwide Program on Participatory Research and Gender Analysis for Technology Development and Institutional Innovation of the CGIAR for financial support; and barley breeders of North Africa, Yemen and Eritrea for providing their data, and several farmers in Syria, Egypt, Morocco, Tunisia, Eritrea and Yemen who shared with us their knowledge of the crop.

## References

Akem, C., Ceccarelli, S., Erskine, W. and Lenné, J. (2000) Using genetic diversity for disease resistance in agricultural production. *Outlook on Agriculture* 29, 25–30.

Almekinders, C.J.M. and de Boef, W. (eds) (2000) *Encouraging Diversity.* Intermediate Technology Publications, London.

Almekinders, C.J.M., Louwaars, N.P. and de Bruijn, G.H. (1994) Local seed systems and their importance for an improved seed supply in developing countries. *Euphytica* 78, 207–216.

Atlin, G.N. and Frey, K.J. (1989) Predicting the relative effectiveness of direct versus indirect selection for oat yield in three types of stress environments. *Euphytica* 44, 137–142.

Atlin, G.N., McRae, K.B. and Lu, X. (2000) Genotype × region interaction for two-row barley yield in Canada. *Crop Science* 40, 1–6.

Berg, T. (1996) The compatibility of grassroots breeding and modern farming. In: Eyzaguirre, P. and Iwanaga, M. (eds) *Participatory Plant Breeding. Proceeding of a Workshop on Participatory Plant Breeding, 26–29 July 1995, Wageningen, The Netherlands.* IPGRI, Rome, pp. 31–36.

Blum, A. (1988) *Plant Breeding for Stress Environments.* CRC Press, Boca Raton, Florida.

Braun, H.J., Pfeiffer, W.H. and Pollmer, W.G. (1992) Environments for selecting widely adapted spring wheat. *Crop Science* 32, 1420–1427.

Braun, H.J., Rajaram, S. and van Ginkel, M. (1996) CIMMYT's approach to breeding for wide adaptation. *Euphytica* 92, 175–183.

Byerlee, D. and Husain, T. (1993) Agricultural research strategies for favoured and marginal areas, the experience of farming system research in Pakistan. *Experimental Agriculture* 29, 155–171.

Calhoun, D.S., Gebeyehu, G., Miranda, A., Rajaram, S. and van Ginkel, M. (1994) Choosing evaluation environments to increase wheat grain yield under drought conditions. *Crop Science* 34, 673–678.

Ceccarelli, S. (1989) Wide adaptation. How wide? *Euphytica* 40, 197–205.

Ceccarelli, S. (1996a) Positive interpretation of genotype by environment interactions in relation to sustainability and biodiversity. In: Cooper, M. and Hammers, G.L. (eds) *Plant Adaptation and Crop Improvement*. CAB International, Wallingford, UK; ICRISAT, Andra Pradesh, India; IRRI, Manila, Philippines, pp. 467–486.

Ceccarelli, S. (1996b) Adaptation to low/high input cultivation. *Euphytica* 92, 203–214.

Ceccarelli, S. (2000) Decentralized-participatory plant breeding: Adapting crops to environments and clients. In: *Proceedings of the 8th International Barley Genetics Symposium, Adelaide, 22–27 October 2000*, Vol. I. Department of Plant Science, Waite Campus, Adelaide University, Adelaide, South Australia, pp. 159–166.

Ceccarelli, S. and Grando, S. (1997) Increasing the efficiency of breeding through farmer participation. In: *Ethics and Equity in Conservation and Use of Genetic Resources for Sustainable Food Security, Proceeding of a Workshop to Develop Guidelines for the CGIAR, 21–25 April 1997, Foz de Iguacu, Brazil*. IPGRI, Rome, pp. 116–121.

Ceccarelli, S., Erskine, W., Grando, S. and Hamblin, J. (1994) Genotype × environment interaction and international breeding programmes. *Experimental Agriculture* 30, 177–187.

Ceccarelli, S., Grando, S. and Booth, R.H. (1996) International breeding programmes and resource-poor farmers: crop improvement in difficult environments. In: Eyzaguirre, P. and Iwanaga, M. (eds) *Participatory Plant Breeding. Proceeding of a Workshop on Participatory Plant Breeding, 26–29 July 1995, Wageningen, The Netherlands*. IPGRI, Rome, pp. 99–116.

Ceccarelli, S., Grando, S., Tutwiler, R., Baha, J., Martini, A.M., Salahieh, H., Goodchild, A. and Michael, M. (2000) A methodological study on participatory barley breeding. I. Selection phase. *Euphytica* 111, 91–104.

Ceccarelli, S., Grando, S., Amri, A., Asaad, F.A., Benbelkacem, A., Harrabi, M., Maatougui, M., Mekni, M.S., Mimoun, H., El-Einen, R.A., El-Felah, M., El-Sayed, A.F., Shreidi, A.S. and Yahyaoui, A. (2001) Decentralized and participatory plant breeding for marginal environments. In: Cooper, H.D., Spillane, C. and Hodgink, T. (eds) *Broadening the Genetic Base of Crop Production*. CAB International, Wallingford, UK; FAO/IPGRI, Rome, pp. 115–136.

Chaudhary, R.C. and Ahn, S.W. (1996) International network for genetic evaluation of rice (INGER) and its *Modus operandi* for multi-environment testing. In: Cooper, M. and Hammers, G.L. (eds) *Plant Adaptation and Crop Improvement*. CAB International, Wallingford, UK; ICRISAT, Andra Pradesh, India; IRRI, Manila, Philippines, pp. 139–164.

CIMMYT (2000) *CIMMYT in 1999–2000. Science and Sustenance.* CIMMYT, Mexico, D.F.

Cleveland, D.A. and Soleri, D. (2002) Indigenous and scientific knowledge of plant breeding: similarities, differences, and implications for collaboration. In: Sillitoe, P., Bicker, A. and Pottier, J. (eds) *Participating in Development: Approaches to Indigenous Knowledge.* Routledge, London.

Cleveland, D.A., Soleri, D. and Smith, S.E. (2000) A biological framework for understanding farmers plant breeding. *Economic Botany* 54, 377–390.

Cooper, M. (1999) Concepts and strategies for plant adaptation research in rainfed lowland rice. *Field Crops Research* 64, 13–34.

Duvick, D.N. (1996) Plant breeding, an evolutionary concept. *Crop Science* 36, 539–548.

Engledow, F.L. (1925) The economic possibilities of plant breeding. In: Brooks, F.T. (ed.) *Report of the Proceedings of the Imperial Botanical Conference.* pp. 31–40.

Falconer, D.S. (1952) The problem of environment and selection. *American Naturalist* 86, 293–298.

Falconer, D.S. (1981) *Introduction to Quantitative Genetics,* 2nd edn. Longman Group, London.

Farrington, J. (1996) Farmers' participation in agricultural research and extension: lessons from the last decade. *TAA Newsletter* June 1996 (9–10), 15.

Grando, S., von Bothmer, R. and Ceccarelli, S. (2001) Genetic diversity of barley: use of locally adapted germplasm to enhance yield and yield stability of barley in dry areas. In: Cooper, H.D., Spillane, C. and Hodgink, T. (eds) *Broadening the Genetic Base of Crop Production.* CAB International, Wallingford, UK; FAO/IPGRI, Rome, pp. 351–372.

Grisley, W. (1993) Seed for bean production in Sub-Saharan Africa, issues, problems, and possible solutions. *Agricultural Systems* 43, 19–33.

Hardon, J.J. and de Boef, W.S. (1993) Linking farmers and breeders in local crop development. In: de Boef, W., Amanor, K., Wellard, K. and Bebbington, A. (eds) *Cultivating Knowledge. Genetic Diversity, Farmer Experimentation and Crop Research.* Intermediate Technology Publications, London, pp. 64–71.

Harlan, J.R. (1992) *Crops and Man,* 2nd edn. American Society of Agronomy and Crop Science Society of America, Madison, Wisconsin.

Hayes, H.K. (1923) Controlling experimental error in nursery trials. *Journal of the American Society of Agronomy* 15, 177–192.

Itoh, Y. and Yamada, Y. (1990) Relationships between genotype × environment interaction and genetic correlation of the same trait measured in different environments. *Theoretical and Applied Genetics* 80, 11–16.

Joshi, A. and Witcombe, J.R. (1996) Farmer participatory crop improvement. II. Participatory varietal selection, a case study in India. *Experimental Agriculture* 32, 461–477.

Kornegay, J., Beltran, J.A. and Ashby, J. (1996) Farmer selections within segregating populations of common bean in Colombia: crop improvement in difficult environments. In: Eyzaguirre, P. and Iwanaga, M. (eds) *Participatory Plant Breeding. Proceeding of a Workshop on Participatory Plant*

*Breeding, 26–29 July 1995, Wageningen, The Netherlands*. IPGRI, Rome, pp. 151–159.

Lakew, B., Semeane, Y., Alemayehu, F., Gebre, H., Grando, S., van Leur, J.A.G. and Ceccarelli, S. (1997) Exploiting the diversity of barley landraces in Ethiopia. *Genetic Resources and Crop Evolution* 44, 109–116.

Martin, G.B. and Adams, M.W. (1987) Landraces of *Phaseolus vulgaris* (*Fabaceae*) in Northern Malawi. II. Generation and maintenance of variability. *Economic Botany* 41, 204–215.

Maurya, D.M., Bottral, A. and Farrington, J. (1988) Improved livelihoods, genetic diversity and farmer participation: a strategy for breeding in rainfed areas of India. *Experimental Agriculture* 24, 311–320.

Rhoades, R. and Booth, R. (1982) Farmer-back-to-farmer: a model for generating acceptable agricultural technology. *Agricultural Administration* 11, 127–137.

Rosielle, A.A. and Hamblin, J. (1981) Theoretical aspects of selection for yield in stress and non-stress environments. *Crop Science* 21, 943–946.

Schnell, F.W. (1982) A synoptic study of the methods and categories of plant breeding. *Zeitschrift für Pflanzenzüchtung* 89, 1–18.

Semeane, Y., Lakew, B., Alemayehu, F., van Leur, J.A.G., Grando, S. and Ceccarelli, S. (1998) Variation in Ethiopian barley landrace populations for resistance to barley leaf scald and net blotch. *Plant Breeding* 117, 419–423.

Simmonds, N.W. (1984) Decentralized selection. *Sugar Cane* 6, 8–10.

Simmonds, N.W. (1991) Selection for local adaptation in a plant breeding programme. *Theoretical and Applied Genetics* 82, 363–367.

Singh, M. and Ceccarelli, S. (1995) Estimation of heritability using varietal trials data from incomplete blocks. *Theoretical and Applied Genetics* 90, 142–145.

Sperling, L., Loevinsohn, M.E. and Ntabomvura, B. (1993) Rethinking the farmer's role in plant breeding, local bean experts and on-station selection in Rwanda. *Experimental Agriculture* 29, 509–519.

Sthapit, B.R., Joshi, K.D. and Witcombe, J.R. (1996) Farmer participatory crop improvement. III. Participatory plant breeding, a case study for rice in Nepal. *Experimental Agriculture* 32, 479–496.

Tekle, B., Ceccarelli, S. and Grando, S. (2000) Participatory barley breeding in Eritrea. In: *Proceedings of the 8th International Barley Genetics Symposium, Adelaide, 22–27 October 2000*, Vol. III. Department of Plant Science, Waite Campus, Adelaide University, Adelaide, South Australia.

Thiele, G., Gardner, G., Torrez, R. and Gabriel, J. (1997) Farmer involvement in selecting new varieties: potatoes in Bolivia. *Experimental Agriculture* 33, 275–290.

Tilman, D. (1998) The greening of the green revolution. *Nature* 396, 211–212.

van Eeuwijk, F.A., Cooper, M., DeLacy, I.H., Ceccarelli, S. and Grando, S. (2001) A vocabulary and grammar for the analysis of multi-environment trials, as applied to the analysis of FPB and PPB trials. *Euphytica* 122, 477–490.

Ward, P.J. (1994) Parent-offspring regression and extreme environments. *Heredity* 72, 574–581.

Weikai Yan, Hunt, L.A., Qinglai Sheng and Szlavnics, Z. (2000) Cultivar evaluation and mega-environment investigation based on the GGE biplot. *Crop Science* 40, 597–605.

Weltzien, E., Smith, M.E., Meitzner, L.S. and Sperling, L. (1999) *Technical and Institutional Issues in Participatory Plant Breeding – Done from a Perspective of Formal Plant Breeding. A Global Analysis of Issues, Results and Current Experience*. Working Document No. 3. CGIAR, Systemwide Program on Participatory Research and Gender Analysis for Technology Development and Institutional Innovation, Cali, Colombia.

Witcombe, J.R. (1999) Do farmer-participatory methods apply more to high potential areas than to marginal ones? *Outlook on Agriculture* 28, 43–49.

Witcombe, J.R. and Joshi, A. (1996) Farmer participatory approaches for varietal breeding and selection and linkages to the formal seed sector. In: Eyzaguirre, P. and Iwanaga, M. (eds) *Participatory Plant Breeding. Proceeding of a Workshop on Participatory Plant Breeding, 26–29 July 1995, Wageningen, The Netherlands*. IPGRI, Rome, pp. 57–65.

Witcombe, J.R., Petre, R., Jones, S. and Joshi, A. (1999) Farmer participatory crop improvement. IV. The spread and impact of a rice variety identified by participatory varietal selection. *Experimental Agriculture* 35, 471–487.

Zeigler, R.S. (1997) Implementing farmer participatory plant breeding: a research management perspective. *New Frontiers in Participatory Research and Gender Analysis for Technology Development. Proceedings of an International Seminar on Participatory Research and Gender Analysis for Technology Development, 1996, Cali, Colombia*. CIAT, Cali, Colombia.

Zimmermann, M.J.O. (1996) Breeding for marginal/drought prone areas in northern Brazil. In: Eyzaguirre, P. and Iwanaga, M. (eds) *Participatory Plant Breeding. Proceeding of a Workshop on Participatory Plant Breeding, 26–29 July 1995, Wageningen, The Netherlands*. IPGRI, Rome, pp. 117–122.

# Index

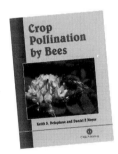
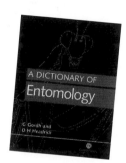
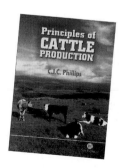